COMPUTATIONAL SEISMOLOGY

Edited by
V. I. Keilis-Borok

O. Yu. Shmidt Institute of Terrestrial Physics
Academy of Sciences of the USSR
Moscow, USSR

Translated from Russian by Stephen B. Dresner

Translation Edited by
EDWARD A. FLINN
Technical Director
Alexandria Laboratories
Teledyne Geotech
Alexandria, Virginia

CONSULTANTS BUREAU · NEW YORK–LONDON · 1972

This volume (with the exception of two papers) is comprised of articles appearing in Vychislitel'naya Seismologiya, Volumes 1-5, published in Moscow by Nauka Press in 1966, 1967, 1968, and 1971. The exact source of each article appears in a footnote on the opening page of that article. The articles have been revised and corrected by the editor for this edition.

ВЫЧИСЛИТЕЛЬНАЯ СЕЙСМОЛОГИЯ

VYCHISLITEL'NAYA SEISMOLOGIYA

Library of Congress Catalog Card Number 76-140827

ISBN 978-1-4684-8817-3 ISBN 978-1-4684-8815-9 (eBook)
DOI 10.1007/978-1-4684-8815-9

A Division of Plenum Publishing Corporation
227 West 17th Street, New York, N.Y. 10011

United Kingdom edition published by Consultants Bureau, London
A Division of Plenum Publishing Company, Ltd.
Davis House (4th Floor), 8 Scrubs Lane, Harlesden, London, NW10 6SE, England

COMPUTATIONAL SEISMOLOGY

VYCHISLITEL'NAYA SEISMOLOGIYA

ВЫЧИСЛИТЕЛЬНАЯ СЕЙСМОЛОГИЯ

PREFACE

This book contains selections from Volumes I-V of the series "Computational Seismology," which was initiated a few years ago by the Academy of Sciences of the USSR. Volume V was still in preparation when the translation was begun, and the translations of papers from it were made from manuscripts. Most of the authors are members of the Department of Computational Geophysics of the Institute of Physics of the Earth, Moscow.

The series is dedicated to theoretical and computational aspects of the analysis of seismological data. The present state of this field is typical of our times. The rapidly increasing flow of information is already too vast to be processed or even comprehended in a traditional way. This has forced analysis to be performed on random samples of data, and has led to attempts to reestablish the integrity of the information by computerization.

In making such attempts in seismology, we faced a situation that is perhaps also typical: direct computerization of the traditional approach to information processing can destroy the remnants of integrity rather than reestablish it. This is due to the fact that the traditional methods of seismology are of necessity designed to shorten computations by limiting the goals – by narrowing the data sample and by introducing *a priori* hypotheses. The success of such methods is greatly dependent on intuition and common sense in the choice of these limitations and in the evaluation of the results.

Results that are supported by broad intuition can be more general and precise than a rigorous formal analysis would imply. Striking examples are Jeffreys' model of the inner core boundary, the Gutenberg–Richter magnitude scale, and Gamburtsev's models of the crust.

Computerization of such limited methods of data processing leads to a deluge of new results and at the same time to a decline in the value of these results, since the available intuition and common sense do not grow as fast as computational possibilities.

Moreover, for some strange reason, computers usually create a spirit of haste, though they are intended to provide time for meditation. In computerizing seismology, therefore, one must first generalize the methods and then make them more rigorous mathematically. All relevant data must be processed jointly. Insofar as is possible, *a priori* hypotheses should be avoided. Particular attention must be given to exact formulation of the problem, to questions of uniqueness and stability, to the confidence limits of the results, etc. This general approach is required in solving the main problems of modern seismology, which are by definition general problems. This approach has other advantages. It simplifies the logical structure of seismology – removes the contradiction between the simple physical and geological goals and the complicated introversive nature of the methods. It makes the interpretation of seismological data one of the most promising areas of application of modern mathematics, as classical theoretical geophysics was at its dawn.

In the introductory paper of this collection, I have tried to make the above considerations more specific. They seem trivial, until one looks at what is happening in the area of computer applications. I do not claim that they are consistently followed in the papers presented here, although they were written by one closely connected group. This is because the general approach is difficult and probably also because of the spirit of haste mentioned above.

This collection is divided into four parts. Part I describes a system of algorithms for automation of seismological surveying: determination of epicenter, identification of pP and sP, determination of hypocenter, identification of secondary phases, and the fault-plane solution. The phase identification method was developed in cooperation with the International Seismological Centre in Edinburgh.

Part II contains papers on direct problems: propagation of seismic waves in a given earth model. These problems arose during work on the inverse problem, which is described in Part III.

The first three papers of Part III involve the inversion of seismological observations into earth structures. The last four papers are concerned with problems of uniqueness and stability. Among other things, they give a method for inversion of travel-time curves in the presence of a low-velocity zone.

Part IV deals with seismological applications of statistical methods: recognition of multiparameter patterns with application to focal depth estimation, an investigation of regularities in earthquake recurrence, and a game theoretical approach to estimation of earthquake risk.

The selection of papers for this collection, as well as additional editing for the English translation, was made in cooperation with Professor L. Knopoff at UCLA during our joint work on the International Upper Mantle Project.

I am extremely grateful to Professor Edward A. Flinn who has made a brilliant scientific edition of the translation.

I am also very grateful to Paul Kiefer, Kent Cummings, and Stephen B. Dresner of Plenum Publishing Corporation for their cooperation and patience.

V. I. Keilis-Borok

CONTENTS

INTRODUCTION SEISMOLOGY AND LOGIC†

V. I. Keilis-Borok

1. Introduction

1.1. About 1300 seismic stations are in operation in the world. Most of the 250,000 earthquakes they record annually are very weak and are detected only by a few nearby stations, but a few tens of strong earthquakes are recorded each year by the greater part of the world's stations. On the average, about a million three-component records – three million curves – are obtained each year. Along with physical and geological information, these comprise the initial data for working out the principal problems of seismology: determination of the earth's density and velocity structure, the study of seismically dangerous zones, and detection of underground nuclear explosions.

Seismology is such an active science that every year produces three or four refined earth profiles (curves of physical parameters versus depth), several maps with refinements of boundaries of seismically dangerous zones, and one or two refinements of criteria for identification of underground explosions. Thus millions of curves are reduced to tens. In the process of this reduction a number of intuitive judgments must be made: first, when the existing rules do not permit a single unique decision (for example, in detecting a signal in a background of noise); when a limited sample must be selected for analysis out of a great mass of initial data.

1.2. Ten or fifteen years ago we were to a certain extent able to cope with the processing of initial data using intuitive methods of sampling. But the situation has changed since then: a five- to tenfold increase in accuracy of the analysis of initial data has been demanded, and the volume of the data has increased correspondingly by about two orders of magnitude. It has become insufficient to process only a small intuitively selected set of data. Thus the problem has arisen of how to process much more data. Large-scale use of digital computers is the evident solution.

1.3. Work in this direction has led to a situation that probably everyone concerned with the automation of analysis of a vast flow of data has encountered in one way or another: namely, that the main problem lies not in the direct computerization (developing algorithms, programming, etc.) but in the logical ordering of the data analysis – selecting a small number of basic standard operations, and reducing the diversity of specific problems to various combinations of these operations.‡

This is possible in principle because the solutions of most of the diverse problems of seismology are based on the same comparatively few physical phenomena – mainly the laws of seismic wave propagation. Methods for solving these problems are at present highly specialized, but this is due primarily to the fact that in the precomputer period special simplified procedures had to be developed for each problem, and these have often ballooned into whole branches of seismology. With computers this specialization is no longer justified. Thus the logical ordering of our analysis comes down to

---

† Translated and adapted from Vychislitel'naya Seismologiya, No. 4, Moscow, Nauka (1968), pp. 317-350. Revised version of a report given at a symposium on the results of the International Geophysical Year organized in honor of the 100th anniversary of the National Academy of Sciences of the United States [36].

‡ This does not imply conformity: for example, a similar reform in writing, the replacement of thousands of hieroglyphs by some twenty-six letters, greatly increased the possibilities of really individual thought.

reestablishing a direct relationship between the initial physical phenomena and the final seismological conclusions; in other words, to step away from the mere apparatus of seismology back to its physical roots. With logical selection of the standard operations, therefore, the results of each operation will have an independent physical meaning, so that the physical content of the analysis will be free of the artificial complications introduced by the particular technique selected for interpretation of the data.

A strictly logical ordering of analysis could greatly facilitate our use of computers (the computerization of seismology would otherwise hardly be possible), and sometimes it might even make it possible for us to get by without computers at all. Without such systematic ordering any use of computers is completely impractical – there would result mass production of unwieldy and difficult-to-compare individual conclusions, which would only bring chaos to the general state of the problem.

Thus, that which at first glance would appear to be the auxiliary problem merely of introducing computers is seen to be in reality the problem of rearranging all of seismology. This conclusion may seem at first to be an exaggerated one, so we shall therefore discuss the need for this kind of rearrangement more concretely.

1.4. A first example is the relationship between the determination of the profile of the upper mantle, and the science of seismology as a whole.

The upper mantle is the part of the earth that begins at the bottom of the crust and extends to a depth of approximately 1000 km. Until recently, about six competing versions of the velocity profile of the upper mantle were seriously considered. Naturally the problem arose of characterizing the entire set of possible profiles that satisfy the available observations: this problem can be reduced mathematically to finding the minimum of some function in a multidimensional space (see below, Section **3.5**).

In Yanovskaya's study of Lehman's travel-time curve for the P-wave in Europe (i.e., the travel time of this wave as a function of distance to the source) [28], no less than 115 profiles (out of 2000 randomly selected) were found for which the standard deviations and maximum deviations of the observed travel times and the calculated one were small – less than 1 sec and 1.5 sec, respectively. A similar ambiguity has arisen each time this problem has been studied since then [3, 37].

Nevertheless this result does contain some unambiguous information: the profiles found lie within a limited band of possible velocities. The remaining ambiguity, however, is very substantial – it is even unclear whether or not a low-velocity zone exists.

Additional data can be used to reduce this ambiguity: the amplitudes of body waves, especially in the shadow zone, triplications of travel-time curves, surface wave dispersion, more accurate corrections for regional travel-time anomalies, etc. As a rule, however, new methods must be developed to measure these data with adequate accuracy.

To use the body wave amplitudes, for example, we must construct detailed amplitude–distance curves individually for various wave periods; to use surface-wave dispersion we must learn to separate the higher modes [13], and to refine the travel-time curves we must determine more accurately the hypocenter locations, detect triplications of the travel-time curves, and determine local corrections for each station.

Thus the levels of almost all branches of seismology must be raised for the sake of the upper mantle alone; and this is not an unfortunate trait only of the upper mantle, for the same thing can be said for any serious problem of seismology. For example, all branches of seismology were represented in a list of the research projects necessary to improve techniques of detection and identification of underground nuclear explosions [30]. And everything in this list is also necessary – in a somewhat different manner – for prediction of earthquakes, studying the structure of the earth, etc.

1.5. This situation has serious consequences. In practice it is impossible to develop one area and at the same time raise the levels of understanding of all related areas. Indeed, it is possible only to make lightning raids on them – in each case to bypass the principal difficulties and devise methods which are narrowly specialized and often rule of thumb approaches. It happens as a consequence that the very same topic in different areas is studied by different methods (for example, observations of long-period surface waves in studying the earth's structure, and in studying source mechanism).

What, as a result, has thus become of seismology as a whole? During the classical period the different branches were naturally confined to specific objects of study: crust, mantle, core, geography of earthquake occurrence, etc.; these branches were more or less clearly separated. Then they began to expand and overlap; however, they did not merge, because each area had already become overgrown with a thick encrustation of its own apparatus and its own simplified ideas about neighboring areas. This led to a fragmentation of seismology along the very same internal boundaries.

Thus, we now have a seismology of seismicity, two seismologies of the structure of the upper mantle (based on body and surface-wave observations), two seismologies of magnitudes, and so on. Each of these is itself split up, and seismology is now to a great extent occupied with clarifying the relationships between its own branches, (for example, the matching of different magnitude measurements with one another and with earthquake energies, the reconciliation of structural models derived from different samples of data, etc.). With slight exaggeration it might be said that seismologists are little by little settling onto different planets; the inhabitants of each planet consider theirs to be the center of the entire system, and they justify the consequent discrepancies by a host of complicated explanations resembling the epicycles of Ptolemaic astronomy.

This fragmentation could in time transform seismology into a merely descriptive science. It would obviously be wrong to reinforce this disconnection by creating specialized algorithms and programs for each of the historically separated branches of seismology: as we have already indicated, this would lead only to proliferation of incompatible results.

The following simple conclusion suggests itself: since each of these seismologies revolves around one and the same center – consistent analysis of the same observations – we could define this very analysis as the center of the entire system. In other words, it seems advisable to introduce a research structure that corresponds not to the individual topics, since these are closely interrelated, but to the natural sequence of analysis of seismic observations. This sequence is described in the following section.

2. Analysis of Seismological Data Flow

The analysis of seismological data is divided into four stages (Fig. 1).

I. Analysis of Records from a Single Station

I.1. Detection of signal in noise background (microseisms or other signals).

I.2. Measurement of parameters of individual signals or groups of signals: arrival time, maximum amplitude, apparent period, polarization, duration of oscillation.

I.3. Preliminary identification of signals – identification of each detected signal as one or another known wave type.

The initial operation is **I.1.** It is essentially intuitive. Operations **I.2** seem formalized, but actually many of them are done by eye. They are almost never done all at once: we return to the same record many times to study different problems. Signal identification (**I.3**) is performed on the basis of the parameters measured in operations I.2. Criteria for identification of various waves are used here in fixed form; from time to time they are refined in the later stages III and IV.

The intuitive recognition of waves also takes part in signal identification. The significance of its role is very great because up to now only a few cri-

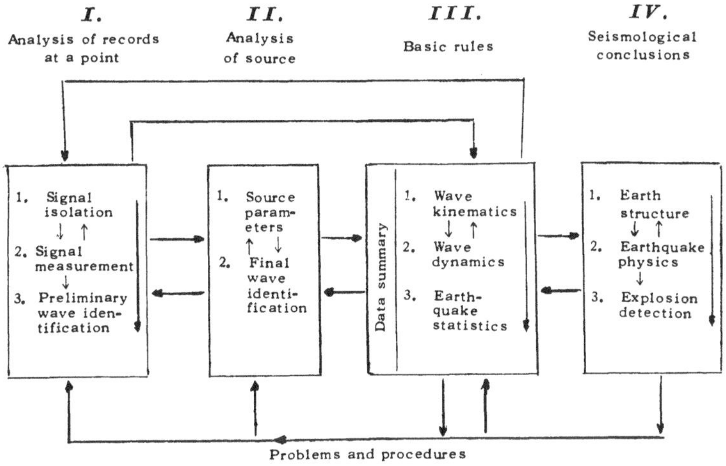

Fig. 1. Basic stages of interpretation of seismic observations.

teria of signal identification – based on arrival time – have been quantitatively formulated.

With the introduction of magnetic tape recording, high-speed computers, and large arrays, spectrum estimates and phase velocity have also begun to be determined in stage I.

As we have already indicated, at the output of this stage we have on the order of a million records a year. From one to twenty individual signals can be isolated on each record. A minimum of two numbers (the arrival time and an estimate of its amplitude on one of the components) are measured on each record, which gives a total output of about ten million numbers a year.

These data are probabilistic in nature, especially since almost all of them are determined visually (and to a considerable extent intuitively). It is obvious that all conclusions based on these data will also be probabilistic: in principal we have the right to seek not a unique solution of a problem, but only its distribution function within a set of possible solutions. The neglect – often forced – of this simple consideration is partially compensated for by averaging together a great quantity of data, but it still has far-reaching consequences, which compelled Jeffreys as early as in the 1930's to write, "It is astonishing that experimenters will spend months in making a series of observations, and grudge a day or so to present the results in an intelligible form" [9]. With today's volume and complexity of data processing, the statistically meaningful representation of results is much more time-consuming, but at the same time much more necessary.

II. Analysis of Source

II.1. Determination of source parameters: epicenter, focal depth, mechanism, energy or magnitude, and sometimes a rough estimate of source dimensions.

II.2. Refined identification of signals.

The source parameters **II.1** are determined by joint analysis of the data of **I.2** from many stations, using the averaged relations obtained in stage III – primarily travel-time and amplitude curves. At the same time, residuals – the deviations of each individual measurement from these relations – are determined for some waves. Residuals are a result of measurement errors and inexact knowledge of the relations.

Preliminary signal identification **I.3** is not necessary here, in principle. But it greatly simplifies the problem, since it makes it unnecessary to sort out a great number of variants. Without computers, therefore, operation **I.3** is practically unavoidable, although some possible variants can be missed.

The source parameters give effective additional criteria for signal identification. Therefore, final identification **II.2** follows **II.1**. This creates strong feedback: if errors are revealed in preliminary identification (**I.3**), we must reexamine **II.1**, and sometimes even return to stage **I**, to find new phases or make additional measurements.

The characteristics of 15,000 to 20,000 seismic events are determined each year, so that the output of this stage is 100,000 to 150,000 numbers. The operations in stages I and II are done continuously as observations arrive. In the next two stages results are generated by sporadically separate impulses, as a sufficient amount of data is accumulated.

III. Constructing Empirical Relations

The following basic relations are studied:

III.1. Seismic-wave kinematics: travel time as a function of epicentral distance and focal depth (travel-time curves); surface-wave velocity as a function of period (dispersion curves); the periods of the earth's free oscillations.

III.2. Seismic-wave dynamics: wave amplitude or (less frequently) spectrum as a function of distance and depth of source (amplitude curves); attenuation of the earth's free oscillations.

III.3. Statistical relations in seismicity: the distribution of hypocenters with various energies.

These operations are done more or less in parallel by averaging the data obtained in stages II and I. The common preliminary operation is to form a summary of data. Unfortunately at the present time only summaries of travel times and data on foci are being systematically compiled. The other summaries are compiled selectively, according to the specific problem. The annual output is several thousands of numbers (allowing for the graphs and maps), and one or two thousand words.

Some of the relations obtained in stage III are used in stages I and II. These are chiefly travel-time and amplitude curves. Their refinement mainly involves taking into account local or regional corrections. The corrections are determined by statistical analysis of the above-mentioned residuals. When the next refinement of these relations appears at the output of stage III, improvements of methods are fed back to stages I and II. Results obtained in the intervals between successive refinements are naturally not equivalent.

IV. Seismological Conclusions

IV.1. The earth's structure: the distribution of seismic-wave velocities, density, and absorption coefficient.

IV.2. Earthquake physics: seismic zoning, investigation of the conditions for and causes of earthquakes; estimation of seismic risk.

IV.3. Source identification.

The annual output of stage IV is up to thousands of numbers and the same amount of words. In this stage, assignments are developed for the preceding stages and data are exchanged with other areas of geophysics and with geology. This exchange applies to the structure of all of geophysics and will not be considered here.

At present, the solutions of problems of stage IV are often nonunique, due to inaccurate and incomplete data [3, 28, 45]. Different authors can arrive at different solutions, simply because they encounter them at random. This is especially likely when they use different data samplings. This is evidently the cause of the many differences of opinion about the structure of the upper mantle. At the same time, the geophysical data grow more numerous and more diverse, and the time approaches when it can be said, with only slight exaggeration, that a data sample can be found to support any given hypothesis.

On the other hand, the development of theory widens so much the possibilities for explaining facts that with the proper training any given data sample could be made consistent with a sufficiently complicated hypothesis.

In particular, a danger of circular reasoning is created. For example, in the classical period the concept of a two-layer crust was based on the identification of P* as a head wave. This identification was based solely on the assumption of a two-layer crust, which was eventually shaken only by the intuitive uneasiness of Gutenberg, confirmed later by Tatel and Tuve. Of course, such cycles always appear to be especially consistent.

So far, this nonuniqueness has been resolved by change and intuition. In fact, it is obvious that the entire set of solutions must be investigated in each problem. This was best formulated by J. W. Tukey: "Far better an approximate answer to the right question, which is often vague, than an exact answer to the wrong question, which always can be made precise" [46]. And to narrow this set, all available data must be analyzed jointly. The volume of such data is too great, and the data can hardly be processed especially for a particular problem. However, most of the data analysis in stages I-III can be made uniform for most if not all problems, and such analysis can be completely computerized. This requires a reasonable complex of programs. Some have been created in recent years; in particular, in this series of publications, we can mention descriptions of programs for automatic signal detection [21, 22, 29] and measurement [15] (stage I), for studying the coordinates and mechanism of foci and phase identification [5, 26, 40, 43] (stage II), for determination of seismic profiles [4], for estimation of seismic risk, and for classification of complex objects [24, 25, 46] (stage IV).

3. Some Possibilities for Seismology as a Whole

3.1. Let us summarize the preceding exposition.

a. The initial data required for solving today's problems of seismology are becoming so numerous and varied, and their analysis must be so complicated and accurate, that the portion of data that can be analyzed is actually growing smaller. As a result, only more and more highly specialized problems can be solved, and this at the cost of such a differentiation of methods that the results are incompatible; thus the solution of basic problems which are of complex nature, becomes more and more difficult. The way out lies in using computational mathematics and computers themselves to unite the different branches of seismic data analysis on the basis of their common physical sources (Section 1).

b. Such sources are the empirical relations in seismic-wave propagation and in the seismicity of the earth. Analysis of data in the different areas of seismology can be reduced to refinement of these relations and can therefore be done by the unified scheme described in Section 2. As can be seen from Fig. 1, the data flow branches only in the last (fourth) stage where these relations are used differently in different problems.

c. This scheme implicitly underlies seismic research. But the same relations are refined differently in connection with different problems, which leads to the difficulties described in Section 1.

d. An analysis ordered in accordance with Section 2 would consist of a reasonable number of operations, which would allow its computerization.

It is possible to formalize, and consequently computerize, the greater part of the operations, including those that have been done intuitively or even have been treated as a subject of scientific research.

3.2. The manner in which seismological research could be reorganized is clear from 3.1: there must be created a system for standard routine processing of observations by the unified scheme described in Section 2 that is common for all areas of seismology. This system will encompass the first three stages of analysis, up to and including the refinement of empirical relations. Then the emphasis of research could be shifted to stage IV – the direct investigation of basic problems.

This system could perform the following principal functions:

Stage I: determination of kinematic parameters: the arrival times of all types of waves, and (for magnetic recording) the arrival times of the various periods of surface waves, as well.

Determination of dynamic parameters: visual characteristics of wave groups (maximum amplitude, apparent period, duration) for film and paper recording and the spectrum, energy, and polarization of individual waves and wave groups for magnetic recording. With network or array observations, the apparent wave velocities and surface-wave dispersion must also be determined.

I do not think that the measurements in this stage should be automated completely, but they should be completely formalized and should include an error estimate.

Stage II: Computer determination of hypocenters with estimation of confidence regions; energy determination for all waves for which the appropriate scales exist, automatic identification of arrivals, with determination of residuals.

Stage III: periodic refinement of the basic relations by statistical processing of the data accumulated in stages I and II: in wave kinematics (travel-time and dispersion curves), and in seismicity (maps of hypocenters, the seismic activity, seismic risk, etc.).

To facilitate feedback coupling, i.e., the refinement of the specifications for a system on the basis of the data produced by it, the routine processing must include some essentially standard operations of stage IV: determination of a set of earth structures that correspond to the refined relations in wave propagation, and statistical analysis of the relationship between the various characteristics of seismicity, tectonics, and other geophysical fields.

The data produced by this system will be the standard elements of which the solution of all seismological problems is made up. Take, for example, amplitude spectra. If we compare them at different stations for the same epicentral distance, we can define more accurately the dependence of the spectrum upon local conditions – the principle of microseismic zoning. If the same spectra are grouped so that their dependence upon distance is brought into prominence, the procedure for determining source energy can be improved. This same dependence, along with the dispersion of the higher modes of surface waves [16], underlies the study of waveguides. The dependence of the high-frequency part of the spectrum on source depth makes it possible in principle to estimate focal depth within the crust [1, 41]. The relationship between the spectra of different waves is one of the main criteria of explosion identification.

3.3. Of course, some properties of seismic waves and seismicity have always provided a basis for solving seismological problems. In the solution of any problem with which seismology can cope at the present time, some of the properties listed in **3.2** are always determined, and the individual fragments of the same basic relations are constructed from them. But the system proposed here could give quite new results. In the first place, as soon as the properties listed in **3.2** begin to be systematically defined in a standardized and comparable form, today's problems in seismology could be solved much more simply and at a higher level.† Secondly, and this is the main point, it will be possible to join the separate fragments of the above-mentioned relations and obtain more general and accurate relations – a basis for solving more general and complex problems.

3.5. The proposed system could eventually develop the following additions to the relations known today:

(1) For each empirical value, it could give not only its averages but also other statistical characteristics: the distribution function or at least the confidence intervals.

(2) For seismic-wave kinematics and dynamics, it could estimate the effect of the horizontal inhomogeneity of the earth. For example, travel-time curves would be defined not as $t(\Delta, h)$ but as $t(h, \lambda_0, \varphi_0, \lambda, \varphi)$, where t is travel time, h is focal depth, and λ_0, φ_0 and λ, φ are the coordinates of the focus and of the station, with proper discretization.

(3) For seismic-wave kinematics:

(a) the possible amplitude of each wave would be taken into account in the summary of travel-time curves: the travel-time curves of sufficiently large secondary waves (reflected, converted, etc.) would be added, and the travel-time curves of those phases which are clearly impossible to observe would be eliminated [16];

† Every seismologist would easily agree with this, imagining that he systematically receives a bulletin containing these data.

(b) dispersion curves would be plotted for the first five or six surface-wave modes (only one or two are systematically studied today).

(4) For seismic-wave dynamics:

(a) the visual characteristics of each wave group would be plotted against epicentral distance, focal depth, earthquake energy, and, in accordance with (2), the coordinates of the epicenter and the station. Now, besides the early classical summary of Gutenberg and Richter [34], we have chiefly only amplitude curves or (what is the same) magnitude scales, and these only for two or three wave groups;

(b) in addition to the usual amplitude curves, the amplitude spectrum of each wave or group of superposed waves could be plotted as a function of the same factors.

(5) In the analysis of seismicity, it has so far been difficult to indicate parameters that would describe it in a sufficiently complete and compact way. It seems advisable not to terminate the routine processing contemplated in 4.2 with the compilation of earthquake catalogs but here, too, to make maps of the density of hypocenters of various energy ranges over consecutive time intervals, and also maps of the seismic risk. It is still unclear which more compact relations could then be formulated.

3.6. The general goal is to replace the fundamental classical relations (Jeffreys–Bullen travel-time curves and the Gutenberg–Richter summary of seismicity [8]) with new relations that are more complete and accurate. This is necessary for the solution of the basic problems of modern seismology and is therefore today the main problem of processing seismic observations.

This conclusion is dictated by the experience of the past. Indeed, the scheme described in Section 2 follows from the very nature of seismology and was implicitly outlined in the preclassical period at the beginning of the 1920's. Then too, in solving each problem it was necessary to go through the scheme completely: begin with the collecting of observations, then determine the epicenters, etc. Afterwards, the above-mentioned classical relations were obtained. The foundation of classical seismology was thereby created, defining for many years ahead a general ordered basis for developing its problems: clarification of the basic outlines of the earth's structure, rough energy classification of earthquakes, and the geography of large earthquakes. The processing of primary data became standard and the emphasis of research was shifted from it to direct solution of these problems. Many particular problems – for example, determination of hypocenters and, later, magnitudes – ceased completely to be topics for scientific work and were given over to routine survey operations.

Then new problems gradually arose: refinement of the structure of the Earth in accordance with the needs of geology, seismic zoning and earthquake prediction, and, somewhat later, detection of nuclear explosions. The accuracy of the classical relations became more and more inadequate. Regional, spectral, and other types of corrections† were needed, and seismology returned to a highly complicated analogy of the preclassical situation: the solution of each problem begins with a special collecting and processing of primary data. The natural way out of this is also analogous: to determine new, more complete, and accurate relations.

3.7. History does not repeat itself entirely, however. The difference now is that we can formalize and automate a considerable part of the entire analysis, i.e., a considerable portion of the entire area of science, down to the final conclusions. Such deep penetration of formalized operations, along with their diversity, complex interrelationship, and interweaving with intuitive operations, changes the appearance of seismology. After a long interval, it is again in direct contact with the most advanced areas of mathematics. Perhaps we shall see the return of that golden age when seismology enriched mathematics with fundamental problems and was in turn ennobled by it.

LITERATURE CITED

1. Andrianova, Z. S., V. I. Keilis-Borok, A. L. Levshin, and M. G. Neigauz (1965), Surface Love Waves, Moscow, Nauka. [English translation: Seismic Love Waves, Consultants Bureau, New York (1967).]
2. Azbel', I. Ya., V. I. Keilis-Borok, and T. B. Yanovskaya (1966), "Method for joint interpretation of travel-time and amplitude curves in upper mantle studies," in: Computational Seismology, No. 2, pp. 3-45.
3. Valyus, V. P., A. L. Levshin, and T. M. Sabitova (1966), "Joint interpretation of body and surface waves for one region of Central Asia," in: Computational Seismology, No. 2, Moscow, Nauka, pp. 95-103.
4. Valyus, V. P., "Determining seismic profiles from a set of observations," this volume, p. 114.

† Almost all of today's observation processing comes down to finding corrections.

5. Keilis-Borok, V. I., S. S. Mebel', I. I. Pyatetskii-Shapiro, L. Yu. Vartanova, and T. S. Zhelankina, "Computer determination of earthquake focal depth," this volume, p. 16.
6. Vil'kovich, E. V., A. L. Levshin, and M. G. Neigauz (1966), "Love waves in a vertically inhomogeneous medium (allowance for sphericity, parameter variations, and absorption)," in: Computational Seismology, No. 2, Moscow, Nauka, pp. 130-149.
7. Gotsadze, O. D., V. I. Keilis-Borok, I. V. Kirillova, S. D. Kogan, T. I. Kukhtikova, L. N. Malinovskaya, and A. A. Sorskii (1957), "Investigation of earthquake mechanism," Geofiz. Inst. Akad. Nauk SSSR, No. 40, p. 166.
8. Gutenberg, B., and C. F. Richter (1954), Seismicity of the Earth and Related Phenomena, Princeton, Princeton University Press.
9. Jeffreys, H. (1952), The Earth: Its Origin, History, and Physical Constitution, Cambridge, Cambridge University Press.
10. Keilis-Borok, V. I., I. L. Nersesov, and A. M. Yaglom (1962), Estimating the Economic Effect of Earthquake-proof Construction, Moscow, Izd. Akad. Nauk SSSR.
11. Buné, V. I., M. V. Gzovskii, and K. K. Zapolskii (1960), "Methods for detailed study of seismicity," Inst. Fiziki Zemlii Akad. Nauk SSSR, No. 9, Chap. 6, p. 176.
12. Keilis-Borok, V. I., 1960, "The level of seismic energy and the level of the predominant frequency of earthquakes," Inst. Seismostoikogo stroitel'stva i seismologii Akad. Nauk Tadzh. SSR, 6:33-39.
13. Keilis-Borok, V. I., 1967, "On using computers to study the structure and development of the earth," Izv. Akad. Nauk SSSR, Ser. Geol., No. 1, pp. 3-11.
14. Keilis-Borok, V. I., and A. S. Monin (1959), "Magnetoelastic waves and the boundary of the earth's core," Izv. Akad. Nauk SSSR, Ser. Geofiz., No. 11, pp. 1529-1541.
15. Keilis-Borok, V. I., and L. N. Malinovskaya (1966), "On one relationship in the development of strong earthquakes," in: Memorial Collection for Academician G. A. Gamburtsev, Moscow, Nauka, pp. 88-98.
16. Kilinchuk, L. M., and T. B. Yanovskaya (1968), "The amplitude of fundamental seismic waves. I. Waves propagated in the earth's mantle," in: Computational Seismology, No. 4, pp. 226-251, Moscow, Nauka.
17. Kolesnikov, Yu. A., L. N. Malinovskaya, and Yu. V. Sladkov (1968), "Spectral calibration of seismic stations," in: Computational Seismology, No. 4, Moscow, Nauka, pp. 287-316.
18. Levshin, A. L. (1964), "Love waves and the low-velocity zone in the upper mantle," Izv. Akad. Nauk SSSR, Ser. Geofiz., No. 11, pp. 1595-1607.
19. Lyubimova, E. A. (1966), Research into the Earth's Thermics, author's abstract of doctoral dissertation, Institute of Physics of the Earth, Moscow.
20. Magnitskii, V. A. (1965), Internal Structure and Physics of the Earth, Moscow, Nedra. [English translation: NASA Technical Translation TTF-395, Washington, D.C., 1967.]
21. Naimark, B. M. (1966), "Algorithm for detecting a seismic signal against a background of microseisms," in: Computational Seismology, No. 1, Moscow, Nauka, pp. 5-9.
22. Naimark, B. M. (1967), "Collection of 'correlator' programs for processing seismic observations," in: Computational Seismology, No. 3, Moscow, Nauka, pp. 118-244.
23. Neigauz, M. G., and G. V. Shkadinskaya, "Method for calculating surface Rayleigh waves in a vertically inhomogeneous halfspace," this volume, p. 88.
24. Pisarenko, V. F., and T. G. Rautian (1966), "Statistical classification by certain criteria," in: Computational Seismology, No. 2, Moscow, Nauka, pp. 150-182.
25. Pisarenko, V. F., and T. G. Rautian, "Effect of station and source factors on accuracy of seismic parameter determination," this volume, p. 189.
26. Keilis-Borok, V. I., L. G. Pavlova, I. I. Pyatetskii-Shapiro, P. T. Reznyakovskii, and T. S. Zhelankina, "Computer determination of earthquake epicenters," this volume, p. 16.
27. Richter, C. F. (1958), Elementary Seismology, San Francisco, W. H. Freeman and Co.
28. Yanovskaya, T. B. (1963), "Determination of the velocity distribution in the upper mantle as an inverse mathematical problem," Izv. Akad. Nauk SSSR, Ser. Geofiz., No. 8, pp. 1171-1178.
29. Yanovskaya, T. B. (1966), "Program for calculation of travel-time and amplitude curves of body waves in a layered medium," in: Problems of Quantitative Study of Seismic Wave Dynamics, 8:87-92.
30. Berkner, L. V. (ed.) (1959), The Need for Fundamental Research in Seismology: Report of the Panel of Seismic Improvement, Washington, D.C., U.S. Department of State.

31. Bolt, B. A. (1960), "The revision of earthquake epicenters, focal depths, and origin times using a high-speed computer," Geophys. J. Roy. Astr. Soc., 3:433.
32. Brune, J. N., J. E. Nafe, and J. Oliver (1960), "A simplified method for the analysis and synthesis of dispersed wave trains," J. Geophys. Res., 65:287-304.
33. Ewing, M., W. S. Jardetzky, and F. Press (1957), Elastic Waves in Layered Media, New York, McGraw-Hill Book Co.
34. Gutenberg, B., and C. F. Richter, On Seismic Waves: Gerlands Beitr. z. Geophysik, 43:56 (1934); 45:280-360 (1935); 47:73-131 (1936); 54:94-136 (1939).
35. Reference omitted.
36. Keilis-Borok, V. I. (1964), Seismology and Logic, in: (H. Odishaw, ed.), Research in Geophysics, Vol. 2, Cambridge, MIT Press.
37. Yanovskaya, T. B., and I. Ya. Azbel' (1964), "The determination of velocities in the upper mantle from the observations of P waves," Geophys. J. Roy. Astr. Soc., 8:313.
38. Keilis-Borok, V. I., and T. B. Yanovskaya (1967), "Inverse problems of seismology (structural review)," Geophys. J. Roy. Astr. Soc., 13:223-234 (see also this volume, p. 109).
39. Keilis-Borok, V. I., and L. N. Malinovskaya (1964), "One regularity in the occurrence of strong earthquakes," J. Geophys. Res., 69:3019-3024.
40. Knopoff, L., 1960, "Analytical calculation of the fault plane problem," Publ. Dominion Obs. Ottawa, 24:309-315.
41. Levshin, A. L., M. G. Neigauz, and T. M. Sabitova (1965), "Spectra of Love waves and the depth of the normal source," Geophys. J. Roy. Astr. Soc., 9:253-260.
42. Lehmann, I., 1955, "The times of P and S in northeastern America," Ann. di Geofisica, 8:351-370.
43. Moltshan, G. M., V. F. Pisarenko, and N. A. Smirnova (1964), "Some statistical methods of detection of signal in the presence of noise," Geophys. J. Roy. Astr. Soc., 8:319.
44. Proc. IEEE, Special Issue on Nuclear Test Detection: 53(12):1965.
45. Bullard, E., and W. Penney (conveners), "A discussion of recent advances in the technique of seismic recording and analysis," Proc. Roy. Soc. Lond., 290A:287-476 (1966).
46. Tukey, J. W. (1962), "The future of data analysis," Ann. Math. Stat., 33:1-67.

PART I

AUTOMATION OF SEISMOLOGICAL SURVEYS

COMPUTER DETERMINATION OF EARTHQUAKE EPICENTERS†

V. I. Keilis-Borok, L. G. Pavlova, I. I. Pyatetskii-Shapiro, P. T. Reznyakovskii, and T. S. Zhelankina

O. Yu. Shmidt Institute of Physics of the Earth, Academy of Sciences, USSR
(Presented by Academician E. K. Fedorov, December 26, 1962)

1. This paper describes a method for determining epicenters on a general-purpose digital computer. The method is applicable when the initial approximation is unknown and some of the observations contain large errors (as is the case in seismological surveying). We judge from rather extensive experience with our program that when the necessary minimum quantity of data is available, failures (cases in which no solution is found) are very rare, and significant errors are virtually eliminated.

A method for the solution of analogous problems was developed in [1] and, later, by a number of other workers [5, 6].

The initial data used in this method are the arrival times of P waves at epicentral distances of up to 105° and the arrival time of PKP_1 waves at distances greater than 110°. Cases of erroneous identification are discarded in the course of computations. Arrivals in the shadow zone from 105° to 110° are considered unreliable and are not used. The travel-time curve is assumed to be the same for the entire earth. The only allowance for the earth's departure from radial symmetry is that calculations are made in geocentric coordinates.

2. Formulation of the problem: the arrival times t_k of P or PKP_1 waves at m_0 seismic stations, are given (k is the number of the station) and the coordinates of these stations, λ_k (longitude) and φ_k (latitude), are known. The travel-time table $g = g(\Delta, h)$ is known, which determines the travel time g as a function of epicentral distance Δ and focal depth h. We shall find the time of occurrence of the earthquake t and the coordinates of the hypocenter λ (longitude), φ (latitude), and h.

We find (t, h, λ, φ) from the condition that the mean-square residual is sufficiently small

$$\bar{\psi} = \left\{ \frac{1}{m'} \sum_{i=1}^{m'} f^2 k_i \right\}^{1/2} \tag{1}$$

where $f_k = t + \psi_k$; $\psi_k = g(\Delta_k, h) - t_k$; and Δ_k is the epicentral distance for the k-th station. Stations for which the $|f_k|$ are overly large, as well as stations in the shadow zone, are excluded from the sum in (1).

The solution involves the following steps: first, we find the minimum of the function $\bar{\psi}$ as determined using all stations except those in the shadow zone. The method for finding the minimum for a particular set of stations k is described below. All f_{k_i} and Δ_{k_i} are calculated at the minimum point. Then we find the minimum of the new function $\bar{\psi}$. As before, it is determined using formula (1), but this time the sum includes only max $(m_0/C_4, C_1)$, stations with minimum $|f_{k_i}|$, and also stations

† Adapted from Doklady Akademii Nauk SSSR, 151(2):323-325 (1963).

for which

$$|f_{k_i}| < C_2 + C_3\bar{\psi} \tag{2}$$

(C_i are given constants; in routine practice $C_1 = 5$, $C_4 = 4$).

If the minimum of the new function $\bar{\psi}$ is less than the minimum of the preceding function, we calculate new f_{k_i}, again determine which stations should be excluded from the sum, and again find the minimum of $\bar{\psi}$. At this point, the stations excluded earlier may be included in subsequent calculations, provided their residuals have become sufficiently small. This process is continued as long as the successive minima of $\bar{\psi}$ decrease. We assume that a solution has been found when the minimum of $\bar{\psi}$ is less than a given number A. Otherwise, we assume that a local minimum of $\bar{\psi}$ that does not coincide with the hypocenter has been found. Then we add some large arbitrary numbers to the latitude and longitude of this minimum (114° in our usual practice) and seek the minimum again from this point, which is essentially arbitrarily determined. The number of iterations of this procedure is limited. The coordinates of the minimum thus found are used as an initial approximation for repetition of the entire process, with a decrease in the constants C_2 and C_3 (i.e., with a more rigid requirement on the residuals f_{k_i}). In practice we select one of two epicenters corresponding to two sets of C_2 and C_3 by the following rule: let $\bar{\psi}_1$ and $\bar{\psi}_2$ ($\bar{\psi}_1 < \bar{\psi}_2$) be the $\bar{\psi}$ values for these two epicenters and let m_1' and m_2' be the corresponding m' values. We take, as the solution, the epicenter corresponding to $\bar{\psi}_1$ if $\psi_2 - \psi_1$ is greater and $(m_1' - m_2')/m$ is smaller than some threshold. Otherwise, we take the epicenter with the larger m_2'. In routine practice we take first $C_2 = 1.0$, $C_3 = 1.5$, and then $C_2 = C_3 = 1.0$.

The following iteration is used to find the minimum of the function $\bar{\psi}$ for a fixed set of stations. We replace $\bar{\psi}$ near the n-th approximation by a quadratic form and use the minimum of this form as the (n+1)-th approximation.

Despite the apparently restricted applicability of this method, it allowed the coordinates of the epicenter to be determined precisely, regardless of how poor the initial approximation was. It is interesting to note that the time expended depends little upon the accuracy of the initial approximation. This is demonstrated by the results of calculation of several hundred epicenters from two initial approximations: from an arbitrary point near the equator (h = 0, $\lambda = \varphi = 0.1°$) and the vicinity of the true epicenter. The initial data were those usually used in the seismological survey of the USSR. The difference in the number of iterations for 55% of the earthquakes was less than 10 and almost always was less than 100. For ~15% of the earthquakes, this difference was even negative, due to serious errors in t_k and an unfavorable distribution of stations (in these cases, the iterations deviate from the true solution until the erroneous t_k are discarded).

As an example of the convergence of the iterations, Table 1 shows the successive iterations in the determination of the epicenter of an earthquake in California on July 21, 1952 (h = 0-18 km, λ = 110° W, φ = 35°N).

TABLE 1

Number of iteration	ψ	λ	φ	m'
Reckoned from equator				
0	431.3	0.1°E	0.1°N	64
1	360.7	171.4°E	73.2°S	64
2	311.7	75.2°W	70.4°S	56
3	260.9	78.7°W	59.2°S	66
4	281.5	157.8°W	42°S	65
5	268.4	75.9°W	10.2°S	66
6	199.1	140.9°W	10.9°S	66
7	66.5	118.4°W	26.1°S	67
8	15.6	121.8°W	32.5°N	65
9	3.5	119.5°W	34.67°N	66
10	2.8	119.2°W	34.9°N	66
16	1.16	119.02°W	34.98°N	58
17*	1.02	119.0°W	34.96°N	55
21	0.903	118.9°W	34.95°N	52
Reckoned from station closest to epicenter				
0	10.8	119.7°W	34.4°N	66
1	3.12	119.4°W	34.7°N	66
2	2.8	119.15°W	34.9°N	66
3	2.79	119.16°W	34.9°N	66
6	1.16	119.0°W	34.9°N	58
8*	1.02	119.0°W	34.9°N	55
9	0.938	118.9°W	34.9°N	53
12	0.903	118.9°W	34.9°N	52

*Decrease of C_3.

3. The accuracy of the results may be characterized as follows: on the basis of the data used by the seismological survey of the USSR, it was almost always possible to obtain $\bar{\psi} < 1.5$-2. The failures were about 5%; these can, for the most part, be attributed to inadequate data. The use of a com-

puter increased the number of routinely determined epicenters by 30-50%, because there was no longer a need for a good initial approximation or the S phase. The accuracy of determination of the true hypocenter is indirectly characterized by the distance of the epicenter, which we calculated from the data in the bulletins of the USCGS and ISS. In 55% of the cases this distance is less than 0.3°; in 92% of the cases, it is less than 1°. For the most part, the difference can be attributed to the use of different sets of stations.

Experience has shown that very distant epicenters can be determined from data for a small number of stations if the t_k values are sufficiently accurate (within reasonable limits). For example, the epicenter of the "Blanca" explosion in Nevada was determined with an error of 10 km on the basis of observations from only seven stations.

The focal depth is determined with an error of less than 50 km for 60-70% of the earthquakes. This percentage is unexpectedly high, but it is not high enough for the pP and sṖ phases to be dispensed with.

Our experience in the routine use of the described program is analyzed separately in [4].

The authors wish to express their sincere appreciation to N. V. Kondorskaya, who furnished data from the seismological survey of the USSR, and to S. Z. Mebel' and T. P. Starova, who assisted in the preparation of the data and in the analysis of the calculation results.

LITERATURE CITED

1. Bolt, B. A. (1960), "The revision of earthquake epicenters, focal depths, and origin times using a high-speed computer," Geophys. J. Roy. Astr. Soc., 3:433.
2. Gel'fand, I. M., and M. L. Tseytlin (1962), "On some methods of control of complicated systems," Uspekhi Matematicheskikh Nauk, 17:3.
3. Jeffreys, H. (1936), "On travel times in seismology," Publ. Bur. Centr. Seism. Int., 14A: 1-86.
4. Kondorskaya, N. V., T. S. Zhelankina, S. S. Mebel', and L. Yu. Vartanova (1966), "Some results of computer application in the generalization of seismological data" [in Russian], Computational Seismology, No. 1, Moscow, Nauka, pp. 31-53.

COMPUTER DETERMINATION OF EARTHQUAKE FOCAL DEPTH†

V. I. Keilis-Borok, S. S. Mebel', I. I. Pyatetskii-Shapiro, L. Yu. Vartanova, and T. S. Zhelankina

1. Introduction

This paper is a continuation of [1], which described a method of determining a distant hypocenter from the arrival times of P waves.‡

As is well known, the accuracy of seismic observations and theoretical travel-time curves is insufficient for determining focal depth from these arrival times alone (the case when observations from close stations are available is an exception). That is why the depth of a distant focus is best determined from the arrival times of pP and sP waves; their delay relative to P is more sensitive to focal depth than to epicentral distance. The arrival times of these waves can be interpreted by the same method of the first arrivals: by minimization of the mean-square residual. This was first done in [2]. The question arises of identifying pP and sP among the group of observed arrivals. Manual identification has obvious disadvantages.

This paper describes an iterative process in which the pP and sP phases are identified, the focal depth is determined, and the epicenter is refined. The problem is formulated and the main idea for its solution is presented in this section.

The problem of determining focal depth has yet to be rigorously formulated. We have attempted to formulate it as follows:

The times of the first few arrivals (up to five) at a certain number of stations, and the travel-time curves of the waves of the P group are given. It is required to find the optimal phase identification and the corresponding epicenter and focal depth.§ Optimality is defined as follows. For each fixed focal depth h, from the observed data we can find:

1) The epicenter, the origin time, the epicentral distances of each station, the number of stations $\overline{m}$ at which the residual of the first arrivals is sufficiently small, and the mean-square residual $\overline{\psi}$ at these stations. These data are determined from the first-arrival times by the program described in [1].

2) The anticipated arrival times of pP and sP waves (from the travel-time curve and epicentral distances). The number of given arrivals that are sufficiently close to the anticipated we shall denote by s (farther along, S will be defined not as the number but as the total weight of the arrivals).

We shall let n be the number of arrivals in s that can be identified as pP or sP. The number n does not coincide with s, since not all of the phases in s can be identified simultaneously (for example, several observed arrivals can be rather close to one calculated arrival). We shall let σ be the mean-square residual of these n arrivals (below, σ will be converted to the variance of the depths determined from each identified arrival). Thus, we can estab-

† Translated from Vychislitel'naya Seismologiya, No. 1, Moscow, Nauka (1966), pp. 10-30.

‡ Here and below, PKP_1, $pPKP_1$, and $sPKP_1$ waves are understood to be P, pP, and sP waves, respectively.

§ It is not possible to require only that the phases be identified and the focal depth be found, since this problem cannot always be solved uniquely: different depths may correspond to different phase identification.

lish a correspondence between each focal depth h, each set of specific h phase identifications, and the five parameters m(h), $\bar{\psi}$(h), s(h), n(h), and σ(h). An optimal phase identification is one in which m is greater and ψ and σ are smaller than specified limits and n is maximal. Here, a choice is made between two maxima of n: for h < 100 km and h > 100 km. The h value that corresponds to the second maximum is taken as a solution only when the second maximum exceeds the first by a given value (the earthquake is considered normal if this does not result in great inconsistencies). This value is defined as a function of the overall form of the seismograms, i.e., of the relative number of stations at which the recording of a given earthquake is, in the opinion of the interpreters, typical for h > 100 km. If a majority of interpreters choose h > 100 km, this value is reduced.

Such a formulation of the problem apparently corresponds to the formulation that is tacitly understood in manual processing. Of course, this problem does not always have a unique solution.

Different focal depths can correspond to approximately equivalent combinations of parameters – for example, identical or slightly different n. All possible variants of phase identification cannot be considered with manual processing when we stop at the first variant encountered that agrees with the overall form of the recording. The parameter s is apparently taken into account intuitively: it is natural to seek the phases pP, sP where the chances for their identification are greater.

With computer processing, all possible versions must be considered. The problem is solved as follows. All arrivals except P are in turn taken as pP and sP, and the corresponding focal depth is determined each time. Then the distribution of these values on the h axis is studied (for $0 \leq h \leq 790$ km). In this case, s(h) and n(h) are defined as the number of values encountered in some interval whose center is at h. Two modifications of the method are described below. In the first, the distribution of only s(h) over the entire h axis for a fixed epicenter is studied, and n(h) is considered only in two depth intervals: in that for which s(h) is maximal and in the interval $0 \leq h < 100$ km. Allowance for the parameter s compensates for possible errors due to rejecting the true pP and sP phases. In the second modification, the n(h) values for all possible h (in practice, in 12 intervals from 0 to 790 km) are considered. It would probably be better to combine both modifications: to consider all depth intervals and to take into account s(h) together with the other parameters. But sorting of the depth intervals would take too much computer time, so we use mainly the first modification.

2. Procedure

I. Initial Data

The following initial data are used:

a) The times t_{ik} of the first five arrivals on the recordings of the seismic stations (i is the arrival number and k is the station number).

b) The coordinates of the epicenter λ and φ; the focal depth $h = h_f$, which is given or found in determining this epicenter from the first arrivals t_{1k}; the origin time t; the epicentral distance Δ_k; the residuals f_k, i.e., the differences between the calculated (for given λ, φ, t, and h_f) and observed t_{1k}; and the mean-square residual $\bar{\psi}$. It is obvious that $\tau_{ik} = t_{ik} - t - t_{1k} - f_k$ is the delay of the i-th arrival relative to the calculated (corresponding to given h, λ, φ, and t) arrival of P.

These data are determined from t_{ik} by the program described in [1]. Here a slight addition is made to this program: two epicenters are determined by this program for two pairs of C_3 and C_4 values under the condition

$$|f_k| < C_3 + C_4\bar{\psi}. \quad (1)$$

Here and below, C_q are constants that enter the calculation scheme proposed in this paper. They are numbered in accordance with [3]. The numerical values of C_q are given in Table 1.

When condition (1) was satisfied, the arrival t_{1k} was used for final determination of the epicenter.

Let m_0 be the number of stations for which t_{1k} are given, m_1 and m_2 the numbers of stations not in the shadow zone whose residuals satisfy condition

TABLE 1. Numerical Values of Constants Used in Experimental Calculation

q	C_q	q	C_q	q	C_q
3	1	12	0.3	20	2
4	1.5	13	1	21	0.2
4*	1	14	0.003	22	1
5	3.5	15	0.5	23	2
7	1	16	1.5	24	0.05
8	1.5	17	0.5	25	2
9	163	18	1	26	0.2
10	2	19	0.333	27	0.01
				28	1
				29	5

*For second epicenter.

TABLE 2. Intervals for Unreliable Zones

h/R	Waves			
	pP	pPKP$_1$	sP	sPKP$_1$
0.005	22—23°; 67—80°	110—140°; 146—155°	10—25°; 65—80°	110—140°; 146—155°
0.010	22—24, 65—80	— 153—161	10—25; 60—75	110—112; 153—161
0.020	21—28; 53—72	111—116; 162—167	25—35; 50—60	115—120; 166—173
0.040	30—33; 48—60	116—121; 168—172	35—50 —	121—127; 174—179
0.060	30—37; 43—52	119—125; 174—177	35—60 —	128—132 —
0.070	30—45; 45—48	124—129; 177—180	32—38; 55—70	132—138 —
⩾0.090	36—55 —	127—134 —	25—33; 72—87	110—112; 139—145

(1) for the first and second epicenters, respectively, and let $\bar{\psi}_1$ and $\bar{\psi}_2$ be the $\bar{\psi}$ values for these epicenters. We added to the program automatic selection of one of the two epicenters by the following rule. We take the epicenter that corresponds to the smaller of $\bar{\psi}_1$ or $\bar{\psi}_2$, if

$$|\bar{\psi}_1 - \bar{\psi}_2| > C_{16} \text{ and } \frac{m_1 - m_2}{m_0} < C_{17}. \quad (2)$$

If even one of these inequalities is not satisfied, we take the epicenter with the greater m_i. When $\bar{\psi}_1 > C_5$ and $\bar{\psi}_2 > C_5$, the epicenter is assumed to be undetermined.

c) The travel-time curves $g^{(1)}(\Delta, h)$ and $g^{(2)}(\Delta, h)$ are the delays of the pP and sP waves, respectively, relative to the P wave. The values of $g^{(1)}(\Delta, h)$ and $g^{(2)}(\Delta, h)$ were taken from [4, 5].

Besides the travel-time curves, also given are $j^{(1)}(\Delta, h)$ and $j^{(2)}(\Delta, h)$, which are the numbers of phases that arrive, according to the Jeffreys–Bullen tables [5], before the pP and sP waves, respectively.

According to [5], not more than two phases arrive before pP. We shall, however, seek pP among the second, third, and fourth arrivals, since the preceding phases may have been erroneously identified.

d) Unreliable zones are given for pP, pPKP$_1$, sP, and sPKP$_1$, in which these waves can be confused with other waves (chiefly, with P$_c$P, PP, and PPP). We tried two methods for specifying the unreliable zones.

In the first method, those combinations (h, Δ) for which the difference between the arrival times of a given and any other wave (except the first arrival) is greater than 5 sec but less than 10 sec are considered unreliable. The intervals of unreliable zones as given by the first method are shown in Table 2. The line with the closest h/R value is used for intermediate values.

With the second method, those combinations (τ_{ik}, Δ) are considered unreliable for which τ_{ik} can, according to [4, 5], correspond not only to pP and sP waves but also to PP, PPP, and P$_c$P waves for any focal depth. The delay of these waves relative to P is shown in Fig. 1 as a function of Δ. The zones of intersection of the curves are unreliable.

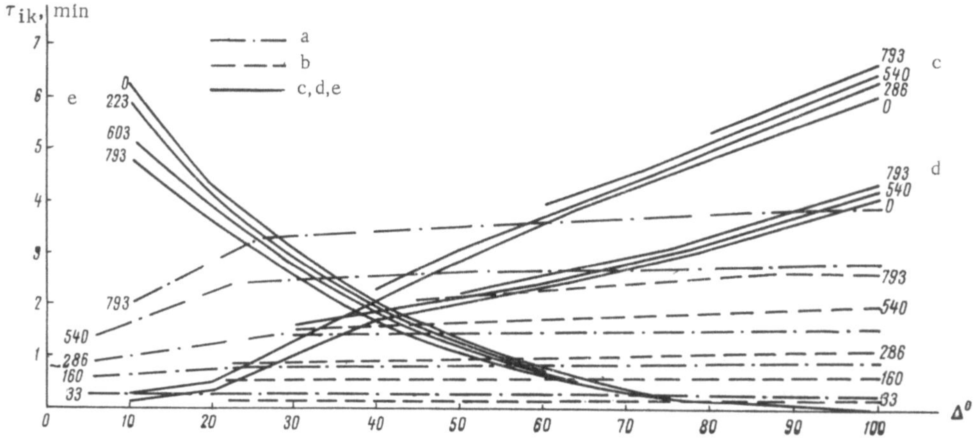

Fig. 1. Intersection of different travel-time curves of the P group: a) sP; b) pP; c) PPP–P; d) PP–P; e) P$_c$P–P. The parameters are focal depth.

The latter method is cruder than the former (it does not allow for the dependence of the unreliable zones upon h), but in return it eliminates the possibility of erroneous but internally consistent identification of an entire group of PP, PPP, or P_cP waves.

e) A preliminary estimate can be made for some stations – that the focal depth h is greater or less than 100 km. This estimate is based on the interpreter's impression of the overall pattern of the seismogram and may be purely intuitive. We assume that in the opinion of most interpreters:

$$h < 100 \text{ km, if } K_1 - K \geqslant C_{18} + C_{19}(K_1 + K); \quad (3)$$

$$h > 100 \text{ km, if } K - K_1 \geqslant C_{18} + C_{19}(K_1 + K). \quad (4)$$

where K_1 and K are the numbers of stations at which h < 100 km or h > 100 km, respectively, is estimated. If

$$|K - K_1| \leqslant C_{18} + C_{19}(K_1 + K), \quad (5)$$

it is assumed that the interpreters are not in agreement. Then only the corrective opinion of a significant majority is used.

II. First Modification of the Method

We shall seek the pP and sP waves among the second through fifth arrivals (at times t_{ik}, for $i \geq 2$), assuming that t_{1k} corresponds to P or PKP_1. If

$$-(C_7 + C_8\bar{\psi}) > f_k > -C_9, \quad (6)$$

we seek pP and sP among the first four arrivals.

A. Preliminary Selection of h Interval. The possible h values lie in the range from 0 to (H + 0.12R), where R is the radius of the mantle and H is the crustal thickness. We divide this range into $(14 - C_{10})$ overlapping intervals:

$$0 \leqslant h \leqslant 0.01R + H,$$
$$H + 0.01R\,(q - 2) \leqslant h \leqslant H + 0.01R(q - 2 + C_{10}),$$

where q = 2, 3, . . . , $(14 - C_{10})$ is the interval number.

For each t_{ik} (i = 2-5 or 1-4, depending upon whether or not condition (6) is satisfied) we determine two solutions, using the travel-time curves $g^{(1)}$ and $g^{(2)}$: $h = h_{ik}^{(1)}$ and $h = h_{ik}^{(2)}$. The former value is determined assuming that t_{ik} is the pP arrival; the latter, that it is the sP arrival. Then we consider the distribution of $h_{ik}^{(l)}$ ($l = 1, 2$) over these intervals. A correspondence is established between each interval and the sum of the statistical weights S of the depths $h_{ik}^{(l)}$ that fall into the given interval. We denoted this weight by $C_{11,i-j}^{(l)}$ (the number $(i - j - 1)$ is the difference between the actual arrival numbers and those found from the travel-time curve). The following values were selected for the calculation:

$C_{11,\,i-j}^{(l)}$	0.7	0.8	0.9	1	0.9	0.8	0.7
$i-j$	−2	−1	0	1	2	3	4

If Δ_k and $h_{ik}^{(l)}$ fall into an unreliable zone, the weight is reduced by a factor of C_{12}. If at the same station $h_{ik}^{(1)}$ and $h_{ik}^{(2)}$ are in the same h interval and $i_1 < i_2$, the weight of the smaller of these two depths is increased by a factor of C_{13}. In addition the weight of $h_{ik}^{(l)}$ in the first interval $0 \leq h \leq (H + 0.01R)$ is doubled.

Thus each arrival "votes" twice with a variable weight for each of the h intervals. We find the interval for which the sum of the weights S is maximal. Let us denote the number of this interval by q_{max}. If $q_{max} = 1$ (normal depth), we consider another interval q' for which S is greater than for all other intervals except q_{max}.

B. Phase Identification and Focal Depth Determination in Individual Intervals. Let us identify the phases for the intervals q = 1 and $q = q_{max}$. If $q_{max} = 1$ but condition (4) is satisfied, i.e., the majority of interpreters vote for q > 1, then the phases are identified for the intervals q = 1 and q = q'. The phases are identified in the q-th interval in the following way determining the mean-square value $h_m^{(q)}$ of all $h_{ik}^{(l)}$ that lie in this interval and the corresponding standard deviation $\sigma_m^{(q)}$. Then select the $h_{ik}^{(l)}$ for which

$$|h_{ik}^{(l)} - h_m^{(q)}| < C_{14} + \sigma_m^{(q)} C_{15}. \quad (7)$$

Among $h_{ik}^{(1)}$ and $h_{ik}^{(2)}$ we find the closest to h_m. Let these be $h_{i_1k}^{(1)}$ and $h_{i_2k}^{(2)}$. Then t_{i_1k} and t_{i_2k} are considered the pP and sP arrivals provided $i_1 < i_2$ (i.e., the assumed pP arrives earlier than the assumed sP). But if $i_1 \geq i_2$ we check whether the following three conditions are satisfied:

1) $|h_{i_1k}^{(1)} - h_m^{(q)}| < |h_{i_2k}^{(2)} - h_m^{(q)}|$. (8)

2) Among the $h_{ik}^{(1)}$ that satisfy (7) there are those for which

$$i < i_2. \quad (9)$$

The number of the closest of them to h_m is denoted by i_3.

TABLE 3. Phase Identification

Condition satisfied			Phase number for wave	
1	2	3	*pP*	*sP*
Yes	Yes	Yes	i_1	i_4
		No	i_3	i_2
	No	Yes	i_1	i_4
		No	i_1	—
No	No	Yes	i_1	i_4
		No	—	i_2
	Yes	Yes	i_3	i_2
		No	i_3	i_2

3) Among the $h_{ik}^{(2)}$ that satisfy (7) there are those for which

$$i > i_1. \quad (10)$$

We let i_4 be the number of the closest of them to h_m.

The phases are identified in accordance with Table 3, which shows arrival to be identified with pP and sP for every possible combination of inequalities (8)-(10). We keep the identification of at least one phase to which corresponds the depth closest to h_m. Of course, this is not the only method of identification: for example, the variance of $h_{ik}^{(l)}$ could have been minimized, the weight of the phases could have been taken into account, etc.

Thus, as a result of step B we find the sP and pP phases assuming that

$$h \leqslant H + 0.01R.$$

In addition, if $q_{max} > 1$ or condition (4) is satisfied, we have the second version under the assumption that

$$H + 0.01R\,(q' - 2) \leqslant h \leqslant H + 0.01R\,(q' - 2 + C_{10}).$$

C. Selection of the h Interval. For q = q' let us consider the inequalities

$$n_{q'} \geqslant C_{20} + C_{21}m_0; \quad (11)$$

$$n_{q'} \geqslant C_{22}m_1 + C_{23}; \quad (12)$$

$$\sigma_{cq'} < C_{24}, \quad (13)$$

and for q = 1

$$n_1 \geqslant C_{25} + C_{26}m_0, \quad (14)$$

where n_1 and $n_{q'}$ are the numbers of identified phases in the first interval and in the interval q', respectively.

The rule for selection of the depth interval is given in Table 4. Depending upon whether inequalities (3)-(5) and (11)-(14) are satisfied and upon the value of q_{max}, we take one of the two solutions:

$$\text{I} \qquad h < H + 0.01R;$$

$$\text{II} \qquad H + 0.01R\,(q' - 2) \leqslant h \leqslant H + {} + 0.01R\,(q' - 2 + C_{10}).$$

D. Matching Focal Depth to Epicenter. In steps B and C, the focal depth $h_a^{(q)}$ and the residuals $h_{ik}^{(l)} - h_a^{(q)}$ are determined simultaneously with the phase identification. In this step, a correspondence is established between $h_a^{(q)}$ and the focal depth estimated using first arrivals only.

Let us compare the found $h_a^{(q)}$ with the focal depth h_f given in determining the epicenter. If

$$|h_f - h_a^{(q)}| \leqslant C_{27}, \quad (15)$$

the problem is solved. The values obtained earlier are assumed for the coordinates of the epicenter, distances, azimuths, and residuals. When inequal-

TABLE 4. Selection of Depth Interval

q_{max}	Inequality satisfied					Interval for h
	(3)	(4)	(5)	(11)–(13)	(14)	
1	—	No	—	—	Yes	I
					No	—
	No	Yes	No	Yes	—	II
				No	Yes	I
				No	No	—
>1	Yes	No	No	Yes	—	II
				No	Yes	I
				No	No	—
	No	Yes	No	Yes	—	II
				No	Yes	I
				No	No	—
	No	No	Yes	Yes	—	II
				No	Yes	I
				No	No	—

Note. In the fifth column, "yes" means that all inequalities (11)-(13) are satisfied and "no" means that at least one of them is not. If inequality (4) is satisfied, C_{22} in condition (12) is reduced by a factor of C_{28}. If inequality (3) is satisfied, C_{22} in condition (12) is increased by a factor of C_{28}.

ity (15) is not satisfied, these are recalculated using a new fixed depth $h_f = h_a^{(q)}$, and the problem is solved from the beginning. This cycle is repeated not more than C_{29} times.

III. Second Modification of Method

Selection of the h interval is most critical as far as the uniqueness of the results is concerned. Ambiguity in the selection of pP and sP among the phases corresponding to this interval is much less important if the dimensions of the intervals are not too great.

The method of interval selection described above (step C) is not unique and has an obvious disadvantage: the result of the voting in step A may be determined not by the arrivals that are later identified as pP or sP in step B, but by arrivals that will be rejected in step B. In this case, the h interval for which the maximum number of arrivals would be identified in step B may be lost for consideration. Of course, this situation is unlikely in practice, but in order to avoid it altogether, we can seek directly the depth interval in which the greatest number of phases is identified. This can be done by the following method, which is very similar to that described in Part 2 of this section.

1) We determine the epicenter (and the entire group b of initial data) for a set of h_f values – for example, for

$$h_f = KC_{30},\ K = 0,\ 1,\ 2\ldots;\ C_{30} = 50 - 100\ \text{km}.$$

2) We find $h_{ik}^{(1)}$ and $h_{ik}^{(2)}$ for each h_f, as in step A.

3) For each h_f we perform an operation similar to step B, except that we seek the focal depth h_a in the interval $|h_a - h_f| \leq C_{30}$.

4) For each h_f we perform an operation similar to step D: we check the condition $|h_a - h_f| > C_{31}$. If it is satisfied, we repeat 1), 2), and 3) with the new h_f, which is equal to the found h_a; if not, we move to 5).

5) We compare the results of 1), 2), and 3) for various h_f and choose the optimal version.

Of course, $S(h_a)$ may be taken into account in the optimality criteria along with $n(h_a)$, $\sigma_a(h_a)$, $f_i(h_a)$, and $\bar{\psi}(h_a)$. This method was tested by printing out the n, σ_a, m, and $\bar{\psi}$ values for h_f = 33 km + K·0.01R, where K = 0, 1, 2, . . . , 12. We preferred the first modification, for the reasons discussed in Section 3.

IV. Description of Program

The described method was programmed as an addition to the program described in [1] for determining epicenters.

Besides the data obtained with the program for determining epicenters (group b), the following data are printed out:

The number of phases identified as pP or sP, the mean h, h_a, σ_r, and σ_a.

The arrival times of the phases that follow P. The phase identification is indicated in the code of these arrivals.

The residuals $h_{ik} - h_a$ for the identified phases.

The distances Δ_k. Cases are noted when condition (6) is satisfied, i.e., when pP and sP were sought among the first through fourth arrivals.

3. Experimental Calculation

The results of an experimental calculation for several tens of earthquakes are described in this section. The calculations were made with both modifications of the method, with both methods of specifying the unreliable zones, and for various values of the constants C_{12}, $C_{11,i-j}^{(l)}$. The number of earthquakes studied was approximately one order of magnitude smaller than that required to evaluate reliably the effectiveness of the method and to optimize the entire set of constants. Additional material is now being gathered by the seismological survey of the USSR.

I. Data

The initial data were taken from the seismological survey of the USSR. Data on 121 earthquakes for which there were observations by not less than eight stations in the seismological bulletin of the USSR were selected. The data on most of these earthquakes were confirmed by the USCGS or BISS bulletins. For 50 of the earthquakes there were observations for near and distant stations. The depth was estimated as follows: the epicenter and the origin time were determined from the first arrivals at the near stations, and separately at the distant stations for various h_f. The h_f value for which the closest t, λ, and φ values were obtained was chosen as the focal depth. Most of the experiments were made with 54 earthquakes whose depths were known most reliably; for 44 earthquakes, it was confirmed by matching the data of near and distant stations, and the data of various bulletins were matched for 10 earthquakes. The calculation was made without the voting of interpreters, since data from past years were used.

II. First Modification

For the 121 earthquakes, calculations were made by the first modification with the first method of specifying unreliable zones. The values in Table 1 were used for the constants. According to the bulletin data, 66 of the earthquakes were normal, 48 were intermediate, and the remaining seven were deep.

Three or more phases were identified for 95 earthquakes, five or more for 78, and 10 or more for 38. Accordingly, in 84, 68, and 36 cases, the depths determined agreed with those given in the bulletin with an accuracy of ± 20 km (more accurate agreement is unlikely, since the computer can identify more phases). The maximum number of identified phases was 27 (of 84 given at 23 stations).

Let us discuss further the constants that govern the calculation. The values of the constants in Table 1 were, in essence, guessed. For 54 of the above earthquakes, the calculation was repeated with $C^{(l)}_{11,i-j}$, equal to 0 and 1 with $i - j \neq 1$. These constants determine the weight reduction when the actual number of the arrival i does not agree with that given in the travel time curve. In addition, the calculation was repeated with $C_{12} = 0$ and $C_{12} = 1$ (this constant determines the weight reduction in the unreliable zones). The corresponding changes in the results are given in Table 5.

As can be seen, the calculation is rather stable with respect to the constants in question; there is still insufficient data to improve them.

The decrease in C_{14} and C_{15} from inequality (7) reduces σ_a, which is the variance of the depths determined from individual identified phases, but in return it also reduces the number of these phases.

Figure 2 shows the scatter of the depths for the earthquakes for which five or more phases were

TABLE 5. Effect of Changes of Constants on Accuracy of h Determination

Accuracy change	Number of cases for			
	C_{12}		$C^{(l)}_{11,\,i-j}$	
	1	0	1	0 (for $i-j=1$)
Improvement	1	3	—	1
Worsening	10	1	6	7
No change	42	49	47	45

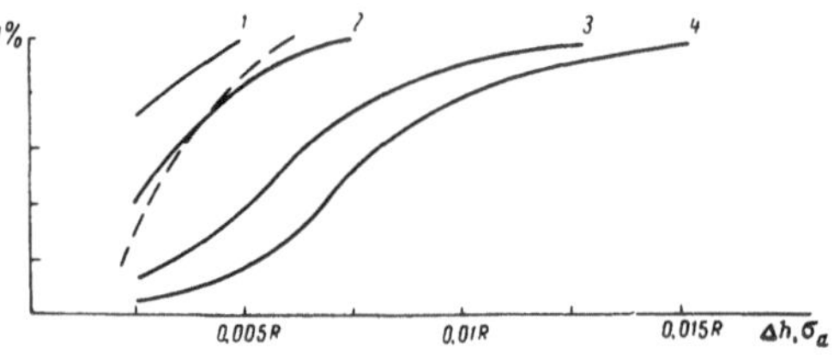

Fig. 2. Scatter of depth determined from individual phases: 1) α = 60%; 2) α = 70%; 3) α = 90%; 4) α = 100%.

identified. The dashed line indicates the distribution function for σ_a for 78 earthquakes, and the solid lines give the distribution function of the depth interval Δh for 26 earthquakes. The parameter α is the percentage of identified phases that give a focal depth in this interval. If this situation is confirmed by a larger amount of material, the accuracy of determining h_a can be substantially increased by reducing C_{14} and C_{15}.

III. Second Modification. Uniqueness of Solution

Calculations by the second modification allow us to evaluate the uniqueness of the solution. For this, we compare σ_a, $\bar{\psi}$, and n for various intervals $h_f - 0.01R \leq h_a \leq h_f + 0.01R$. The calculations were made using the first method for specifying unreliable zones and with the constants given in Table 1.

Four examples of the results are shown in Table 6. The coordinates of the epicenter and the origin time correspond to the indicated h_a. The asterisks indicate h_a that were selected by the first modification of the method. In the first example, the optimal version was selected uniquely, in the second, two similar versions are possible, and in the third and fourth examples, the selection was essentially ambiguous (importance need not be attached to an increase in $\bar{\psi}$ by 1 sec or to a decrease in n by 2).

For the third example (see Table 6), the first modification with the first method of specifying unreliable zones did not give a solution: cycling occurred in step D.

Now let us compare the results obtained by both modifications. The maximum n in five of 37 cases did not fit into the intervals where the h selected by the first modification lay. In three of these cases, the maxima of n were rather distinct ($\Delta h \leq 33$ km for $n \geq \bar{n} - 2$), but they fit into other depth intervals, unlike the maximum of s. It is still unclear whether an interval that corresponds to only one of the maxima (n or S) can be considered reliable.

TABLE 6. Comparison of Parameters for Various h_a Intervals

h_f	ψ, sec	m	n	σ_a, km	h_a, km	h_f	ψ, sec	m	n	σ_a, km	h_a, km
Earthquake of 11 August 1953; t_0 = 3 h, 32 min, 17 sec; φ = 38°N; λ = 20°E; h normal; m_0 = 21; 76 phases given						Earthquake of 16 July 1955; t_0 = 7 h, 07 min, 13 sec; φ = 37°.5N; λ = 27°E; h normal; m_0 = 18; 61 phases given					
0	1.09	20	16	8	17	0	1.6	21	5	11	39
33	1.01	20	16	8	17*	33	1.2	20	6	15	35
97	0.97	20	6	21	88	97	0.9	18	6	16	79*
160	1.29	20	6	8	136	160	0.9	16	6	25	151
223	1.74	19	3	14	234	223	1.6	17	5	23	229
286	1.79	18	8	24	315	286	2.2	17	4	20	273
350	2.24	17	7	19	347	350	0.6	11	4	16	363
413	1.03	9	7	19	395	413	0.8	14	6	25	400
476	1.52	12	5	15	431	476	0.9	14	8	18	459
540	1.82	12	2	26	553	540	1.0	14	4	15	536
603	1.67	11	3	14	623	603	0.4	15	2	32	597
666	2.01	11	2	7	619	666	1.9	16	2	17	646
730	1.01	8	1	6	740	730	1.8	15	4	17	770
Earthquake of 6 April 1956; t_0 = 7 h, 11 min, 36 sec; φ = 36°.5N; λ = 70°.7E; h = 200-250 km; m_0 = 15; 53 phases given						Earthquake of 27 November 1952; t_0 = 7 h, 20 min, 32 sec: φ = 36°.6N; λ = 70°.0E; h = 150-200 km; m_0 = 10; 31 phases given					
0	1.77	13	5	10	43	0	1.47	10	2	6	70
33	2.40	14	4	3	52	33	1.29	10	2	7	69
97	2.05	15	10	14	115	97	1.12	10	7	17	118
160	1.29	15	14	22	186	160	1.39	10	10	19	158*
223	1.07	15	15	16	214*	223	1.93	10	8	22	233
286	0.84	14	8	14	271	286	1.00	9	8	13	255
350	1.16	14	7	10	353	350	1.32	9	3	19	368
413	0.92	13	6	23	370	413	1.61	9	3	24	418
476	0.71	12	6	16	472	476	2.00	9	2	9	431
540	1.05	12	4	11	495	540	1.26	8	2	6	574
603	1.46	12	3	3	647	603	1.52	8	2	6	575
666	1.92	12	3	1	648	666	1.82	8	—	—	—
730	1.94	11	1	0	703	730	2.21	8	—	—	—

IV. Second Method of Specifying Unreliable Zones

Calculations with the second method of specifying unreliable zones were made for 33 earthquakes by both modifications. For all of the earthquakes, $\bar{n} \geq 5$. We tried for the factor C_{12} the values 0.3 (as in Table 1), and also zero and unity.

The change in C_{12} did not affect the results. Both methods of specifying unreliable zones give very similar results. In the first modification, the results were unchanged in 28 cases, improved in four cases, and slightly worsened in one case. In the second modification, the results were unchanged in 29 cases, slightly changed in three, and improved in only one case.

Let us compare both methods of specifying unreliable zones for the earthquake in example 3 of Table 6.

With the first method, two maxima of n were obtained: 35-151 km (n = 6) and 400-459 km (n = 6-8). The more extreme maximum is explained by erroneous identification of eight PP phases as pP. The results given by the second method are shown in Table 8. This error is partially eliminated, but the ambiguity has not been removed.

Conclusion

The main idea of the method is first to identify all of the phases as pP or sP and then to determine the corresponding focal depths. After this, the problem boils down to the simple selection of the matching values. Preliminary experience shows that the problem of determining focal depth in its

TABLE 7. Number of h_a Determinations (with Varying Accuracy) Versus Number of Identified Phases

Number of identified phases	Accuracy of Δh, km			
	$\leqslant 33$	33—100	100—200	> 200
$\bar{n}$	35 (15)	1 (0)	—	1 (0)
$\geqslant \bar{n}-2$	21 (13)	9 (0)	3 (0)	4 (0)

TABLE 8. Comparison of Parameters for Second Method of Specifying Unreliable Zones

h_f	n	σ_a, km	h_a, km	h_f	n	σ_a, km	h_a, km
33	5	11	40	413	5	23	406
97	6	15	79	476	6	11	467
160	5	18	169	540	4	14	536
223	5	14	214	603	2	32	597
286	3	17	287	666	2	13	643
350	6	20	374	730	4	18	764

conventional formulation does not always have a unique solution. For future calculations, therefore, data must be accumulated for developing solution-optimality criteria and for determining the constants. In its present state, however, the method apparently does give more complete or reliable results than manual calculation. It seems advisable to stop with the first modification. As a control, we can make calculations with two different h_f values. For example, besides h_f = 33 km, we must take h_f = 400-500 km, if there is no basis for assuming the earthquake to be clearly normal. If sufficiently high-speed computers are used, the general method indicated at the end of Section 1 seems preferable.

The authors thank N. V. Kondorskaya for assistance in selecting the data of the seismological survey of the USSR and for valuable advice.

LITERATURE CITED

1. Pyatetskii-Shapiro, I. I., T. S. Zhelankina, V. I. Keilis-Borok, L. G. Pavlova, and P. T. Reznyakovskii (1963), "Computer determination of epicenters," Dokl. Akad. Nauk SSSR, 151:323-326.
2. Bolt, B. A. (1960), "The revision of earthquake epicenters, focal depths, and origin times using a high-speed computer," Geophys. J. Roy. Astr. Soc., 3:433.
3. Pyatetskii-Shapiro, I. I., T. S. Zhelankina, V. I. Keilis-Borok, N. V. Kondorskaya, and L. G. Pavlova (1963), "The use of electronic computers in seismological surveys," paper presented at the XIIIth Assembly of the International Union of Geodesy and Geophysics, Berkeley, California.
4. Kondorskaya, N. V. (1956), "Discrimination of the sP wave in shallow earthquakes, and its use in determining focal depth," Tr. Geofiz. Inst. Akad. Nauk SSSR, No. 36, pp. 35-37.
5. Jeffreys, H., and K. E. Bullen (1940), Seismological Tables, London, British Association for the Advancement of Science, Gray-Milne Trust.

AUTOMATIC IDENTIFICATION OF SEISMIC WAVE ARRIVALS†

E. P. Arnold, V. I. Keilis-Borok, A. L. Levshin,
I. I. Pyatetskii-Shapiro, and P. L. Willmore

Introduction

The problem of identification of seismic-wave arrivals consists in indicating which wave corresponds to a signal arrival ("phase") that has been isolated in processing a recording of an earthquake or explosion. The identification of body waves that arrive at teleseismic distances will be considered in this paper. The classical tables [9] list approximately 60 such waves. The number of phases that can be practically distinguished on one recording is, of course, less and only very rarely exceeds 5-10.

The preliminary phase identification that is done in processing recordings is to a considerable extent intuitive and insufficiently reliable. Reliable identification is possible only when the coordinates of the focus are known, or when the interpreter has exceptional intuition, which cannot be relied upon in routine processing.

The P and PKP phases which occur as first arrivals are identified automatically with an electronic computer [5, 6, 7]. Other phases have up to now been identified manually, after determination of the hypocenter. Only in the seismological survey of the USSR have certain phases (pP, sP, pPKP, and sPKP) been identified automatically, simultaneously with refinement of the focal depth [2].

In manual identification, with the hypocenter and origin time known, the anticipated arrival time of each wave is calculated from the travel-time tables. If it is sufficiently close to the observed arrival time of some phase, this phase is identified with the given wave. This procedure is rather tedious and cannot be performed manually for all of the data that require processing at the present time, and the volume of these data is rapidly increasing. Meanwhile, reliable identification of the greatest possible number of phases (not only qualitative identification but also determination of the residual – the difference between the calculated and observed arrival times) is becoming more and more necessary in a number of areas of seismology: determination of focal depth, determination of the energy and mechanism of a source, study of the earth's structure, and discrimination between earthquakes and underground nuclear explosions.

It is becoming necessary to include phase identification in the overall process of automated analysis of data from primary processing of seismograms – from the sorting of these data according to earthquakes to the production of a seismological bulletin [3, 6, 8].

This paper describes an algorithm for automatic phase identification, including residual determination. The algorithm reproduces the corresponding part of manual processing of data gathered from a network of stations at seismological centers. This algorithm is realized in a FORTRAN-IV program formulated by A. L. Levshin when he was at the International Seismological Center at Edinburgh. The center has been using this program since September 1966 in its processing of the data of the world seismological network on a UNIVAC 1107 (beginning with the data for January 1964).

† Translated from Vychislitel'naya Seismologiya, No. 4, Moscow, Nauka (1968), pp. 170-182.

1. Formulation of Problem

Let the following data be known:

1. The parameters of the focus: the origin time t_0, the focal depth h_0, and the latitude and longitude of the epicenter φ_0 and λ_0.

2. The results of primary processing of recordings: the arrival times of the phases t_{in}, the phase number i in order of increasing t_{in} (i = 0, 1, 2, ..., I_n), the station number n (n = 1, 2, ..., N), the clearness characteristic of the phase arrival X_{in} (we used X_{in} = 1 and X_{in} = 2, instead of the usual i and e, for impulsive and emergent phases, respectively); the code (the conventional letter symbol) of the first phase t_{0n} if in determining the epicenter it was confirmed that it corresponded to P, PKP_1, or P_{dif}; and the code B_{in} of all other phases from preliminary identification data. For unidentified phases it was assumed that B_{in} = 0.

3. The list of body waves for which, by particular rules (Section 3) using travel-time tables, the anticipated arrival time at any station can be calculated. Let the code for these waves be A_j, where j = 1, 2, ..., J.

It is required to identify the maximum number of phases t_{in} by establishing a correspondence between them and any one index D_{in}. Of course, B_{in} and D_{in} are the tentative and desired names of the waves; they coincide with particular A_j.

The following rules must be observed in identification.

1. Each observed phase t_{in} corresponds to not more than one wave A_j, and vice versa.

2. The arrival-time residual is sufficiently small:

$$\Delta\tau_{ijn} \equiv |\tau_{in} - g_{jn}| < C_j, \quad (1)$$

where $\Delta\tau_{ijn}$ is the residual of phase arrival time t_{in} if it is identified as A_j, $\tau_{in} = t_{in} - t_0$ is the travel time, which is determined by observation, g_{jn} is the travel time of wave A_j to the n-th station, which is determined from travel-time tables, and C_j is the allowable residual for wave A_j (Fig. 1).

3. There are no other calculated arrivals between the observed and calculated arrival times. In other words, if a phase t_{in} is identified as A_j, the inequality

$$\tau_{in} < g_{sn} < g_{jn} \quad \text{or} \quad \tau_{in} > g_{sn} > g_{jn} \quad (2)$$

is not satisfied for any other waves A_s.

4. First, all impulsive phases (X_{in} = 1) are identified, as if other phases were not distinguished

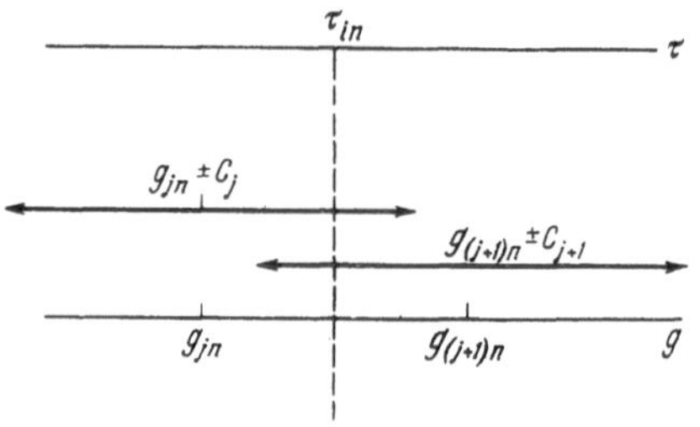

Fig. 1

on the recordings. Then the unclear phases (X_{in} = 2) are identified, but without changing the identification of the impulsive phases. Another rule must be observed here: no impulsive arrival must be located between observed unclear and calculated arrivals. In other words, if a phase t_{in} is identified as A_j and X_{in} = 2, the inequality

$$\tau_{in} < \tau_{sn} < g_{jn} \quad \text{or} \quad \tau_{in} > \tau_{sn} > g_{jn} \quad (3)$$

cannot be satisfied for any t_{sn} with X_{sn} = 1.

5. If an impulsive arrival is identified as S in initial processing at a station (B_{in} = S, X_{in} = 1), the residual of this arrival is sufficiently small, and the epicentral distance is less than 80°, then this identification is unchanged (since the nature of the recording is essential in the identification of S). In other words, if B_{in} = S, X_{in} = 1, $\Delta \leq 80°$, and $(\tau_{in} - g_{jn}) < C_j$ for A_j = S, then D_{in} = S.

2. Identification Method

For a fixed hypocenter, the phases can be identified individually at each station. The simplest identification method is as follows: observing rules 2-5 above, find for each phase t_{in} the wave A_{ji} with the minimum residual, for the entire allowable j, in absolute value $\Delta\tau_{ijn}$. Here, in those cases when the time intervals between adjacent phases are sufficiently great, different phases will correspond to different waves $j_i \neq j_q$ for $i \neq q$. "Competition" between waves will not arise and rule 1 will also be observed, so that the solution $D_{in} = A_{ji}$ can be adopted.

In the presence of competition, the problem is complicated and a more involved method must be used: two adjacent phases are identified simultaneously in a sliding time window. This method consists of the following basic operations.

1. **Selection of the Claimants.** Let a phase with a certain $\tau_{in} = t_{in} - t_0$ be given. We shall find waves A_k and A_l that, according to the travel-time tables, must arrive immediately before and after t_{in}, wherein their travel times

satisfy (1). These waves lay claim to identification with t_{in}. Let A_m ($m = k$ or l) be the wave of A_k and A_l for which the residual is the smaller. If one of the waves is absent, we let its subscript be zero and assume that the other wave is A_m. If both waves are absent ($k = l = 0$), the phase is not identified ($D_{in} = 0$) and is excluded from further consideration.

2. Phase-Pair Identification. Let two successively arriving phases τ_{i1n} and τ_{i2n} ($i_1 < i_2$) be given. Let the subscripts of the corresponding claimants be k_1, l_1, m_1 and k_2, l_2, m_2, respectively. Let d_1 and d_2 be the subscripts of the claimants that we shall select for identification of the phases in question. The solution (selection of d_1 and d_2) is taken according to the diagram in Fig. 2, so that, if possible, both phases are identified. Otherwise, the phase with the smallest residual is identified.

3. Identification of Impulsive Phases ($X_{in} = 1$). Identification begins with the first and second (in time) impulsive phases t_{i1n} and t_{i2n}. The solution is taken according to Fig. 2. For the first phase, the solution is considered final if $d_1 \neq 0$: $D_{i1} = A_{d1}$ and the wave A_{d1} is eliminated from further identification. When $d_1 = 0$, $D_{i1} = 0$: the phase is not identified. When $d_2 \neq 0$, the second phase replaces the first, and the values k_2, l_2, d_2, and i_2 are assigned to the subscripts k_1, l_1, m_1, and i_1. The third (in time) impulsive phase replaces the second, etc. When $d_2 = 0$, $D_{i2} = 0$, and the process of identification begins anew for the third and fourth impulsive phases. The identification diagram is given in Fig. 3.

4. Identification of Emergent Phases ($X_{in} = 2$). Identification is performed as for impulsive phases; in addition the observance of condition (3) is checked.

5. Additional Check. The described procedure has the disadvantage that only two phases

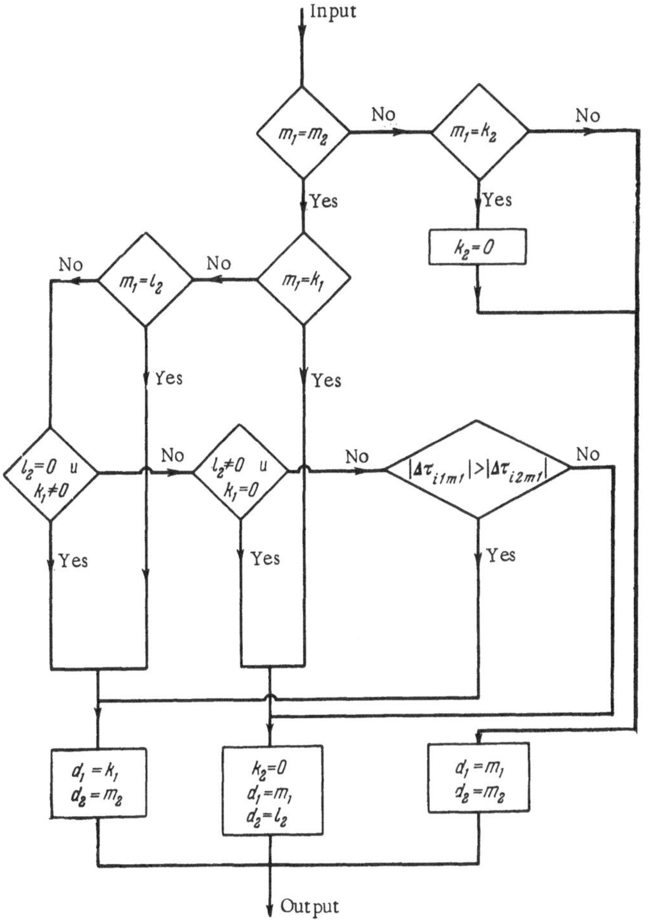

Fig. 2. Solution diagram.

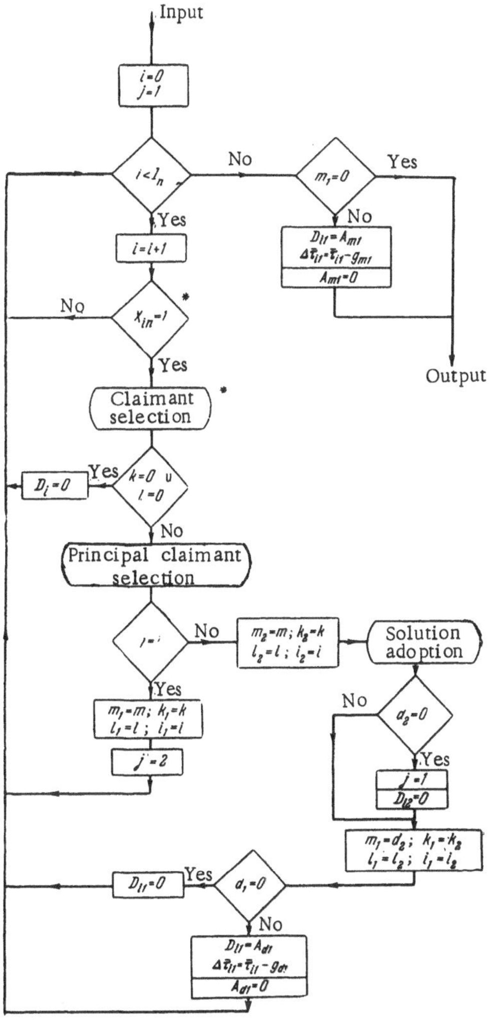

Fig. 3. Sequence of impulsive-phase identifications. The operators or units changed for emergent-phase identification are marked with an asterisk.

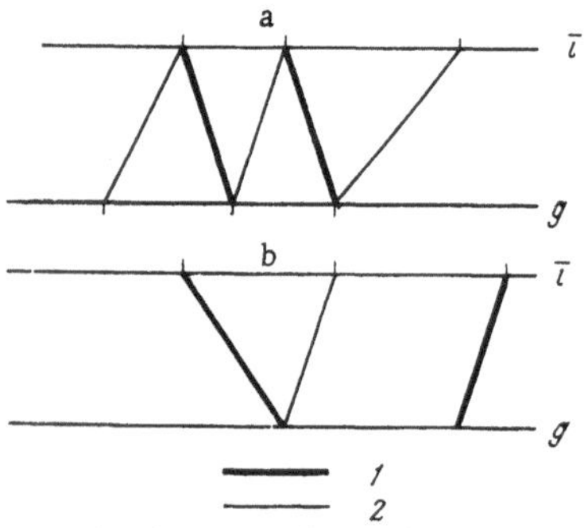

Fig. 4. Theoretically possible cases required in correction to reduce number of unidentified phases (a) or residual (b): 1 and 2 – identification before and after correction, respectively.

are considered simultaneously. In some cases, the maximum possible number of phases will not be identified or the residuals will not be minimal. An example of such cases is given in Fig. 4. Experience shows, however, that these cases are extremely rare (in practice, they are generally not encountered), and we shall ignore them. A subroutine for correcting the identification in cases such as in Fig. 4 was formulated, but its inclusion was considered unnecessary.

3. Calculation of Theoretical Travel Times

A list of theoretical phases A_j for which travel times must be calculated is given in Table 1. To calculate g_{jn}, with given j, n, Δ_n, and h_0, the Jeffreys–Bullen tables [9] are used, with a number of additions and refinements [11]. Differences in the structure of the tables for different types of waves make it necessary to calculate g_{jn} by one of three methods.

1) For P, S, and PKP waves, the travel time at the nodal points of the network of values Δm and h_f is given, where h_f in fractions of the radius of the earth's mantle takes the values −0.0052, 0, 0.1, ..., 0.12. Bilinear interpolation is used to calculate g_{jn}. Note that the P wave cannot enter directly into the

TABLE 1. List of Theoretical Phases*

j	A_j	Branch	Δ_1	Δ_2	Ellipticity	j	A_j	Branch	Δ_1	Δ_2	Ellipticity
1	*PcP*		0	180	+	31	*SKS*	AB	148	130	+
2	*PP*	1	0	180	+	32	*SKP2*	DEF	104	180	+
3	*PP2*	2	180	210	+	33	*ScSPKP*	DF	121	180	
4	*PPP*	1	0	180	+	34	*ScSPKP2*	(A)—C	180	160	
5	*PPP2*	2	180	220	+	35	*ScSPKP3*	A—(C)	206	180	
6	*PKP2*	AB	180	243	+	36	*SKSP*	1	115	180	
7	*PKKP*	DF	198	360		37	*SKSP2*	2	180	205	
8	*PKKP2*	ABC	234	267		38	*ScSP*	1	112	130	
9	*PKKP3*	AB	256	234		39	*ScSP2*	2	140	112	
10	*PcPPKP*	DF	132	180		40	*SKS*	AC	62	133	+
11	*PcPPKP2*	A—(B)	278	180		41	*SKS2*	DF	140	112	+
12	*PKPPKP*	DF	220	360		42	*SKKS*	A—(C)	85	180	
13	*PcS*		0	70	+	43	*SKKS2*	(A)—C	180	243	
14	*PS*		44	147	+	44	*SKKS3*	DF	187	360	
15	*PPS*		44	180		45	*SKKKS*	A—(C)	90	180	
16	*PSS*	1	90	180		46	*SKKKS2*	(A) – C	180	360	
17	*PSS2*	2	180	220		47	*SS*	1	0	180	+
18	*PKS*	DF	104	180	+	48	*SS2*	2	180	214	+
19	*PKS2*	AB	148	130	+	49	*SSS*	1	0	180	+
20	*PKKS*	DF	192	360		50	*SSS2*	2	180	220	+
21	*PKKS2*	BC	214	260		51	*SSP*		44	180	
22	*PKKS3*	AB	226	214		52	*pPKP*	DF	110	180	+
23	*PKPPKS*	DF	215	360		53	*sPKP*	DF	110	180	+
24	*PcSPKP*	DF	127	180		54	*pPKP2*	AB	180	143	+
25	*PcSPKP2*	A—(C)	250	180		55	*sPKP2*	AB	180	143	+
26	*ScS*		0	100	+	56	*pP*		0	105	+
27	*SP*		44	147	+	57	*sP*		0	105	+
28	*ScP*		0	70	+	58	*sS*		0	106	+
29	*SSP*	1	90	180		59	*S*		0	106	+
30	*SSP2*	2	180	220		60	*PKP*	DF	110	180	+

* Δ_1 and Δ_2 are the distances determining the existence interval of the wave. The "+" sign in the "ellipticity" column indicates phases for which ellipticity corrections have been made.

list of A_j. The PKP wave enters into the list only when the wave P_{dif} is the first arrival.

2) For waves whose first reflection from the free surface is near the focus (pP, pS, sS, pPKP, sPKP, etc.), the delay relative to the corresponding fundamental wave (P, S, PKP, etc.) is given at the nodal points of a similar network. The time of the fundamental wave and the correction are calculated by bilinear interpolation; their sum gives the travel time g_{jn}. For very small $h_0 < -0.004$, waves of this type are excluded from the list.

3) For the remaining waves, the surface travel-time curve at points with different Δ_m and corrections for the focal depth at nodal points of the network Δ_l and h_t are given. The time for the focus with the same Δ_n but at $h = -0.0052$ is found by linear interpolation of the surface travel-time curve. Then the correction for the focal depth is found by bilinear interpolation; the difference between the first and second values gives the travel time g_{jn}.

In all three cases, if the tabular time is not given at at least one of the nearest network nodes to the given point Δ_m and h_t, the wave is eliminated from the list for the given station (extrapolation of the travel time is forbidden).

Corrections for Ellipticity. For a number of fundamental waves (see Table 1), corrections for the earth's ellipticity are made in the calculated travel times. The corrections are calculated by approximate formulas from [1, 10]. The heights of the stations relative to the mean sphere are known. The heights of the epicenter and of the points of intermediate reflection (for PP, PPP, PS, SS, etc.) are determined from the coordinates of the epicenter and of the stations. The corrections do not exceed 1.5-2 sec for longitudinal waves or 2.5-3 sec for transverse waves.

4. Some Results

We have analyzed the results of phase identification by the described method for 661 earthquakes of January 1964 – the first month for which data obtained by this method were published in the ISC Bulletin [12]. On the basis of our limited preliminary experience, it was assumed that $C_j = 7.5$ sec for all waves. Only phases with $\Delta \geq 25°$ were identified, since there were no reliable overall travel-time curves for local phases. For 207 of these earthquakes, recordings were made at distances greater than $\Delta \geq 25°$, and computer identification was used for them. Two typical examples are shown in Tables 2 and 3. (All of the results are contained in [12]. They deserve special analysis.) Here, we shall make only a general comparison with the results of preliminary manual identification, i.e., with B_{in}. It should be noted that for many stations – especially in the USA and Japan – B_{in} was determined not from the individual recordings at the stations but after determination of the focus coordinates at regional seismological centers. We identified 640 phases with $B_{in} = 0$ (i.e., phases which were not identified beforehand). The pP, sP, and pPKP phases were found among them most often, which is of great value for refining focal depth. The PcP, PP, and even the exotic PKPPKS and ScSPKP were also found.

In addition, for 626 phases, the preliminary identification of the station was replaced by a new one, which gave a smaller residual (this does not include the comparatively rare cases when the preliminary identification was incorrect but $D_{in} = 0$). Besides the correction of obvious errors, certain interesting changes were noted. Very often, for example, PS and PPS were replaced by SP and SPP. When the source was sufficiently deep, this was not a formality and was due to the greater intensity of S waves.

TABLE 2. Earthquake No. 287 (Kuril Islands, 15 January 1964; $02^h23^m48^s$; 150.8°E; 45.4°N; $h_0 = 0.013R$)

Station	Δ_n°	τ_{in}, sec	X_{in}	B_{in}	Δt_{ijn}, sec*	D_{in}	$\Delta\tau_{ijn}$, sec
Blue Mountains	61.2	625	*e*			*pP*	0.8
Eureka	65.3	654	*e*			*PS*	−1.7
Goldstone	67.4	681	*i*	*pP*	15.9	*PcP*	0.8
Boulder City	68.2	683	*e*			*PcP*	−0.5
Tonto Forest	71.5	691	*i*	*S*	−541.3	*pP*	0.8
Tucson	73.2	699	*e*			*pP*	−0.9
Moxa	78.0	718	*e*			*PcP*	−6.9
		734	*e*			*sP*	2.5

*Residuals for station identification.

TABLE 3. Earthquake No. 298 (south of Honshu, 15 January 1964; $21^h36^m05^s$; 141.1°E; 29.2°N; h_0 = 0.0054R). Sampling for 77° < Δ < 89°

Station	Δ_n°	τ_{in}, sec	X_{in}	B_{in}	Δt_{ijn}, sec	D_{in}	$\Delta\tau_{ijn}$, sec
Oroville	77.2	727	e			pP	0.4
		1294	i	S	2.9	S	2.9
Blue Mountains	77.5	730	e			pP	2.0
Byerly.	77.5	721	i			PcP	0.8
		730	i			pP	2.0
		886	i	PP	−0.6	PP	−0.6
		1317	i	S	23.1	ScS	−1.6
		1615	i	SS	18.8		
		1831	i	SSS	37.5	PKKP	1.6
Roxsburg	78.6	1303	i	S	−2.7	S	−2.7
Mount Hamilton	78.1	734	e			pP	2.3
Reno	78.4	725	i	pP	−8.0	PcP	1.0
Skalstugan	79.1	1312	i	S	1.2	S	1.2
Paraiso	78.8	737	e			pP	1.6
Upsala	79.9	1317	i	S	−2.1	S	−2.1
Trieste	79.4	739	e			pP	0.4
Boulder City.	83.6	755	i	pP	−5.6	pP	−5.6
		762	i	S	−595	sP	−5.9
		937	i	PP	0.3	PP	0.3
Skoresbezund	79.9	1326	i	S	7.6	SKS	−1.8
Simferopol	80.8	1034	i	PP	119.6		
		1328	i	S	−0.9	S	−0.9
		1330	i	SKS	−5.3	SKS	−5.3
		1364	i	sS	3.2	sS	3.2
Eureka	81.0	749	i	pP	1.8	pP	1.8
		916	i	PP	−0.3	PP	−0.3
Pasadena	82.1	1317	i	S	28.1	sS	−3.8
Kishinev	82.7	1347	i	S	−1.2	S	−1.2
Kongsberg	82.9	745	i	PcP	1.8	PcP	1.8
		930	i	PP	−0.7	PP	−0.7
		1352	i	S	2.3	S	2.3
Warsaw	83.5	1357	i	S	0.9	S	0.9
L'vov	83.6	743	i	PcP	−3.4	PcP	−3.4
		1357	i	S	−0.1	S	0.1
		1402	i	SS	13		
		1423	i	PS	3.8	PS	3.8
Salt Lake City	83.0	759	i			pP	1.3
Gothenburg	83.5	1357	i	S	0.8	S	0.8
Copenhagen	84.8	764	i	pP	−2.5	pP	−2.5
		1366	i	S	−3.0	SKS	3.6
		1370	i	S	1.0	S	1.0
		1387	i	sS	−13.9		
Pruhonice	88.1	1395	i	S	−5.0	SKKS	4.4
		1462	i	PS	−11.1	SP	−2.1
		1486	i	PPS	−15.9	SSP	−6.5
		1747	i	SS	−4.5	SS	−4.5
Moxa	88.9	794	e			sP	0.7
		975	i	PP	−4.2	PP	−4.2
		1010	e				
		1391	i	S	−16.6	SKS	2.2
		1407	i	S	−0.6	S	−0.6
		1467	i	PS	−15.8	SP	−6.9
		1763	i	SS	0.1	SS	0.1

A certain percentage of the changes are formal replacements of one phase by another with a slightly smaller residual, close in time and often actually interfering. This is unavoidable in a method based on purely kinematic data. With sufficiently complete calculations of the intensities of various waves [4], it is possible to completely or partially eliminate from the travel-time curves the clearly weak waves or, by introducing weights, to give preference to the stronger waves in selecting the claimants. On the whole, the trial results are satisfactory.

The authors thank D. M. McGregor and A. Fluendy, of the International Seismological Centre, for their great help in this work, and N. V. Kondorskaya and S. S. Mebel' of the Institute of Physics of the Earth, Moscow, for their assistance in supplementing the travel-time tables.

LITERATURE CITED

1. Bullen, K. E. (1953), An Introduction to the Theory of Seismology, Cambridge, Cambridge University Press.
2. Keilis-Borok, V. I., S. S. Mebel', I. I. Pyatetskii-Shapiro, L. Yu. Vartanova, and T. S. Zhelankina, "Computer determination of earthquake focal depth," this volume, p. 16.
3. Keilis-Borok, V. I., "Seismology and logic," this volume, p. 1.
4. Kilinchuk, L. M., and T. B. Yanovskaya (1968), "The intensities of fundamental seismic waves," Computational Seismology, No. 4, Moscow, Nauka, pp. 226-251.
5. Keilis-Borok, V. I., L. G. Pavlova, I. I. Pyatetskii-Shapiro, P. T. Reznyakovskii, and T. S. Zhelankina, "Computer determination of earthquake epicenters," this volume, p. 13.
6. Arnold, E. P., and P. L. Willmore (1966), "A progress report on the International Seismological Research Centre," paper presented at the Fourth International Symposium on Geophysical Theory and Computers, Cambridge, England.
7. Bolt, B. A. (1960), "The revision of earthquake epicenters, focal depths, and origin times using a high-speed computer," Geophys. J. Roy. Astr. Soc., 3:433.
8. Fluendy, A., D. M. McGregor, and P. L. Willmore (1966), "The collection of seismic station readings," Proc. Roy. Soc. Lond., 290A: 461-466.
9. Jeffreys, H., and K. E. Bullen (1940), Seismological Tables, London, British Association for the Advancement of Science, Gray-Milne Trust.
10. Jeffreys, H. (1935), "On the ellipticity correction in seismology," Mon. Not. Roy. Astr. Soc., Geophys. Supp., 4:271-274.
11. Jeffreys, H., and M. Shimshoni (1964), "The times of pP, sS, sP, pS," Geophys. J. Roy. Astr. Soc., 8:324.
12. International Seismological Centre, Bulletin for January, 1964, Edinburgh, 1967.

COMPUTER DETERMINATION OF EARTHQUAKE MECHANISM†

V. I. Keilis-Borok, V. F. Pisarenko,
I. I. Pyatetskii-Shapiro, and T. S. Zhelankina

This paper describes a method of computer determination of a source equivalent to a seismic focus. The initial observations are the signs of the first arrivals of various body waves.

The method is suitable for use in an automated seismological survey. It is realized in a program that is an extension of the program package for the seismological survey [1, 4, 8].

1. Formulation of Problem

Let us introduce the following symbols (the terms used for fault-plane solutions are indicated below in quotation marks):

λ and φ, the longitude and latitude of the epicenter;

λ_k and φ_k, the coordinates of the station, where k is the station number;

Δ_k, the epicentral distance;

A_k, the azimuth from the epicenter to the station;

$\bar{A}_k$, the azimuth from the station to the epicenter;

e, n, and z, the axes of a rectangular coordinate system, which are directed to the east, north, and upward, respectively; the origin coincides with the epicenter.

r, i, and q are spherical coordinates: i is the angle between the radius vector r and the z axis, measured from the z axis; q is the angle between the projection of r onto the plane en and the n axis, measured from the n axis ("r azimuth"). i_k^m and q_k^m ("coordinates of arbitrary observation point") are the i and q values for the tangent to the seismic ray at the hypocenter (for "rectified seismic ray"); k is the seismic station index; and m indicates the wave type (P, S, sP, etc.). I_c and Q_c are the i and q values for the C axis of the source that is used as the model of the focus.

The sources which we shall consider here are characterized by not more than two axes x and y, which are perpendicular to one another. Sometimes the two auxiliary axes z (perpendicular to x and y) and f (bisectrix of x and y) are introduced. Thus, C can take the values x, y, z, and f.

E_k^m, N_k^m, Z_k^m, a_k, b_k, d_k, α_k, β_k, δ_k are parameters that take values of ± 1; E_k^m, N_k^m, Z_k^m have the same sign as the arrivals at the NS, EW, and Z components of a recording of a wave m at a station k.

a_k and α_k, b_k and β_k, and d_k and δ_k have the same sign as the first arrivals, respectively, of longitudinal, transverse SH, and transverse SV waves emitted by the source ("signs of displacements on rectified rays").

$a_k(i, q)$, $b_k(i, q)$, and d_k are determined from observed E_k^m, N_k^m, and Z_k^m at the points $i = i_k^m$ and $q = q_k^m$. The method of determining a_k, b_k, and d_k from observations is described in detail in [5]. Note in particular that Z_k determines a_k; E_k^s and N_k^s together determine one of the parameters b_k or d_k.

α_k, β_k, and δ_k are calculated theoretically by formulas that depend upon the type of source used as the focal model (expansion center, dipole, simple force, etc.). These formulas are given for various sources in [5]. α_k, β_k, and δ_k are functions of i and

† Translated from Vychislitel'naya Seismologiya, No. 5, Moscow, Nauka (1971), pp. 3-27.

q and of I_c and Q_c. If the sign of any of the parameters E_k^m, N_k^m, Z_k^m, a_k, b_k, and d_k is not observed, this parameter is assumed to be equal to zero. α_k, β_k, and δ_k are zero on the "nodal surfaces," which are described in detail in [5]. These are the surfaces formed by the rectified rays on which the displacements in the corresponding wave are zero.

We shall consider the observed parameters E_k, N_k, and Z_k, and thereby a_k, b_k, and d_k, to be random values. Let p(u) be the probability of correct determination of the parameter u from observations. The probability that the sign of the parameter will be the opposite of the true sign, due to random noise, is $1 - p(u)$.

Sometimes we may assume that p(u) is known from previous experiments, and sometimes we may only assume that in practice the probability p(u) is less than some value.

Now let us formulate the three problems (a, b, and c) that will be considered in this paper.

Given are:

1. The coordinates of the hypocenter λ, φ, and h and of the station λ_k and φ_k.

2. The empirical tables of the functions $i_k^m = i^m(\Delta_k, h)$.

3. The observed parameters E_k^m, N_k^m, and Z_k^m, together with the corresponding probabilities p. Only some of these parameters might be given for each station k and wave m.

4. The functions α_k, β_k, and δ_k, which correspond to certain types of sources (i.e., possible models of the focal mechanism). Only the functions α_k might be given.

It is required:

a. To check the hypothesis that the observations (data of 3) agree with a given type of model.

b. Of two types of models, to select the one that better agrees with the observations.

c. To find a confidence region for the axes of the focal model (that is, for the angles I_x, Q_x, I_y, and Q_y) for a given type of model.

These problems can be solved independently. The solution is divided into two parts.

I. From the data in 1 and 2, we find i_k^m and q_k^m (i.e., the directions of the rectified rays), and then from the data in 3 we find a_k, b_k, and d_k at the points $i = i_k^m$ and $q = q_k^m$ (i.e., the signs of the arrivals on the rectified rays are found from the observed signs of the arrivals). The method for this is described in detail in [5].

II. Problems a, b, and c are solved by comparing the derived estimates a_k, b_k, and d_k with the theoretically calculated α_k, β_k, and δ_k. The method is described in Section 2.

2. Probabilistic Solution of Problems

A probabilistic formulation of the above problems is given in this section and a maximum-likelihood method is proposed for their solution.

To simplify the symbolism we shall write the solutions in the signs of longitudinal waves a_k. Everything will remain valid when a_k and α_k, respectively, are understood as the observed and calculated signs of any waves.

Problem a. We shall assume that the signs of $a_1, \ldots, a_M$ from M stations are known. We must statistically check the hypothesis H_1 that these observations agree with the selected focal model.

Let $\bar{I}_x$, $\bar{Q}_x$, $\bar{I}_y$, and $\bar{Q}_y$ be the true directions of the axes of the model. Since the axes are perpendicular, it is sufficient to consider any three angles – for example, $\bar{I}_x$, $\bar{Q}_x$, and $\bar{I}_y$. At the k-th station, the theoretical sign must be $\alpha_k(\bar{I}_x, \bar{Q}_x, \bar{I}_y)$. Let us assume that due to the presence of noise this theoretical sign is distorted to the opposite sign with probability $(1-p)$ and remains correct with probability p.

We shall assume that distortions at the various stations occur independently. Then $a_1, \ldots, a_M$ are independent random values. The probabilities that a_k is +1 or − 1 have the form

$$\mathcal{P}\{a_k = +1\} = p\,\frac{1+\alpha_k(\bar{I}_x, \bar{Q}_x, \bar{I}_y)}{2} + (1-p)\,\frac{1-\alpha_k(\bar{I}_x, \bar{Q}_x, \bar{I}_y)}{2},$$

$$\mathcal{P}\{a_k = -1\} = 1 - \mathcal{P}\{a_k = +1\}.$$

Two cases of checking hypothesis H_1 may be considered. In the first case, it is assumed that $p > 0.5$ is known *a priori* ; in the second, p is considered unknown. In the first case we proceed as follows. Since the true $\bar{I}_x$, $\bar{Q}_x$, and $\bar{I}_y$ are unknown, we find their estimates I_x^*, Q_x^*, and I_y^* by the maximum-likelihood method. Let

$$\pi_k(I_x, Q_x, I_y) = p\,\frac{1+\alpha_k(I_x, Q_x, I_y)}{2} + (1-p)\,\frac{1-\alpha_k(I_x, Q_x, I_y)}{2}. \quad (1)$$

Then the likelihood function has the form

$$L(I_x, Q_x, I_y) = \prod_{k=1}^{M} [\pi_k(I_x, Q_x, I_y)]^{\frac{a_k+1}{2}}\,[1-\pi_k(I_x, Q_x, I_y)]^{\frac{1-a_k}{2}}. \quad (2)$$

The values I_x^*, Q_x^*, and I_y^*, which give the maximum of the function L, are the maximum-likelihood estimates:

$$L(I_x^*, Q_x^*, I_y^*) = \max_{I_x, Q_x, I_y} L(I_x, Q_x, I_y). \qquad (3)$$

It is easy to see that we can use any combination of I_x^*, Q_x^*, and I_y^* for which the number of sign coincidences between α_k and a_k ($k = 1, \ldots, M$) is maximal. Since the stations are arranged discretely $L(I_x, Q_x, I_y)$ is a step function and, thus, there always exists a region (sometimes disconnected) of values (I_x, Q_x, I_y) in which the maximum L is reached. For this entire region, the number of coincidences of sign pairs $\alpha_k(I_x, Q_x, I_y)$ and a_k is the same. Let R be this number. Thus, when our model has the best orientation, of the M stations R will have signs that coincide with the theoretical and $M-R$ will have signs that disagree.

The number R/M is the estimate for p, and our hypothesis is checked by determining whether or not a deviation of this estimate from p can be explained by accidental variance. If M is great (say $M > 20$), we can use a normal approximation for the binomial distribution that the random value R obeys, providing H_1 is valid.

In this case, $(R - Mp)/\sqrt{Mp(1-p)}$ has an approximately normal distribution. Therefore, if we find from tables of the normal distribution the quantile λ_α of the level α (i.e., a number such that $1-\alpha$ lies within the interval $-\infty, -\lambda_\alpha$ with 100% probability), we can use the following rule for checking the hypothesis:

When

$$\frac{|R - Mp|}{\sqrt{Mp(1-p)}} > -\lambda_\alpha \qquad (4a)$$

the hypothesis is rejected.

When

$$\frac{|R - Mp|}{\sqrt{Mp(1-p)}} \leqslant \lambda_\alpha \qquad (4b)$$

it is assumed that the data contradict the hypothesis. If the hypothesis H_1 is valid, we shall err with this checking method only with a probability of $\alpha \cdot 100\%$.

If $M < 20$ we can use exact formulas for the binomial distribution (see [2], for example); for given M, p, and significance level α they give the lower $R_1(\alpha, M, p)$ and upper $R_2(\alpha, M, p)$ limits within which the observed number of coincidences R must lie. If the number goes beyond these limits, we reject our hypothesis. If the hypothesis is valid, we shall err only with a probability of $\alpha \cdot 100\%$.

Let us return to the second case, when p is not assumed to be known *a priori*.

The question of whether or not a given set of observations $a_1, \ldots, a_M$ agrees with a given theoretical model is greatly dependent upon p. If, for example, $p = 0.5$, all models are equivalent, since any combination of signs among $a_1, \ldots, a_M$ has the same probability 2^{-M}. It is natural, therefore, to assign a boundary value p_0 (say, $p_0 = 0.6$ or $p_0 = 0.7$) and replace the check of hypothesis H_1 by a check of the new hypothesis H_2: $p > p_0$. If the hypothesis $p > p_0$ is refuted, the focal model under study must be rejected. The check of hypothesis H_2 is essentially the same as in the preceding case, except that we must test a unilateral hypothesis for the number of coincidences R. In other words, we must select a quantile λ_α and when

$$\frac{R - Mp_0}{\sqrt{Mp_0(1-p_0)}} < \lambda_\alpha \qquad (5a)$$

refute H_2, or when

$$\frac{R - Mp_0}{\sqrt{Mp_0(1-p_0)}} > \lambda_\alpha \qquad (5b)$$

adopt H_2. The probability of error when H_2 is adopted is, in this case, $\alpha \cdot 100\%$.

If $M < 20$, from the binomial distribution tables we must find only the lower limit $R_1(\alpha, M, p_0)$ and reject H_2 when $R < R_1$ and adopt H_2 when $R > R_1$.

We should make one comment: the estimate R/M for the parameter p is the maximum-likelihood estimate, but it may be biased toward the high side relative to the true value of p. This shift is especially appreciable when the number of observations is small (say, when $M < 10$). For example, if $M = 2$ and the stations are not very far from one another, a dipole model will always give agreement between the theoretical and observed signs, if the axes have the required orientation (the observed signs can be arbitrary here). Thus, in this case, we always have $R/M = 1$, which gives an overestimate for the true p.

For $M > 10$, however, and especially when the stations surround the center more or less uniformly, this bias is obviously insignificant. Moreover, since we use only a discrete network in seeking the best orientation of the source, the shift is toward the decreasing side. Experiments with a known answer gave satisfactory results; they are described in Section 4.

Problem b. Given are the observed signs of the waves $a_1, \ldots, a_M$ and two focal models, between which we must choose the one that better corresponds to these signs. We shall use a likelihood ratio to test these two models. When the focal mechanism has the orientation (I_x, Q_x, I_y), let the first and second models give the signs a_k^1 and a_k^2, respectively, at the k-th station.

Then the likelihood function for the i-th model will have the form

$$L^{(i)}(I_x, Q_x, I_y) = \prod_{k=1}^{M} [\pi_k^{(i)}]^{\frac{a_k+1}{2}} [1-\pi_k^{(i)}]^{\frac{1-a_k}{2}}, \qquad (6)$$

where $\pi_k^{(i)}$ is given by formula (1), and i = 1, 2.

Let $L^{(1)}(I_x, Q_x, I_y)$ reach its maximum at the point (I_x^*, Q_x^*, I_y^*), and (I_x, Q_x, I_y) at the point $L^{(2)}$ $(I_x^{**}, Q_x^{**}, I_y^{**})$. Then, the logarithm of the likelihood ratio will have the form

$$\ln \frac{L^{(1)}(I_x^{\bullet}, Q_x^{\bullet}, I_y^{\bullet})}{L^{(2)}(I_x^{\bullet\bullet}, Q_x^{\bullet\bullet}, I_y^{\bullet\bullet})} = \sum_{k=1}^{M} \left[\frac{a_k+1}{2} \ln \frac{\pi_k^{(1)}(I_x^{\bullet}, Q_x^{\bullet}, I_y^{\bullet})}{\pi_k^{(2)}(I_x^{\bullet\bullet}, Q_x^{\bullet\bullet}, I_y^{\bullet\bullet})} + \right.$$
$$\left. + \frac{1-a_k}{2} \ln \frac{1-\pi_k^{(1)}(I_x^{\bullet}, Q_x^{\bullet}, I_y^{\bullet})}{1-\pi_k^{(2)}(I_x^{\bullet\bullet}, Q_x^{\bullet\bullet}, I_y^{\bullet\bullet})} \right]. \qquad (7)$$

If this expression is greater than zero, the first model is considered valid; otherwise the second model is assumed valid.

In other words, the delection rule may be formulated as follows: for each model, we seek the orientation for which the number of coincidences between theoretical and observed signs is maximal. Then we compare these coincidence numbers and select the model with the greater number.

Of course, the logarithm of the likelihood ratio may be compared with some constant other than zero, which changes the relationship of the error probabilities. We shall explain how these probabilities can be calculated. Let us assume that the probability p is known. Let the maximum number of coincidences for the first model be $K_1(I_x^*, Q_x^*, I_y^*)$, and for the second model $K_2(I_x^{**}, Q_x^{**}, I_y^{**})$. The criterion based on the likelihood ratio is, as we have already seen, equivalent to the following criterion:

When $$K_1(I_x^{\bullet}, Q_x^{\bullet}, I_y^{\bullet}) - K_2(I_x^{\bullet\bullet}, Q_x^{\bullet\bullet}, I_y^{\bullet\bullet}) > K \qquad (8)$$

we take the first model, and when

$$K_1(I_x^{\bullet}, Q_x^{\bullet}, I_y^{\bullet}) - K_2(I_x^{\bullet\bullet}, Q_x^{\bullet\bullet}, I_y^{\bullet\bullet}) \leqslant K \qquad (9)$$

we take the second model, where K is some threshold constant.

Let us show how to find the probability of error of the first kind μ, i.e., the probability that inequality (8) is satisfied, when the second model is valid.

Let M_1 be the number of theoretical disagreements of the signs

$$a_k^{(2)}(I_x^{\bullet\bullet}, Q_x^{\bullet\bullet}, I_y^{\bullet\bullet}) \quad \text{and} \quad a_k^{(1)}(I_x^{\bullet}, Q_x^{\bullet}, I_y^{\bullet}),$$

where k = 1, 2, . . . , M.

Then each station at which the theoretical signs agree will make a zero contribution to the value of

$$\psi = K_1(I_x^{\bullet}, Q_x^{\bullet}, I_y^{\bullet}) - K_2(I_x^{\bullet\bullet}, Q_x^{\bullet\bullet}, I_y^{\bullet\bullet}), \qquad (10)$$

while a station at which the theoretical signs disagree will make a +1 contribution with probability 1 − p and a −1 contribution with probability p (since it is assumed that the second model is valid). Thus, ψ is a binomially distributed random number, consisting of M_1 terms, each of which is +1 with probability 1 − p and −1 with probability p. The expected value and the variance of ψ are $M_1(1-2p)$ and $4M_1p(1-p)$, respectively. Hence, it is easy to find the probability μ that inequality (8) is satisfied, providing the second model is valid. For this, we must use either a normal approximation for ψ (at $M_1 > 20$) or exact binomial distribution tables [2]. For example, if $M_1 > 20$, then

$$\mu = \Phi\left(\frac{-K - M_1(2p-1)}{2\sqrt{M_1p(1-p)}}\right), \qquad (11)$$

where $\Phi(x)$ is a normal distribution function. The probability of an error of the second kind ν, i.e., the probability that inequality (9) is satisfied, providing the first model is valid, is found in a similar way. The only difference is that each of the M_1 terms of ψ will be +1 with probability p and − 1 with probability 1 − p. When $M_1 > 20$, we have

$$\nu = \Phi\left(\frac{K - M_1(2p-1)}{2\sqrt{M_1p(1-p)}}\right). \qquad (12)$$

<u>Problem c.</u> We shall assume that (I_x, Q_x, I_y) are random numbers. We have no *a priori* information about the preferable (I_x, Q_x, I_y) values. Therefore, it is natural to assume that the *a priori* distribution of (I_x, Q_x, I_y) is uniform over the entire region of possible (I_x, Q_x, I_y). We shall obtain the confidence interval for the true angles (I_x, Q_x, I_y) from the *a posteriori* distribution.

Let Ω be the range of possible variation of (I_x, Q_x, I_y). Then the conditional distribution of the observed signs $a_1, \ldots, a_M$, provided that the orientation of the axes of the center is given by the angles I_x, Q_x, and I_y, is written in the form of (2). The unconditional distribution of $a_1, \ldots, a_M$ is written as

$$\frac{1}{\Omega} \iiint_{\Omega} L(I_x, Q_x, I_y)\, dI_x, dQ_x, dI_y = F(a_1, \ldots, a_M). \qquad (13)$$

The *a posteriori* distribution of (I_x, Q_x, I_y), provided that the observed data are $a_1, \ldots, a_M$, is written as

$$V(I_x, Q_x, I_y \mid a_1, \ldots, a_M) = \frac{L(I_x, Q_x, I_y)}{F(a_1, \ldots, a_M)} = \\ = \frac{L(I_x, Q_x, I_y)}{\frac{1}{\Omega} \iiint\limits_{\Omega} L(I_x, Q_x, I_y)\, dI_x\, dQ_x\, dI_y}. \quad (14)$$

To obtain the confidence region Ω_1 for $(\bar{I}_x, \bar{Q}_x, \bar{I}_y)$, we must select a constant C such that in the region Ω_1 the inequality

$$V(I_x, Q_x, I_y \mid a_1, \ldots, a_M) > C \quad (15)$$

is satisfied, and, in addition,

$$\iiint\limits_{\Omega_1} V(I_x, Q_x, I_y \mid a_1, \ldots, a_M)\, dI_x, dQ_x, dI_y = (1-\gamma)\,100\,\%. \quad (16)$$

Then Ω_1 will be the confidence region for $\bar{I}_x$, $\bar{Q}_x$, and $\bar{I}_y$, with the confidence level $(1-\gamma) \cdot 100\%$.

Computer solution of Problem c was also examined in [10, 12]. The method proposed in [12] was applied in [14]. The main differences in the method proposed here are as follows: the confidence region is sought (in [14], instead, the region in which the number of disagreeing signs was greater by 2 than the minimum possible value is considered). Moreover, another definition of the likelihood function is used in [14]: it is assumed that the weight of the observed sign is proportional to the amplitude of the displacement that would be observed on the rectified ray in a homogeneous medium and, therefore, increases monotonically with distance from the nearest nodal line.

This assumption is unjustified both theoretically (since the amplitude is a function of a host of other factors and is much less stable than the sign of the first arrival) and in the light of practical experience [5]. Finally, another method of finding the maximum of the likelihood function is used in [14].

3. Algorithm and Program

Limitations

The method proposed in Section 2 is applicable to any observations if they can be used to calculate the signs of the displacements on the rectified rays (any of a_k, b_k, or d_k) and to find the directions of these rays (i_k, q_k). It is also applicable to any center model for which the corresponding functions α, β, or δ are given.

The program has the following specific possibilities.

Initial Observations. The signs z_k^p of the arrival of the longitudinal wave P (or PKP) and the probabilities $p(z_k^p)$; the signs e_k^s and n_k^s of the arrival of the transverse wave S (or SKS) and the probabilities $p(e_k^s)$, $p(n_k^s)$; e_k^s and n_k^s must be assigned simultaneously.

These data can be given at any number of stations not over 100. Instead of the coordinates of the stations λ_k and φ_k and of the hypocenter λ, φ, and h, we can specify Δ_k, A_k, and $\bar{A}_k$ directly. Otherwise, Δ_k, A_k, and $\bar{A}_k$ can be calculated by the program in [8].

The limitations on the wave types are due to the fact that the memory contains tables of i_k^m only for m = P, S, PKP, and SKS.

Instead of all of the above data, we can give directly (i_k, q_k) and (a_k, b_k, d_k), together with the corresponding probabilities p_k. These data can be obtained for waves of any type.

The probabilities p_k can take one of five values (in practice, the values 0.7, 0.8, 0.9, 0.95, and 0.99 are given). This limitation is for memory economy.

Focal Models

The limitations on the focal model are determined by the formulas that the program uses to calculate α, β, and δ.

Problems a and b are solved for an expansion model $\alpha = 1$ and a dipole

$$\alpha = 1 \cdot \operatorname{sign}\{\cos(\mathbf{x}, \mathbf{r}_k)\cos(\mathbf{y}, \mathbf{r}_k)\}. \quad (17)$$

Solution of Problem a for an expansion model comes down to counting the number of negative a_k. We shall assume that Problem b is solved also for a single couple and a double couple, since the problem of choosing between these models is often discussed in the literature.

The solution of Problem a for model (17) for known earthquakes should not be considered unnecessary: all of the models were constructed for the case when there are no interfaces near the center. The presence of interfaces (which is highly probable for normal foci) can greatly change the functions α, β, and δ [5, 7].

Problem c is divided into two parts. In the first part, the confidence region for the x and y axes is found from the signs of the longitudinal waves a_k for model [17]. In conventional representations, the x axis is identified with the direction of the movement at the center, and the y axis, with the normal to the fault plane.

Since x and y enter formula (17) symmetrically, longitudinal waves can be used to study only the direction of a pair of axes, one (unknown) of which coincides with x while the other coincides with y. For definiteness in solutions obtained with longitudinal waves, the perpendicular z to x and y ("null vector") and the bisectrix f_k of the x and y axes ("contraction axis") are often considered. The directions of the z and f axes are also calculated and printed out.

In the second part, from the signs of the transverse waves b_k and d_k for

$$\beta = 1 \cdot \operatorname{sign}\{[\cos(e, x)\cos(r_k, n) - \cos(n, x)\cos(r_k, E)]\cos(r_k, y)\}, \quad (18)$$

$$\delta = 1 \cdot \operatorname{sign}\{[\cos(r_k, x)\cos(r_k, z) - \cos(x, z)]\cos(r_k, y)\} \quad (19)$$

(single couple) we determine which of the axes found in the first part corresponds to x and which to y.

Problem c could be solved simultaneously for longitudinal and transverse waves, but in routine processing the signs of longitudinal waves are determined much more reliably, and it is difficult to count on refining the directions of the x and y axes from data on transverse waves.

Algorithm

Here, we shall write an algorithm for Problem c, since it is the most complicated.

The solution of this problem includes determination of the maximum of the likelihood function for a given focal model.

After this maximum has been found, the solution of Problems a and b becomes simple. It will be illustrated with specific examples in the next section.

1. Calculation of a_k, b_k, and d_k at the points $i = i_k^m$ and $q = q_k^m$. The solution of Problem c begins when the number n_a of given a_k is not less than C_{51}.†

Solution of the second part of Problem c (interpretation of b_k and d_k) begins when the number n_s of given b_k and d_k is not less than C_{52}.

1.1. Calculation of i_k^m and q_k^m.

A_k, $\bar{A}_k$, and Δ_k are calculated by the conventional formulas [6]; the calculation units for A_k and Δ_k were taken from the program in [8]. We calculate i_k^m (m = P; PKP; S; SKP) by bilinear interpolation tables [9] for Δ and h, which are in the memory. For all waves, $q_k^m = A_k$.

1.2. Calculation of a_k, b_k, and d_k and the corresponding p_k.

As is obvious, where m = p or PKP,

$$a_k^m = Z_k^m;\ p(a_k^m) = P(Z_k^m).$$

We calculate b_k and d_k from e_k^s and n_k^s according to Table 1, which was set up for $i_k > \pi/2$; when $i_k < \pi/2$, the signs of d should be reversed. It is easy to see that $p(b_k)$ or $p(d_k)$ equals $\frac{1}{2}[e_k^s + n_k^s]$ if positive and negative b_k and d_k are considered *a priori* equally probable. It is also assumed that $b_k = d_k = 0$ when $\Delta_k < C_{37}$, since the transverse-wave sign is determined reliably, due to total internal reflection [7]. This condition is intended for interpretation of teleseismic observations. For near earthquakes, it may be replaced by the reverse condition.

An additional operation reproduces manual processing: the sign of the wave at the conditional point is rejected if it is surrounded by conditional points with the opposite sign for the same wave. Let j be the number of the conditional point and let a_j be given at it. We find the points k for which $a_k \neq 0$ and

$$\cos r_j \cos r_k < C_{38}.$$

The groups of points with

$$q_k > q_j \quad \text{and} \quad q_k \leqslant q_j$$

are examined separately. If there are more than C_{39} stations in both groups and at each group $a_j a_k = -1$ for C_{40} or most of the points, then we assume $a_j = 0$. We proceed similarly with b_j and d_j.

2. Determination of the confidence region for the x and y axes from the signs of a_k.

2.1. We consider all possible directions of x and y.

† Here and below, C_k are given constants; their numbering follows [1, 4, 8].

TABLE 1

ε	N	$0 \leqslant A_k \leqslant 90°$		$90° \leqslant A_k \leqslant 180°$		$180° \leqslant A_k \leqslant 270°$		$270° \leqslant A_k \leqslant 360°$	
		d_k	b_k	d_k	b_k	d_k	b_k	d_k	b_k
+1	+1	−1	0	0	+1	+1	0	0	−1
−1	+1	0	+1	+1	0	0	−1	−1	0
+1	−1	0	−1	−1	0	0	+1	+1	0
−1	−1	+1	0	0	−1	−1	0	0	+1

In order not to consider equivalent combinations of the axes (according to (17), these are the combinations (x, y) and (−x, −y) and also (x, y) and (y, x)), it is more convenient to solve the problem in the space I_z, Q_z, Q_x.

Let Ω be the region of possible I_x, Q_z, Q_x. If we exclude the equivalent combinations of axes, then

$$\Omega = \Omega' + \Omega'', \tag{20}$$

where Ω' is the region

$$0 \leqslant I_z \leqslant \frac{\pi}{2}, \quad 0 \leqslant Q_z \leqslant 2\pi, \quad Q_z + \frac{\pi}{2} \leqslant Q_x \leqslant Q_z + \pi. \tag{21}$$

We obtain Ω'' from Ω' by substituting $-x$ for x, i.e., $Q_x + \pi$ for Q_x and $\pi - I_x$ for I_x (21). In the region Ω', the values of I_z, Q_x, and Q_z move over a rectangular network with steps of C_{46}, C_{47}, and C_{48}, respectively.

2.2. At each node of this network the sines and cosines of I_x, I_y, and Q_y are calculated by the following formulas, which follow from the condition of perpendicularity of the x, y, and z axes:

$$\sin I_x = \frac{\eta}{\sqrt{\nu^2 + \eta^2}}, \quad I_x \leqslant \frac{\pi}{2}, \tag{22}$$

where

$$\nu = \xi \sin I_z; \quad \eta = \cos I_z,$$
$$\xi = (\cos Q_x \cos Q_z + \sin Q_z \sin Q_z).$$

Instead of $\sin I_x = 1$ we take $\sin I_x = 0.99999$. If

$$\sin I_x = 1 \text{ and } 1 - |\sin I_z| < C_{45},$$

then

$$\sin I_y = \sin Q_y = 0, \quad \cos I_y = \cos Q_y = 1,$$
$$\sin Q_y = \frac{\eta_1}{\sqrt{\nu_1^2 + \eta_1^2}}. \tag{23}$$

Here,

$$\nu_1 = \sin I_x \sin Q_x \cos I_z - \sin I_z \cos I_x \sin Q_z,$$
$$\eta_1 = \sin I_x \cos Q_x \cos I_z - \sin I_z \cos Q_z \cos I_z,$$
$$\operatorname{sign}\{\sin Q_y\} = \operatorname{sign}\{\eta_1\}.$$

from

$$\nu_1 \sin Q_y + \eta_1 \cos Q_y = 0$$

we have

$$\operatorname{sign}\{\cos Q_y\} = -\operatorname{sign}\{\nu_1\} \tag{24}$$
$$\sin I_y = \frac{\eta_2}{\sqrt{\nu_2^2 + \eta_2^2}}, \quad I_y \leqslant \frac{\pi}{2},$$

where

$$\nu_2 = \sin I_x [\cos Q_x \cos Q_y + \sin Q_x \sin Q_y], \quad \eta_2 = \cos I_x.$$

Then we calculate I_x, I_y, Q_x, and Q_y at the corresponding points of Ω'' using formulas (22)-(24).

2.3. We calculate α_k at each node of the network for each point at which a_k is given.

If for any point the expression in the brackets on the right of (10) is smaller in absolute value than C_{35}, we assume that $\alpha_k = 0$ for this station. This means that observations at points close to the nodal line are not used.

2.4. We calculate

$$L_a = \prod_{k=1}^{M} \pi_k, \tag{25}$$

where

$$\pi_k = l_a p(a_k) \quad \text{for } a_k\alpha_k = 1,$$
$$\pi_k = l_a [1 - p(a_k)] \quad \text{for } a_k\alpha_k = -1, \tag{26}$$
$$\pi_k = 1 \quad \text{for } a_k\alpha_k = 0,$$

$$l_a = \frac{1}{1 - p_a} \left[\frac{1 - p_a}{p_a}\right]^{\frac{\bar{K}(n_a, p_a)}{n_a}}, \tag{27}$$

$$K_a(n_a, p_a) = n_a\bar{p}_a - C_{36}\sqrt{n_a\bar{p}_a(1 - \bar{p}_a)}, \tag{28}$$

and $\bar{p}_a$ is the mean value of p_k.

Formula (26) differs from (1) in its normalization. The normalization factor l_a is chosen such that L_a has a high probability of being not much less than 1. More precisely, the value ln L_a has a binomial distribution, which when there is a large number of stations can be considered approximately normal. The probability that $L_a < 0$ (i.e., $L_a < 1$) is approximately equal to the probability that some normalized value ξ exceeds the constant C_{36}. Thus, if C_{36} is approximately equal to 2 or 3, the probability that $L_a > 1$ is small.

The following condition is introduced to reduce machine time: if a number less than C_{49} is obtained in multiplying Π_k by formula (25), we assume that $L_a = 0$.

We select C_{49} such that there is a fairly small probability of missing L_a that are greater than some small threshold (or than the computer zero).

2.5. If $L_a \geq C_{41}$, we calculate L_a by the formulas in **2.4** at 26 additional points surrounding the given point:

$$I_z + \frac{\sigma_1}{3} C_{46}, \; Q_z + \frac{\sigma_2}{3} C_{47}, \quad Q_x + \frac{\sigma_3}{3} C_{48},$$

where $(\sigma_1, \sigma_2, \sigma_3)$ take all possible combinations of the values 0, 1, and −1. Thus this step is triply compressed when L_a is sufficiently great.

2.6. The number of negative $\alpha_k a_k$ (globally and individually over each of the quadrants) and the number of negative α_k are calculated simultaneously with L_a at each point.

2.7. Operations 2.4-2.6 are performed for the points in region Ω'', which is obtained by substituting −x for x. In doing this the signs of α_k are reversed, so that there is no need to calculate them by formulas (17) and (22)-(24).

2.8. All L_a greater than C_{41} are summed. Their sum is the approximate value F from formula (13), which is calculated by the rectangle method.

2.9. The L_a that are greater than C_{41} are arranged in decreasing order. The C_{49} maximum L_a are stored together with the angles I_x, Q_x and I_y, Q_y and the number of negative α_k. Here, the angles are rounded off to 1°.

If for a maximum L_a the number of negative $\alpha_k a_k$ is greater than

$$n_a(1-p_a)+C_{50}\sqrt{n_a p_a(1-p_a)}, \qquad (29)$$

the problem does not have a solution.

For three values of the confidence interval $\mu_a^{(i)}$ (i = 1, 2, 3) we find the confidence regions $\Omega_a^{(i)}$. For this, we sum up the stored L_a, beginning with the greatest, until the sum exceeds $F\mu_a^{(i)}$. The set of values (I_x, Q_x, I_y, Q_y) corresponding to the L_a that have entered into the sum is also the desired confidence region Ω_a. If the sum of all stored L_a does not exceed $F\mu_a^{(i)}$, the confidence region is too large and problem does not have a solution.

2.10. Estimate of width of confidence regions.

In 2.9, we found a set of discrete points that indicate the regions $\Omega_a^{(i)}$ with an interval determined by C_{46}, C_{47}, and C_{48}. Now we must estimate the width of each region in cross sections along I_x, Q_x, I_y, and Q_y. If at least one of these cross sections is sufficiently small (i.e., if the confidence region is sufficiently narrow for at least one of these angles), we assume that the problem has a solution.

Let $(I_\nu)_{max}$, $(Q_\nu)_{max}$ and $(I_\nu)_{min}$ and $(Q_\nu)_{min}$ be the maximum and minimum I_ν and Q_ν in the confidence region in question (ν = x or y). The cross section of the region along I_ν is simple to estimate: we check the condition

$$(I_\nu)_{max}-(I_\nu)_{min}-C_{42}<0. \qquad (30)$$

If it is satisfied for at least one of the two ν, the problem has a solution, with the given confidence level.

Estimation of the cross sections over Q_ν is more complicated. It must be done separately for 12 intervals of I_ν,

$$15°l<I_\nu\leqslant 15°(l+1), \qquad l=0,1,\ldots,11. \qquad (31)$$

A condition similar to (30) is checked within each interval:

$$Q_{\nu_1}-Q_{\nu_2}-C_{43}\sin\bar{I}_\nu<0. \qquad (32)$$

Here $\bar{I}_\nu$ is the maximum of the I_ν that correspond to $Q_\nu = Q_{\nu_1}$ and Q_{ν_2}. We determine Q_{ν_2} as follows (see Fig. 1).

If $(Q_\nu)_{max} \leq 180°$ or $(Q_\nu)_{min} \geq 180°$, then $(Q_\nu)_{max} = Q_{\nu_1}$, and $(Q_\nu)_{min} = Q_{\nu_2}$.

If $(Q_\nu)_{min} < 180°$ and $(Q_\nu)_{max} > 180°$, then we calculate $Q^{(1)} = \max Q_\nu$ in the interval 0°-180° and $Q^{(2)} = \min Q_\nu$ in the interval 180°-360°.

If

$$360°+Q^{(1)}-Q^{(2)}<(Q_\nu)_{max}-(Q_\nu)_{min}, \qquad (33)$$

then

$$Q_{\nu_1}=360°+Q^{(1)}, \qquad Q_{\nu_2}=Q^{(2)}.$$

If (26) is not satisfied, then

$$Q_{\nu_1}=(Q_\nu)_{max}, \quad Q_{\nu_2}=(Q_\nu)_{min}.$$

If for at least one ν (32) is satisfied in all intervals (31), the problem has a solution with a given confidence level.

Estimate (32) is approximate. In practice it is fairly effective, but there is no assurance that it could not be simplified.

Other estimates are possible: over the ellipse of concentration of the points that fall into $\Omega_a^{(i)}$; over the cross sections of a convex polyhedron stretched over these points; etc. These estimates are not considered here.

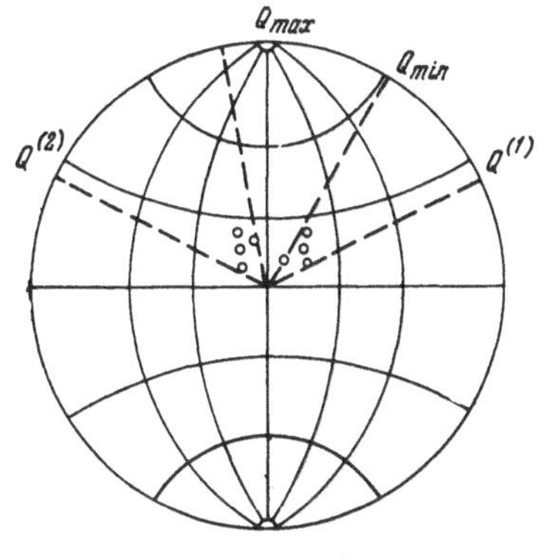

Fig. 1. Determination of confidence-region cross sections.

TABLE 2

k	c_k	k	c_k	k	c_k	k	c_k
34	7	39	3	45	1 — sin 85°	51 *	800
34 *	5	40	0,4	46	15°	$52^{(1)}$	0,85
35	0	41	0,1	47	15°	$52^{(2)}$	0,95
35 *	0	42	30°	48	15°	$52^{(3)}$	0,99
36	1,8	43	30°	49	10^{-10}	—	—
37	0,05	44	0,01	50	1,8	—	—
38	(cos 5°)			51	1000		

*With the index S.

2.11. In each region $\mu_a^{(i)}$, the characteristics of the points with maximum L_a and with extremal I_x, Q_{xi}, I_y, and Q_{yi} are always printed out (the extremal Q_{ν_i} are the Q_{ν_1} and Q_{ν_2} for the interval of (24) in which the left side of (25) is maximal). These characteristics are: L_a, I_x, Q_x, I_y, Q_y, I_b, Q_b (b is the bisectrix of x, y). α_k are negative.

The number of negative ($\alpha_k a_k$) as a whole for this point and separately over the quadrants x, y.

Conditionally printed out are:

The numbers of the stations for which $\alpha_k a_k = -1$; for memory economy, $\alpha_k a_k$ are recalculated before this (for points close to the nodal lines, α_k can differ from those obtained in calculating L, due to rounding off the angles to 1°. But this is not important, since operation 2.3 eliminates these points from consideration);

The numbers of the stations at which $a_k = 0$ according to 1.2, i.e., because the station is surrounded by stations with opposite signs;

The numbers of the stations at which $\alpha_k = 0$ according to 2.3, i.e., because the station is near a nodal line;

A list of all points in the confidence regions or of all stored points with any of the above-mentioned characteristics of each point.

3. Interpretation of b_k and d_k (second part of problem c).

3.1. We consider all points of $\Omega_a^{(3)}$ (the maximum of $\Omega_a^{(i)}$) and the points obtained from them by transposition of x and y. At each point, we calculate the likelihood ratio

$$L_S = \Pi \pi_k (b_k \beta_k,\, p_k^{(b)})\, \pi_k (d_k\, \delta_k,\, p_k^{(d)}).$$

where π_k are calculated by formulas similar to (25)-(28).

The next operations are quite similar to 2.5-2.10 and their description is repeated almost word for word if α_k is replaced by β_k and δ_k and a_k by b_k and d_k, and in the remaining symbols, the subscript "a" is replaced by "S."

In addition, the constant C_{41} is replaced by C_{44}/n_a, where n_a is the number of points that fall into $\Omega_a^{(3)}$.

By solving this part of the problem, we determine three confidence regions within $\Omega_a^{(1)}$, which is the largest of the regions found in operation 2.9..

The print-out is similar to that in 2.11. The station numbers are printed out separately for b_k and for d_k.

4. Examples

Two examples were calculated to test the method. One was a model example with a known "true" answer. The positions of 35 stations in a real earthquake (6 May 1958, at 0416, epicenter 47.8 E, 43.1 N) were taken. The corresponding conditional points for P and S waves were determined. Each point was given a sign Z_k, N_k, or E_k so that from the obtained a_k, b_k, and d_k the "true" x and y axes would be determined practically uniquely. The signs at certain randomly selected stations were reversed, in order to simulate errors. Then the calculation was carried out by the program described above.

The second example was taken from practice. It was a catastrophic earthquake in Alaska (28 March 1964, at 1336, epicenter 147.5 W, 61 N, M = 8.5). The signs Z_k for P and PKP_1 waves at 83 stations were interpreted; the data are given in [3, 9].

Problem c. The calculation was made with the constants in Table 2. The constants C_{46}, C_{47}, C_{48}, and C_{35} were varied as shown in Table 3 which contains a summary of certain calculation results.

Figures 2-10, below, are Wulff stereographic projections [5].

First, let us consider the model example (Figs. 2-8). Figure 2 shows the initial data: the locations of the conditional observation points and

TABLE 3

Version					Result				
Version No.	$C_{46}=C_{47}=C_{48}$	C_{35}	Nos. of stations with erroneous a_k, b_k, d_k	Fig. No.	Number of solutions with $L_a > C_{41}$	Calculation time, min	Number of solutions in $\Omega_a^{(1)}, \Omega_a^{(2)}, \Omega_a^{(3)}$	Nos. of stations with disagreeing a_k, b_k, d_k for L_{max}	L_{max}
1	15	0	a_k: 8, 14, 24, 26, 34	3	54	4	6 13 —	a_k: 3, 8, 10, 14, 26, 34	102
2	9	0	a_k: 8, 14, 24, 26, 34	4	213	15	8 12 37	a_k: 8, 14, 24, 26, 34	108
3	6	0	a_k: 8, 14, 24, 26, 34	5	789	53	23 62 —	a_k: 8, 14, 24, 26, 34	110
4	15	0	a_k: 34	6	163	5	5 8 11	a_k: 20, 34, 36, d_k: 37	101
5	15	0	a_k: 18, 34, 36, 38 b_k: 4 d_k: 5, 7, 9	7	147	25	1 2 4	a_k: 17, 30, 34 b_k: 4, 38 d_k: 5, 7, 9, 18, 23, 37	109
6	9	0	a_k: 4, 5, 8, 11, 13, 14, 24, 26, 29, 33, 34, 38	—	1468	21	475	a_k: 6, 7, 13, 15, 16, 18, 22, 24, 27, 31, 36	104
7	9	sin 2°	a_k: 8, 14, 24, 26, 34	8	3529	45	22	a_k: 1, 3, 4, 8, 10, 14, 20, 23, 24, 26, 32, 34	111

nodal lines. The signs b_k and d_k correspond to a single couple.

Figure 3 shows the results of calculation with the largest step (5° within the confidence region). The pair of points with the same number is a pair of axes, one of which corresponds to x and the other to y. The bisectrices of x and y are omitted here and in subsequent figures. The points are numbered in order of decreasing L_a (operation 2.9), and for identical L_a, in the order in which they are encountered in sorting the points of the network (operation 2.1), i.e., randomly with respect to the relief of L_a.

The L_a values are given in Table 4. Points 1-6 and 1-13 form confidence regions with levels of 85% and 95%, respectively. The 99% level turned out to be too high, according to criteria (23) and (25).

It can be seen from Fig. 3 that Problem c is solved quite effectively. The "true" solution, for which their example was constructed, lies in the 95% confidence region. The maximum L_a is only slightly shifted relative to the "true" solution (due to the discreteness of the network) and the number of disagreeing a_k is only 1 more than in the "true" solution.

The effect of the network step is apparent from a comparison of Figs. 3, 4, and 5 (versions 1, 2, and 3).

In Fig. 4 (3° step), the maximum L_a already included the "true" solution, but the confidence region is not much narrower than that in Fig. 3; it cannot be narrowed with the existing observation system.

Figure 5 shows only the 85% confidence region, from which it can be seen that it is useless to reduce the step, provided there are no exceptional reasons for being interested in details of the form of the confidence region.

The effect of the number of increased a_k's is evident from a comparison of versions 1 and 4 and 5 and 6.

TABLE 4

Version	Numbers of solutions	L_a	Version	Numbers of solutions	L_a
1	1	4.8	5	1	3324.0
	2—6	0.5		2	375.8
	7—13 (17)*	0.06		3—4	42.5
2	1—10	4.8	7	1—4	76.9
	11—19	0.5		5	70.4
	20—37 (48)	0.06		6	50.6
3	1—5	42.5		7, 8	33.3
	6—26	4.8		9, 10	30.5
	27—62 (66)	0.5		11—15	20.1
4	1—3	3324.0		16—20	13.2
	4—9	375.8		21—22 (23)	8.7
	10—11 (12)	42.5			

* The number is given in parentheses of the last solution with the same L_a that did not fall into the confidence region.

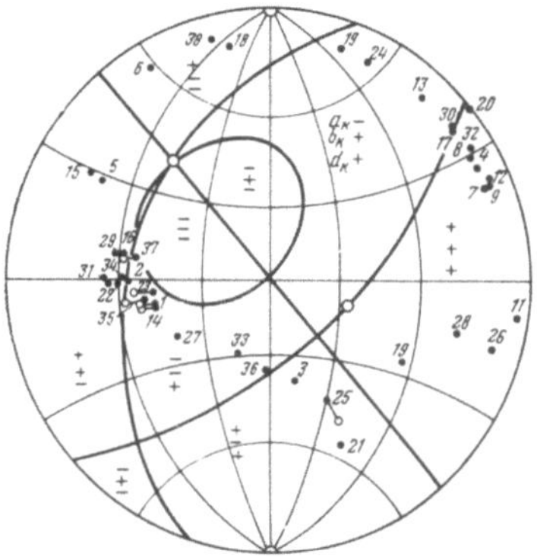

Fig. 2. Initial data in model example.

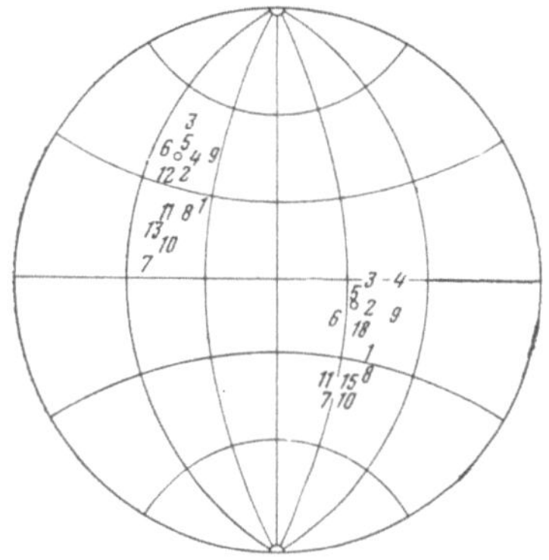

Fig. 3. Version 1.

In version 4 (Fig. 6), one incorrect a_k was given instead of five, but the results differ little from those of version 1 (Fig. 3). Of course, when the number of errors is greatly increased, the confidence regions begin to widen and, ultimately, the problem will not have a solution, according to criteria (23) and (25). For example, when 12 out of 35 signs are incorrect (version 6 in Table 3), the 85% confidence region is still too small.

One defect in the method of selecting the confidence regions can be seen from Fig. 6 and Table 4. Of the points 5-9 with identical L_a, point 5 was in the confidence region $\Omega_a^{(1)}$, so it accidentally turned out to be the first in the list. It would have been more advantageous to include point 8 in $\Omega_a^{(1)}$. This would have reduced the dimensions of $\Omega_a^{(1)}$, and the confidence level would have remained as before. To avoid this, we could, for example, have ordered the points with identical L_a according to increasing distance from the points with higher L_a. We could have drawn a convex polyhedron to the points with one L_a value and then ordered the points with the next lowest L_a according to increasing distance from this polyhedron. It should be noted that conditions (23) and (25) may seem to be violated due to an unfortunate division of the axes into two groups (x and y or q_1 and q_2) in interpreting a_k. This is possible when some of the solutions apply to Ω'_a and some to Ω''_a. These defects are not eliminated, due to the limited size of the computer memory. They can be eliminated manually in each individual case, with conditional storage of all points of $\Omega_a^{(i)}$. This is hardly necessary very often, especially since we are concerned with overestimation of the confidence region.

The effect of the choice of C_{35}. We recall that C_{35} is the sine of the angular distance from the nodal line, within which the observed signs are not used (operation 2.3).

It is obvious that the confidence regions widen as C_{35} is increased, and this is especially true of narrow regions.

From a comparison of Figs. 4 and 8 (versions 2 and 7, which differ only in the value of C_{35}) it is clear that this widening can be appreciable: a 95% confidence region at $C_{35} = 0$ (Fig. 4) is approximately

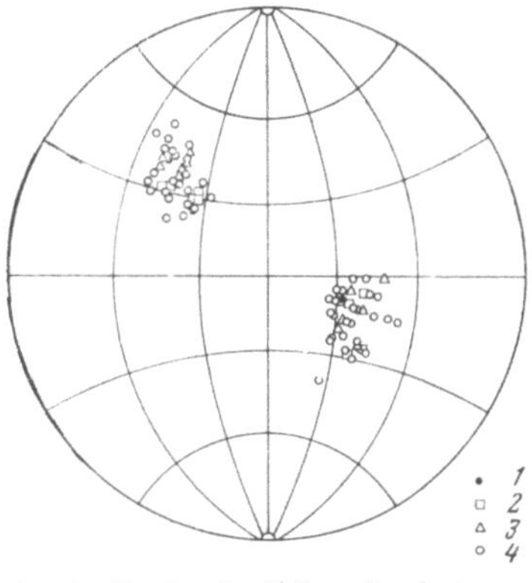

Fig. 4. Version 2. 1) "true" solution; 2-4) solutions in 85%, 95%, and 99% confidence regions, respectively.

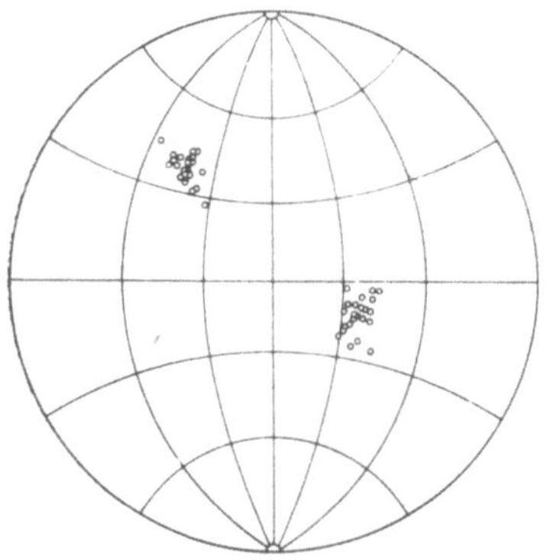

Fig. 5. Version 3. 85% confidence region.

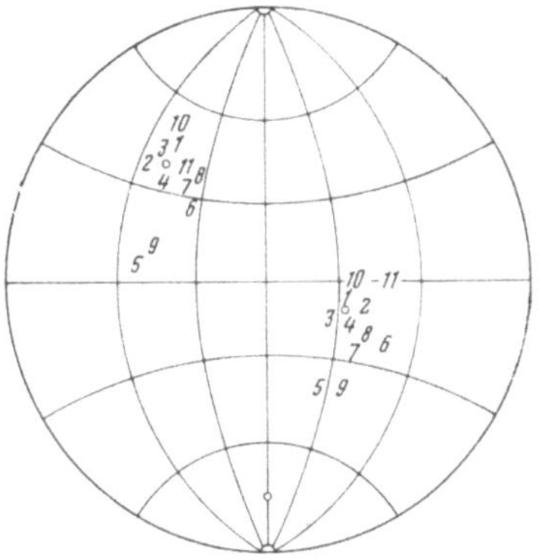

Fig. 6. Version 4.

the same as an 85% region at C_{35} = sin 2° (Fig. 8). To be sure, in actual cases the widening will not be so great, since usually fewer conditional points than in the model example lie near the nodal lines.

The signs of transverse waves b_k and d_k were used in versions 4 and 5. In all cases, as can be seen from Table 3, the maximum of L_s is very sharp; all three confidence regions Ω_b run together and contain only two points each from Ω_a. Of course, when $P(b_k)$ and $P(d_k)$ are reduced, this maximum will spread out. The maxima of L_a and L_s coincided in both cases.

Now let us consider the interpretation of the Aleutian earthquake. The initial data are shown in Fig. 9.

The calculation results are as follows. Three points were discarded, since they were surrounded by points with opposite signs for a_k (condition 2). The function L_a was very flat for the remaining points (Table 5). L_a contained 2657 solutions, 44 of which, corresponding to maximum L_a, are shown in Fig. 10. We see that despite the abundance of observations, it is impossible to obtain a sufficiently narrow confidence region for them. The nodal lines

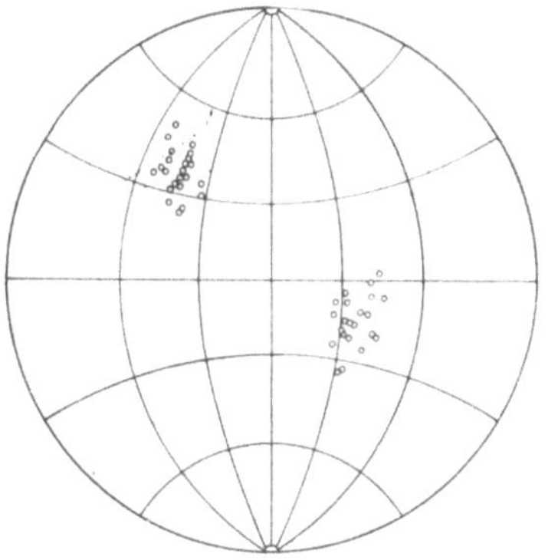

Fig. 8. Version 7, 85% confidence region.

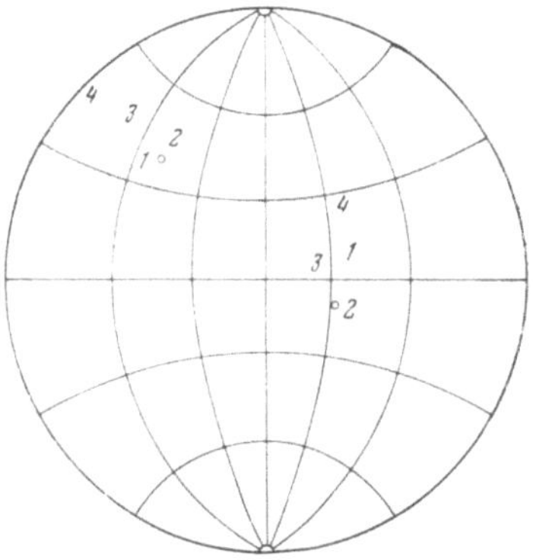

Fig. 7. Version 5. 99% confidence region.

may be drawn quite differently, since the signs are divided by one line. One of the nodal-line versions is shown in Fig. 9.

Problem a. Let us consider the solution of Problem a in the same model example for the first version of initial data (Table 3).

In this version, a_k = +1 at 23 stations and a_k = −1 at 12 stations. The signs of α_k at 31 stations agreed and at four stations they disagreed with the solution corresponding to the maximum of the likelikhood function.

Section 2 describes two cases: when the probability p_k of correct determination of a_k at an individual station is *a priori* known and unknown.

Let us consider the first case. We assume that all p_k are the same and equal p = 0.9.

We test the hypothesis H_0 that the dipole model agrees with such observations. The estimate for p in this case is

$$\frac{R}{M} = \frac{31}{35} = 0.886.$$

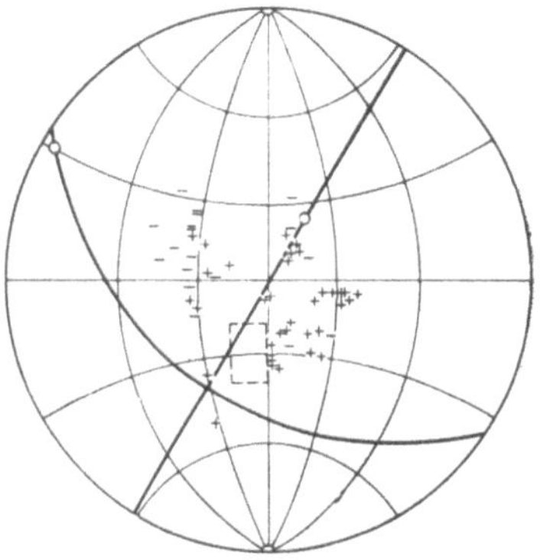

Fig. 9. Initial data for Aleutian earthquake. Inside the rectangular region, there are 23 points with "+" and five points with "−".

TABLE 5

L_a	Number of disagree-ments	Number of solutions	L_a	Number of disagree-ments	Number of solutions
58.6	11	44	0.75	13	400
6.6	12	191	0.085	14	635

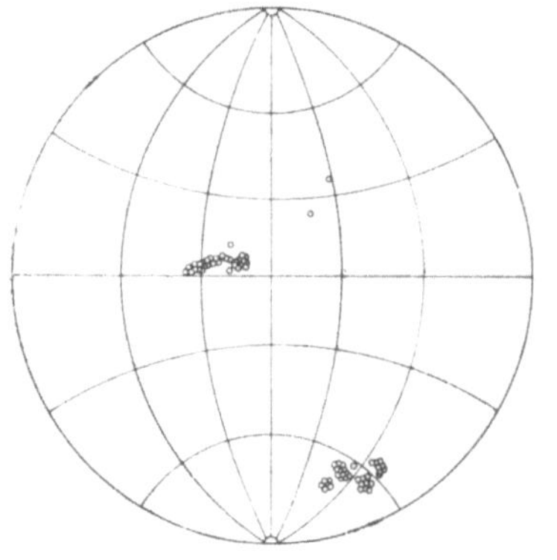

Fig. 10. Solution for Aleutian earthquake (points with maximum L_a).

From formula (4), we set up the value

$$\xi = \frac{R - Mp}{\sqrt{M(1-p)p}} = \frac{31 - 35\cdot 0{,}9}{\sqrt{35\cdot 0.1\cdot 0.9}} = -0.272.$$

We shall use a normal approximation for ξ. Hypothesis H_0 should be adopted, since the significance level of ξ is only about 39% (i.e., the probability that when H_0 is satisfied ξ will take values less than -0.272 is 0.39, which is quite likely and does not provide a basis for rejecting H_0).

Note that in this example we did not obtain an excessive shift in the estimate of P, which we mentioned in Section 2, since four signs were "erroneous" and the best solution also contained four disagreements.

Now let us consider the second case, where p_k is not known *a priori*. Let the threshold value p_0, above which we accept hypothesis H_2, be 0.7. For our example we obtain from formula (5)

$$\frac{R - Mp_0}{\sqrt{Mp_0(1-p_0)}} = \frac{31 - 35\cdot 0.7}{\sqrt{35\cdot 0.3\cdot 0.7}} \approx 2.4.$$

Hypothesis H_2 should be adopted, since we obtained a positive value which for any α exceeds the quantile λ_α.

Problem b. We shall examine the solution of Problem b using the same example as for Problem a.

The two competing models will be a single couple (17) and an expansion focus ($\alpha_k \equiv +1$). Note that the maximum of the likelihood function for the single couple corresponds to four disagreements between observed a_k and theoretical α_k signs: $a_k = +1$ and $\alpha_k = -1$ at two points and vice versa at two points. For the second model, we obtain 23 agreements (the quantity of $a_k = +1$) and 12 disagreements (the quantity of $a_k = -1$).

Using formulas (8) and (9), we formulate

$$K^{(1)}(I_x^{\bullet}, Q_x^{\bullet}, I_y^{\bullet}) - K^{(2)}(I_x^{\bullet\bullet}, Q_x^{\bullet\bullet}, I_y^{\bullet\bullet}) = 31 - 23 = 8.$$

If, for example, the threshold constant "K" in (8) and (9) is zero, we must assume that the couple model is valid. Let us find the probabilities of errors of the first and second kind μ and ν for K = 0. The number of theoretical disagreements M_1 for our models is 12. In order that the difference $K^{(1)} - K^{(2)}$ be greater than the threshold K = 0, it is necessary that of the 12 terms (equal to +1 or −1) that make up $K^{(1)} - K^{(2)}$ not less than seven take +1 (the probability that one of the terms is +1 is 1 − p = 1 − 0.9 = 0.1. Thus.

$$\mu = \sum_{k=7}^{12} C_{12}^k (1-p)^k p^{12-k} = \sum_{k=0}^{5} C_{12}^k p^k (1-p)^{12-k}.$$

From the tables [12] we find that $\mu = 0$ with accuracy to three decimal places.

Similarly, we find that

$$\nu = \sum_{k=6}^{12} C_{12}^k (1-p)^k p^{12-k} \approx 0.$$

If P = 0.7, the same calculations would give $\mu = 0.123$, $\nu = 0.038$.

The threshold K is chosen on the basis of the relative importance of errors of the first and second kind. If it is more important not to miss cases in which the second model is valid (i.e., not to accept false solutions about the validity of the first hypothesis), the threshold K must be greater than zero. The greater K, the lower μ and, accordingly, the greater ν. For example, for K = 3, we obtain

$$\mu = \sum_{k=10}^{12} C_{12}^k (1-p)^k p^{12-k} \leqslant 0.001,$$

$$\nu = \sum_{k=3}^{12} C_{12}^k (1-p)^k p^{12-k} = 0.741.$$

LITERATURE CITED

1. Arnold, E. P., V. I. Keilis-Borok, A. L. Levshin, I. I. Pyatetskii-Shapiro, and P. L. Willmore, "Automatic identification of seismic wave arrivals," this volume, p. 25.
2. Bol'shev, L. N., and N. V. Smirnov (1965), Statistical Tables, Moscow, Nauka.
3. Seismological Bulletin of the Basic Seismological Network of the USSR, March, 1964, Moscow, Izd. AN SSSR, 1965.
4. Keilis-Borok, V. I., S. S. Mebel', I. I. Pyatetskii-Shapiro, L. Yu. Vartanova, and T. S. Zhelankina, "Computer determination of earthquake focal depth," this volume, p. 16.
5. Bessonova, E. N., O. D. Gotsadze, V. I. Keilis-Borok, I. V. Kirillova, S. D. Kogan, T. I. Kukhtikova, L. N. Malinovskaya, I. I. Pavlova, and A. A. Sorskii (1957), "Study of earthquake mechanism," Tr. Geofiz. Inst. Akad. Nauk SSSR, No. 40, p. 166 [English translation: New York, Plenum Press (1960)].
6. Jeffreys, H. (1952), The Earth: Its Origin, History, and Physical Constitution, Cambridge, Cambridge University Press.
7. Malinovskaya, L. N. (1958), "On the dynamic characteristics of transverse waves with total internal reflection," Izv. Akad. Nauk SSSR, Ser. Geofiz., No. 2, pp. 184-195.
8. Keilis-Borok, V. I., L. G. Pavlova, I. I. Pyatetskii-Shapiro, P. T. Reznyakovskii, and T. S. Zhelankina, "Computer determination of earthquake epicenters," this volume, p. 13.
9. International Seismological Center, Bulletin for March, 1964: Edinburgh, 1967.
10. Kasahara, K. (1963), "Computer program for a fault-plane solution," Bull. Seism. Soc. Amer., 53:1-14.
11. Keilis-Borok, V. I. (1958), "The study of earthquake mechanism," Publ. Dominion Obs. Ottawa, 20:279-294.
12. Knopoff, L. (1960), "Analytical calculation of the fault-plane problem," Publ. Dominion Obs. Ottawa, 24:309-315.
13. Ritsema, A. R. (1958), "(i, Δ) curves for body seismic waves of any focal depth," Lembaga Meteorologi dan Geofisik, Verh. 54, Djakarta.
14. Wickens, A. J., and J. H. Hodgson (1967), "Computer re-evaluation of earthquake mechanism solutions, 1922-1962," Publ. Dominion Obs. Ottawa, 33:1-560.

PART II

SEISMIC WAVES

SOME PROBLEMS OF THE USE OF P-WAVE DYNAMICS TO FIND THE VELOCITY PROFILE OF THE UPPER MANTLE†

G. V. Golikova

This paper presents a numerical study of effect on the amplitude-distance curve of body waves caused by a discontinuity in the velocity gradient at a certain depth. The study was made with an elementary model: two layers with different gradients. The results can be used for more complicated models.

1. Standard Working Formulas

Amplitude curves were calculated for velocity profiles approximated by two layers. In each layer, the velocity was represented by the Bullen law

$$v(r) = v_q \left(\frac{r}{r_q}\right)^{-k_q}, \qquad (1)$$

where v(r) is the velocity at distance r from the center of the earth; $q = 1, 2, 3, \ldots, n$; $r_q < r < r_{q-1}$; and k_q is approximately proportional to the velocity gradient.

The calculations were made using the program described in [4].

The wave intensity was calculated by the formula

$$A = \frac{\sqrt{\sin i_0 \cdot \sin i_s} \cdot \prod\limits_j \varkappa^{(1,2)} \prod\limits_i \varkappa^{(2,1)}}{v_0 \sqrt{R\rho \,|\cos i_s|\, \sin \Delta \left|\frac{\partial \tau}{\partial i_s}\right|_{r=R}}}, \qquad (2)$$

where

$$\frac{\partial \tau}{\partial i_s} = 2p_0 \cot i_0 \left(\sum_q \frac{\tan i_q - \tan i_{q-1}}{k_q + 1} - \frac{\tan i_{m-1}}{k_m + 1}\right) \qquad (3)$$

i_s and i_0 are the angles of emergence of the ray from the source and at the earth's surface, respectively; R is the radius of the earth; and $\varkappa^{(1,2)}$ and $\varkappa^{(2,1)}$ are the refractive indices at the interface for a longitudinal wave moving downward and upward, respectively.

All of the factors in expression (2), for the range of epicentral distances that interest us (Δ = 5-25°), vary smoothly and change insignificantly, with the exception of the derivative $\partial\tau/\partial i_s$. This derivative is determined by the nature of the velocity profile and can vary nonmonotonically. Oscillations of the amplitude curve are observed due to variation of this quantity. It can be shown that the derivative $\partial\tau/\partial i_s$ (and, therefore, the amplitude A) is determined by the velocity gradient in the vicinity of the ray minimum.‡ If we exclude the cases when the deepest point of the ray is immediately below one of the boundaries of the second kind, i.e., when i_m are close to $\pi/2$, it is easy to see that the difference ($\tan i_q - \tan i_{q-1}$) in formula (3) is considerably smaller than $\tan i_m$ for all overlying layers. The main contribution to the expression for $\partial\tau/\partial i_s$ is made by the final term $\tan i_{m-1}/k_m + 1$. Thus, we can obtain from formula (3) the following approximate equation:

$$\frac{\partial \tau}{\partial i_s} \approx \frac{2 \cot i_0 \tan i_{m-1}}{\frac{\bar{v}}{r}(k_m + 1)}, \qquad (4)$$

† Translated from Vychislitel'naya Seismologiya, No. 2, Moscow, Nauka (1966), pp. 46-55.

‡ For a sphere, the analog of the velocity gradient dv/dz is $\xi(r) = -r\frac{d}{dr}\left(\frac{v}{r}\right)$, which we shall call the generalized velocity gradient.

where $\bar{r}$ is the distance from the center of the earth to the deepest point of the ray, and $\bar{v}$ is the velocity at that point. In this case, $\bar{v}/\bar{r}(k_m + 1)$ is the generalized gradient $\xi(r)$.

If we ignore the angles i_m that are close to $\pi/2$, then the variation of cot i_0 and tan i_m within the layer can be ignored. Then the amplitude of a refracted wave is approximately proportional to $\sqrt{\xi(r)}$. The segment of the amplitude-distance curve corresponding to the emergence at the surface of rays whose deepest point is in some layer will have an almost constant level which is determined by $\xi(r)$ in that layer.

For a two-layer medium, the amplitude-distance curves $A(\Delta)$ have the following form: the first layer, which has the gradient $(dv/dz)(\partial v/\partial z)_1$, corresponds to a segment of almost constant level (A_1); the layer with the gradient $(dv/dz)_2$ has a similar segment of constant level (A_2), which is either raised $(dv/dz)_2 > (dv/dz)_1$ or lowered $(dv/dz)_2 < (dv/dz)_1$ relative to the first level (A_1). Between these segments is a region of rather sudden amplitude increase or decrease. The approximate equations

$$\frac{\left(\frac{\partial\tau}{\partial i_s}\right)_1}{\left(\frac{\partial\tau}{\partial i_s}\right)_2} \simeq \frac{k_2+1}{k_1+1}; \tag{5}$$

$$\frac{A_1}{A_2} \simeq \sqrt{\frac{k_1+1}{k_2+1}}. \tag{6}$$

follow from relation (4). Otherwise, the level difference between individual segments of the log A curve is determined by the ratio of $\xi(r)$ or $k+1$ for the corresponding layers. Analysis of the formulas shows that the derivative $\partial\tau/\partial i_s$ increases mainly in the layer in which the ray has its deepest point. Therefore, relation (6) is valid also when, for example, the velocity profile contains another layer above these two layers.

2. Models Used for Calculations

The velocity profile of the earth's crust (h_k = 35 km) is considered constant in all cases, since the shape of the amplitude curve at epicentral distances $\Delta > 5°$ is practically independent of crustal structure. The crustal velocity varied linearly – from 6.0 km/sec at the surface to 7.0 km/sec at the Mohorovičić discontinuity, where the jump reached 8.1 km/sec. In the mantle, to a depth of 700 km, the velocity varied continuously. A simple combination of profile elements – two layers with velocity gradients $(dv/dz)_1$ and $(dv/dz)_2$ and a variable boundary between them r – was used for this part of the mantle.

One of eight velocity gradients $(dv/dz)_1$, with values of from 0 to $0.35 \cdot 10^{-2}$ sec^{-1}, was assigned in the first layer. In the second layer, the velocity gradient $(dv/dz)_2$ was given one of 10 values, from 0 to $0.6 \cdot 10^{-2}$ sec^{-1}. The level of the boundary r had eight values, from 6200 to 6125 km. The values of each of the three parameters $(dv/dz)_1$, $(dv/dz)_2$, and r used for the calculations are indicated below:

$\left(\frac{dv}{dz}\right)_1$, sec^{-1}	$\left(\frac{dv}{dz}\right)_2$, sec^{-1}	r, km	$\left(\frac{dv}{dz}\right)_1$, sec^{-1}	$\left(\frac{dv}{dz}\right)_2$, sec^{-1}	r, km
0	0	6025	$0.14 \cdot 10^{-2}$	$0.15 \cdot 10^{-2}$	6125
$0,35 \cdot 10^{-3}$	$0.3 \cdot 10^{-3}$	6050	$0.21 \cdot 10^{-2}$	$0.18 \cdot 10^{-2}$	6150
$0.7 \cdot 10^{-3}$	$0.9 \cdot 10^{-3}$	6075	$0.28 \cdot 10^{-2}$	$0.25 \cdot 10^{-2}$	6175
$0.105 \cdot 10^{-3}$	$0.12 \cdot 10^{-2}$	6100	$0.35 \cdot 10^{-2}$	$0.35 \cdot 10^{-2}$	6200
				$0,45 \cdot 10^{-2}$	
				$0,55 \cdot 10^{-2}$	

Thus, a total of 640 different velocity profiles were calculated. As an example, Fig. 1, a and b, shows five velocity profiles and their corresponding amplitude curves.

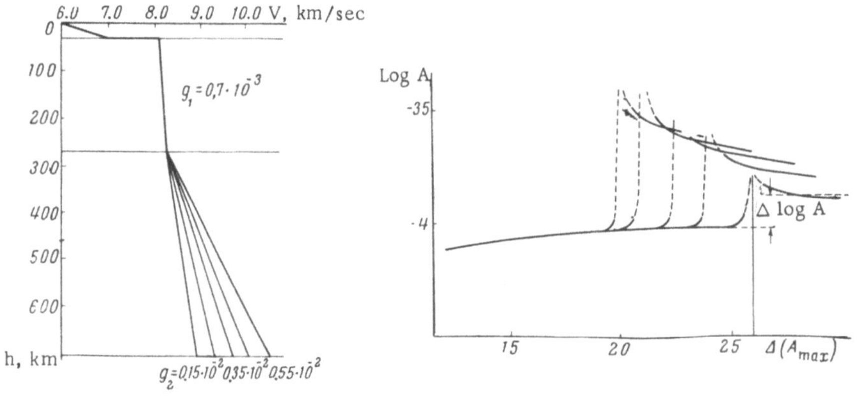

Fig. 1. Example of five velocity profiles (a) and their corresponding amplitude curves (b).

3. Results

As was pointed out above, the amplitude curve for a two-layer medium consists of two almost rectilinear segments of different levels and a region of abrupt amplitude increase or decrease between them. Therefore, the following values can be used to characterize $A(\Delta)$:

a. The level difference between individual curve segments (Fig. 1, b);

b. The epicentral distance corresponding to the region of abrupt amplitude increase or decrease.

The level difference ($\Delta(\log A)$) is determined by the ratio $\xi_1(r)/\xi_2(r)$. A graph of $\Delta(\log A)$ as a function of the ratio $\sqrt{k_2+1}/\sqrt{k_1+1}$ is shown in Fig. 2. It was found that a change in the depth of the boundary at which a velocity-gradient discontinuity occurs does not affect $\Delta(\log A)$. Thus, the graph (Fig. 2) can be used to determine the ratios of the gradients $\xi(r)$ from observed distances in the amplitude-curve level.

Now let us examine the characteristics of the curve $A(\Delta)$ which are related to the absolute values of the gradients themselves and the depth of the boundary at which the gradient discontinuity occurs. Two cases were considered: the gradient in the first layer $\xi_1(r)$ less than and greater than $\xi_2(r)$.

1. In the first case, when $\xi_1(r) < \xi_2(r)$, the epicentral distance Δ corresponding to the maximum on the amplitude curve $\Delta(A_{max})$ was considered. The epicentral distance $\Delta(A_{max})$ is a function of $\xi_1(r)$, $\xi_2(r)$, and r. Since the explicit form of this function was unknown, an attempt was made to determine the function $\Delta = f(\xi_1, \xi_2, r)$ or $\Delta = f(k_1, k_2, r)$ approximately. The unknown function was given as a square paraboloid, i.e., in the expansion of this function into a Taylor series, we limited ourselves to terms of the first and second order

$$\Delta = \Delta_0 + a_1(r - r^0) + a_2(k_1 - k_1^0) + a_3(k_2 - k_2^0) + a_4(r - r^0)^2 + a_5(k_1 - k_1^0)^2 + a_6(k_2 - k_2^0)^2 + a_7(k_1 - k_1^0)(r - r^0) + a_8(k_2 - k_2^0)(r - r^0) + a_9(k_1 - k_1^0)(k_2 - k_2^0). \quad (7)$$

The unknown function was expanded into a series in the vicinity of the point $r^0 = 6100$, $k_1^0 = 0$, and $k_2^0 = 2$. The coefficients of the series $a_1, a_2, \ldots, a_9$, and also Δ_0, were determined by the method of least squares using 240 points (versions of the profile) with the values $\Delta_k (k_1+1)_k$, $(k_2+1)_k$. The coefficients are:

$\Delta_0 = 0.43095$	$a_5 = 0.52022 \cdot 10^{-2}$
$a_1 = -0.88185 \cdot 10^{-3}$	$a_6 = 0.34030 \cdot 10^{-6}$
$a_2 = -0.53484 \cdot 10^{-1}$	$a_7 = 0.11812 \cdot 10^{-3}$
$a_3 = -0.51511 \cdot 10^{-1}$	$a_8 = 0.10823 \cdot 10^{-3}$
$a_4 = -0.76628 \cdot 10^{-6}$	$a_9 = 0.68620 \cdot 10^{-2}$

The coefficients were such that after substitution into series (7), terms one order of magnitude smaller than the preceding were obtained, beginning with the fourth. This confirms the validity of ignoring subsequent terms in the Taylor series.

After substitution of the various (k_1+1), (k_2+1), and r lying at the boundaries of the range of values used, with the aid of the calculated coefficients Δ_0, a_1, a_2, etc., we obtained estimates of Δ that agreed with the calculated values with ± 0.3° accuracy. This confirms once more the validity of representing $\Delta = f(k_1+1, k_2+1, r)$ as the series (7), and allows us to assume that the expansion can be used in a wider range of values of k_1+1, k_2+1, and r.

After a relationship has been established between $\Delta(A_{max})$ and k_1+1, k_2+1, and r, expansion (7) can be used to determine k_1+1 and r. After some transformations, an equation for the parameters k_1+1 and r can be derived from it:

$$(0.5202 \cdot 10^{-2} + 0.6862 \cdot 10^{-2} m)(k_1+1)^2 - [0.8447 \cdot 10^{-1} + 0.5837 \cdot 10^{-1} m - \Delta r (0.1181 \cdot 10^{-3} + 0.1082 \cdot 10^{-3} m)](k_1+1) - 0.13245 \cdot 10^{-2} \Delta r - 0.7662 \cdot 10^{-6} \Delta r^2 + (0.66476 - \Delta_{exp}) = 0. \quad (8)$$

$$r - r_0 = \Delta r; \quad \frac{k_2+1}{k_1+1} = m.$$

This equation contains two unknowns: k_1+1 and $r - r_0$. The value m is determined from the level difference of the curve $\log A(\Delta)$ (Fig. 2). By using

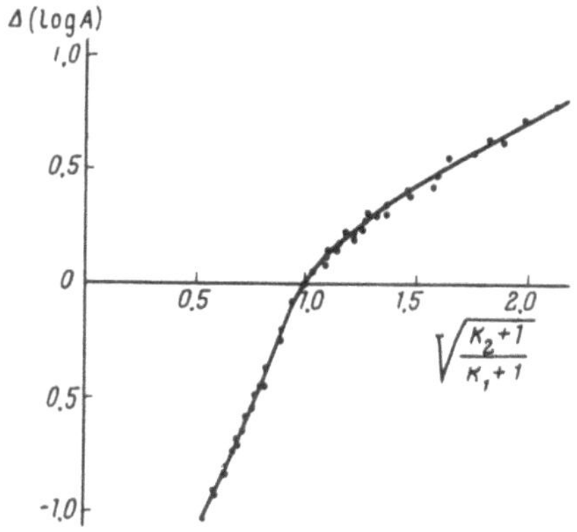

Fig. 2. Graph of $\Delta(\log A)$ versus $\sqrt{k_2+1}/\sqrt{k_1+1}$.

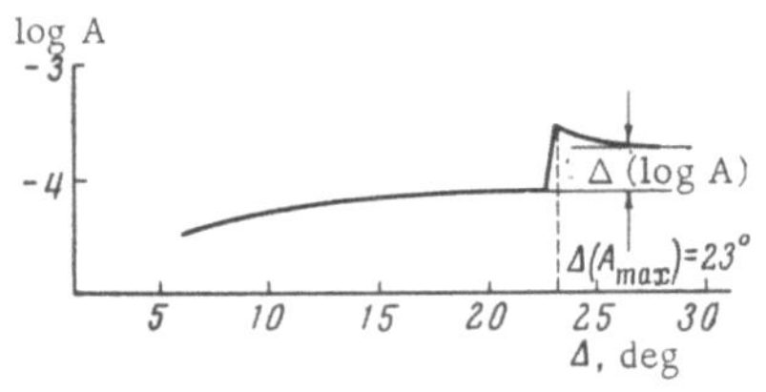

Fig. 3. Initial amplitude curve for two-layer medium.

different values, for example, of the depth of the boundary at which a gradient discontinuity occurs, we can calculate different pairs of k_1+1 and k_2+1. Thus, a unique solution cannot be obtained by interpreting a single amplitude curve. One could be found if we knew the absolute level $A(\Delta)$.

The process of finding the parameters of the velocity profile using only the criteria described above can be illustrated by the following example. The initial amplitude curve corresponded to a two-layer medium with $k_1+1 = 1.268$, $k_2+1 = 3.329$, and $r = 6100$ (Fig. 3). This curve was used to measure the increment Δ (log A) and determine $\Delta(A_{max})$: Δ (log A) = 0.45 and $\Delta(A_{max}) = 23°$. The value m = 2.6 was found with the graph of Δ (log A) as a function of the ratio $\sqrt{k_2+1}/\sqrt{k_1+1}$ (Fig. 2). The corresponding values of the parameters

r	6100	6075	6050	6025	6000	6125	6150	6175	6200
k_1+1	1.26	1.37	1.46	1.54	1.61	1.14	1.005	0.85	0.68
k_2+1	3.32	3.60	3.84	4.06	4.24	3.00	2.64	2.23	1.78

were calculated by Eq. (8), with allowance for the obtained m, for nine proposed boundary depths† at which a gradient discontinuity occurs.

Thus, nine velocity profiles were obtained (Fig. 4).

The corresponding amplitude-distance curves (Fig. 5) were calculated to check the method for these velocity profiles. It was found that all $A(\Delta)$ curves coincided with the initial curve: they had an intensity peak at the same epicentral distance and the same increment Δ (log A) = 0.45. These amplitude curves differ from one another only in absolute level. Unfortunately, however, this cannot be used in practice, because the absolute level of an observed curve cannot be determined.

As can be seen from this example, interpretation of one amplitude curve does not provide a unique solution. However the band of allowed profiles includes a wide range of velocities, and the corresponding travel-time curves must differ greatly from one another. This can be used to reduce the ambiguity. Figure 6 shows the deviation of the travel-time curves for the proposed velocity profiles from the travel-time curve of the initial profile. The minimum deviations were obtained for the velocity profiles with r = 6100, 6125, and 6075 km, i.e., those close to the initial profile.

2. In the case of $\xi_1(r) > \xi_2(r)$, we used the epicentral distance Δ that corresponded to the region of level decrease on the amplitude curve. This epicentral distance is practically independent of $\xi_2(r)$. Therefore the results of the amplitude calculations for the velocity profiles with the gradient relationship $\xi_1(r) > \xi_2(r)$ could be represented as graphs of

† The boundary depth was set within the range of values for which the coefficients of expansion (7) were calculated.

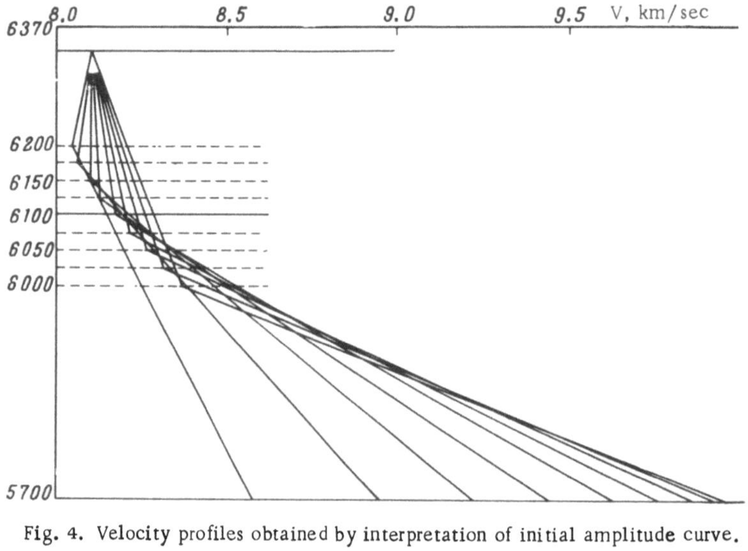

Fig. 4. Velocity profiles obtained by interpretation of initial amplitude curve.

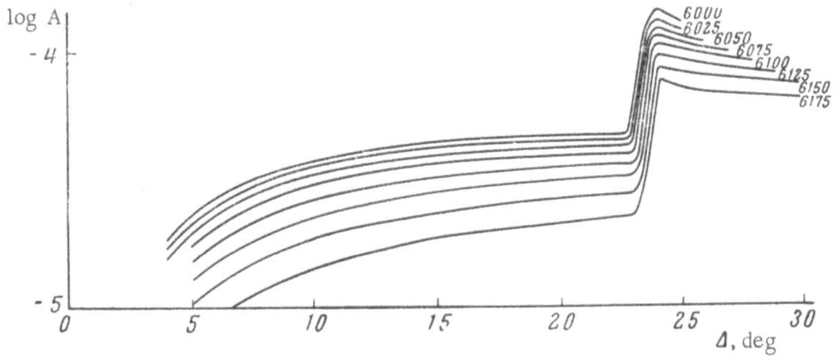

Fig. 5. Amplitude curves for velocity profiles in Fig. 4.

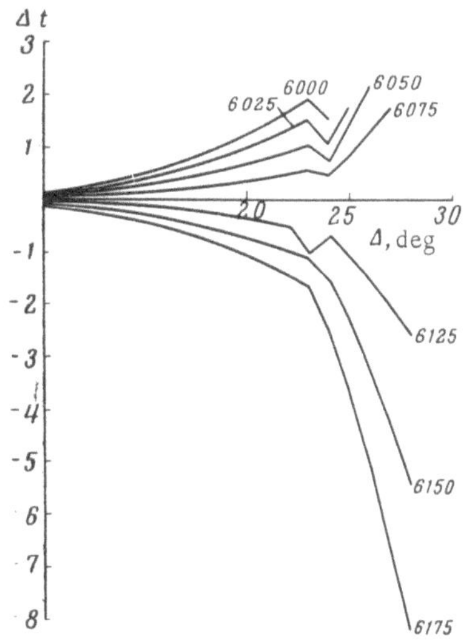

Fig. 6. Deviations of travel-time curves for profiles in Fig. 4 from travel-time curve of initial profile.

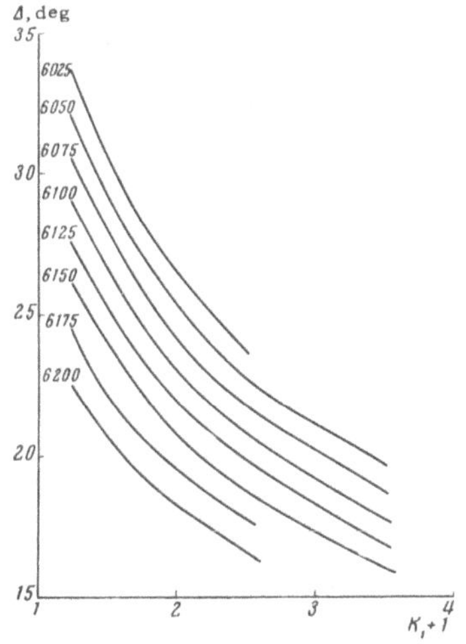

Fig. 7. Graph of epicentral distance Δ versus k_1+1.

Δ versus k_1+1; the parameter of the curves was the boundary depth r (Fig. 7). These graphs can be used to determine the combinations of k_1+1 and r that give the Λ that is obtained from the amplitude curve.

3. When the ratio $\xi_1(r)/\xi_2(r)$ is close to 1, the level differences of the amplitude-distance curves are very slight. Considering that the level differences $|\Delta(\log A)| \geq 0.2$ can be detected from experimental curves, we find that in doing so we can determine gradient ratios that lie outside the interval 1.2-0.9.

Some of the regularities that were found for the two-layer model can be carried over to more complex models. For example, the difference between the levels of the amplitude-distance curve in segments corresponding to layers with different velocity gradients is not a function of the absolute values of the gradients in the layers but only of the ratio of the gradients at the boundary between the layers. It can be shown that this remains valid in the case of a multilayer model. If we divide the amplitude-distance curve into segments with different levels and link to each of these segments a layer in the profile, we can estimate roughly the gradient ratios in the next layers.

LITERATURE CITED

1. Jeffreys, H. (1939), "The times of P, S, and SKS and the velocities of P and S," Mon. Not. Roy. Astr. Soc., Geophys. Suppl., 4:498-533.
2. Vanek, J., and J. Stelzner (1962), "Amplitudenkurven der seismischen Raumwellen," Gerlands Beitr. Geophys., 71:105-119.
3. Yanovskaya, T. B., and I. Ya. Azbel' (1964), "The determination of velocities in the upper mantle from observations of P waves," Geophys. J. Roy. Astr. Soc., 8:313.
4. Azbel', I. Ya., and T. B. Yanovskaya, "Approximation of velocity distributions for calculation of P-wave times and amplitudes," this volume, p. 62.

THE EFFECT OF FOCAL DEPTH IN THE CRUST ON THE WAVE FIELD IN THE INITIAL PART OF A SEISMOGRAM†

G. V. Golikova and T. B. Yanovskaya

Determination of the depth of a source is based on analysis of the first phases of the recording – in particular, on the difference between the arrival times pP – P and sP – P. But the first arrivals can also contain other converted and reflected waves which complicate the wave field in the vicinity of the first arrivals.

Our aim was to find out how the pattern of the first arrivals varied with the depth of the source. We calculated the travel times and amplitudes of waves recorded within 10-25 sec after the first arrival for four source depths: h = 8, 20, 37, and 43 km (Fig. 1). The epicentral distances were the range 5-25°. The types of waves studied depend considerably upon the structure of the crust in the area of the epicenter.‡ The crustal model for the Bukharo-Khivinskii region of Uzbekistan [1] was used for the calculations. The velocity profile is shown in Fig. 1. Thus the conclusions drawn in this paper cannot be considered universal, although some of the theses appear to be of a general nature.

The main waves used in seismology to determine source depth are pP and sP, so it was considered sufficient to calculate the wave field in a time interval that included these waves, i.e., one slightly larger than the sP–P interval. This interval increases with source depth: it was 12 sec at h = 8 km, 15 sec at h = 20, and 20 sec at h = 37 km. The number of different waves (reflected and converted in the vicinity of the source) existing in this time interval reaches 50. We calculated those of these whose amplitude was within at least one order of magnitude of the amplitude of the fundamental waves. In particular, we excluded from consideration the waves that in propagation had undergone more than two conversions at the boundaries of the main velocity discontinuity. Multiply reflected waves were excluded because they did not fall within the time interval in question.

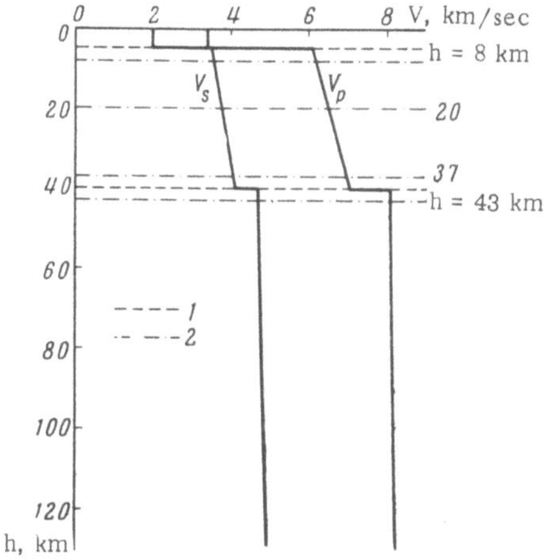

Fig. 1. Velocity profiles for P and S waves: 1) depths of velocity discontinuity; 2) source depths.

The sorting of all possible types of waves for a source located in a given layer can be illustrated by Tables 1 and 2, from which we see how to construct codes for the waves formed when a P or S wave emerges from the source. The tables show

† Translated from Vychislitel'naya Seismologiya, No. 2, Moscow, Nauka, (1966), pp. 83-94.

‡ Waves reflected and refracted in the area of the station can exist in this same time interval. As was shown in [2], in converted waves of the PS type the horizontal component is strong enough to show up on the recording. As calculations in the present paper show, however, these waves have an insignificant vertical component and the main wave field is determined by the superposition of waves formed in the vicinity of the epicenter.

TABLE 1. Source in Crust (5 km < h < 40 km)

$P_2\uparrow$ ($S_2\uparrow$)	$P_1\uparrow$	$P_1\downarrow$ (2) (4)	$P_1\uparrow$	$P_1\downarrow$ (17)
				$S_1\downarrow$
			$S_1\uparrow$	$P_1\downarrow$
				$S_1\downarrow$
		$S_1\downarrow$ (13) (15)	$P_1\uparrow$	$P_1\downarrow$
				$S_1\downarrow$
			$S_1\uparrow$	$P_1\downarrow$
				$S_1\downarrow$
	$S_1\uparrow$	$P_1\downarrow$ (5) (3)	$P_1\uparrow$	$P_1\downarrow$ (18)
				$S_1\downarrow$
			$S_1\uparrow$	$P_1\downarrow$
				$S_1\downarrow$
		$S_1\downarrow$ (16) (14)	$P_1\uparrow$	$P_1\downarrow$
				$S_1\downarrow$
			$S_1\uparrow$	$P_1\downarrow$
				$S_1\downarrow$
	$P_2\downarrow$ (6) (7)			
	$S_2\downarrow$ (19) (20)			

$P_3\uparrow$ ($S_3\uparrow$)	$P_2\uparrow$	$P_1\uparrow$	$P_1\downarrow$ (2) (18)
			$S_1\downarrow$ (13)
		$S_1\uparrow$	$P_1\downarrow$ (5)
			$S_1\downarrow$ (16)
		$P_2\downarrow$ (6) (20)	
		$S_2\downarrow$ (19)	
	$S_2\uparrow$	$P_1\uparrow$	$P_1\downarrow$ (11) (4)
			$S_1\downarrow$ (15)
		$S_1\uparrow$	$P_1\downarrow$ (12) (3)
			$S_1\downarrow$ (17) (14)
		$P_2\downarrow$ (10) (7)	
		$S_2\downarrow$	
	$P_3\downarrow$ (9) (8)		

Note. The arrows indicate the wave direction (upward or downward) and the subscripts give the layer number (1 – sedimentary, 2 – crystalline, 3 – mantle).

TABLE 2. Source in Mantle (h > 40 km)

$P_3\uparrow$ ($S_3\uparrow$)	$P_2\uparrow$	$P_1\uparrow$	$P_1\downarrow$ (2) (18)
			$S_1\downarrow$ (13)
		$S_1\uparrow$	$P_1\downarrow$ (5)
			$S_1\downarrow$ (16)
		$P_2\downarrow$ (6) (20)	
		$S_2\downarrow$ (19)	
	$S_2\uparrow$	$P_1\uparrow$	$P_1\downarrow$ (11) (4)
			$S_1\downarrow$ (15)
		$S_1\uparrow$	$P_1\downarrow$ (12) (3)
			$S_1\downarrow$ (17) (14)
		$P_2\downarrow$ (10) (7)	
		$S_2\downarrow$	
	$P_3\downarrow$ (9) (8)		

*See note in Table 1.

the possible reflections, refractions, and conversions at the boundaries of a velocity discontinuity in the vicinity of the source. Later the wave continues as a direct refracted P.

The waves for which calculations were made are numbered. The upper number refers to a wave that emerges from the source as P; the lower, to a wave that emerges as S. Waves that clearly do not fall into the given time interval are not included in the tables.

When the source is below the Mohorovičić discontinuity, a wave that emerges from the source downward does not undergo reflections or refractions in the vicinity of the source and is therefore omitted from the tables. This direct refracted P wave has the number 1.

All possible waves formed in the vicinity of the source can be sorted using similar tables. We also considered the multiple waves reflected downward from the Mohorovičić discontinuity for the P wave and the strongest waves reflected in the crust. They are denoted by a prime on P or S, according to the type of initial wave: a single prime indicates a singly reflected wave and a double prime a doubly reflected wave. Calculations were also made for waves that underwent reflections and conversions in the vicinity of the station. The crustal structure at reflection points was assumed to be the same as in the area of the station.

The radiation intensity of the source was assumed to be higher by a factor of five for transverse waves than for longitudinal [5]. The radiation of both waves was considered to be uniform in all directions. But in reality the wave field is considerably dependent upon the source radiation pattern, so our conclusions apply only to some averaged first arrival wave pattern.

The detailed structure of the mantle is not important in this problem. For simplicity, it was assumed that the low-velocity zone was absent from the upper mantle. The calculations were made for the vertical component of displacement.

Results

The wave field $\varphi(t, \Delta)$ at a particular epicentral distance is a superposition of different types of waves: $\varphi(t, \Delta) = \sum_k A_k(\Delta) f_k(t - \tau_k(\Delta))$, where $A_k(\Delta)$ is the amplitude of the k-th wave, $f_k(t)$ is the wave form as a function of time, and $\tau_k(\Delta)$ is the relative arrival time [3]. We calculated only $A_k(\Delta)$ and $\tau_k(\Delta)$, and assumed that the wave forms $f_k(t)$

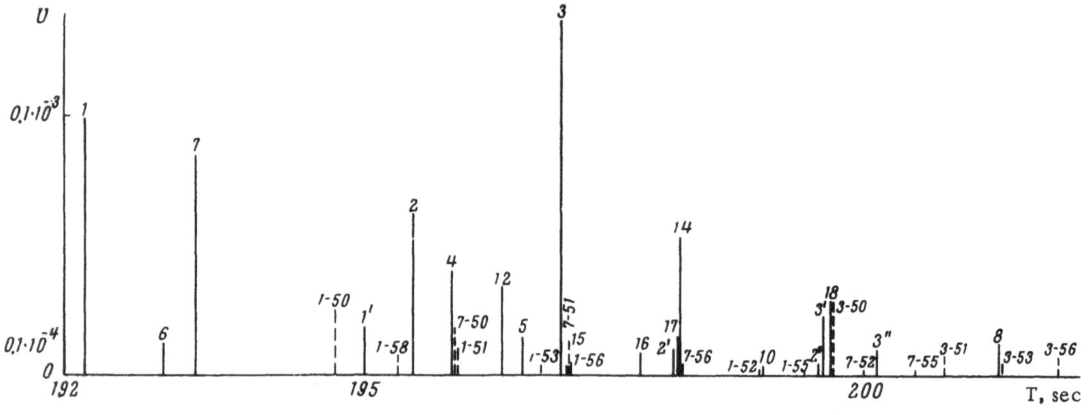

Fig. 2. Amplitudes and delays of waves for h = 8 km and Δ = 14°.

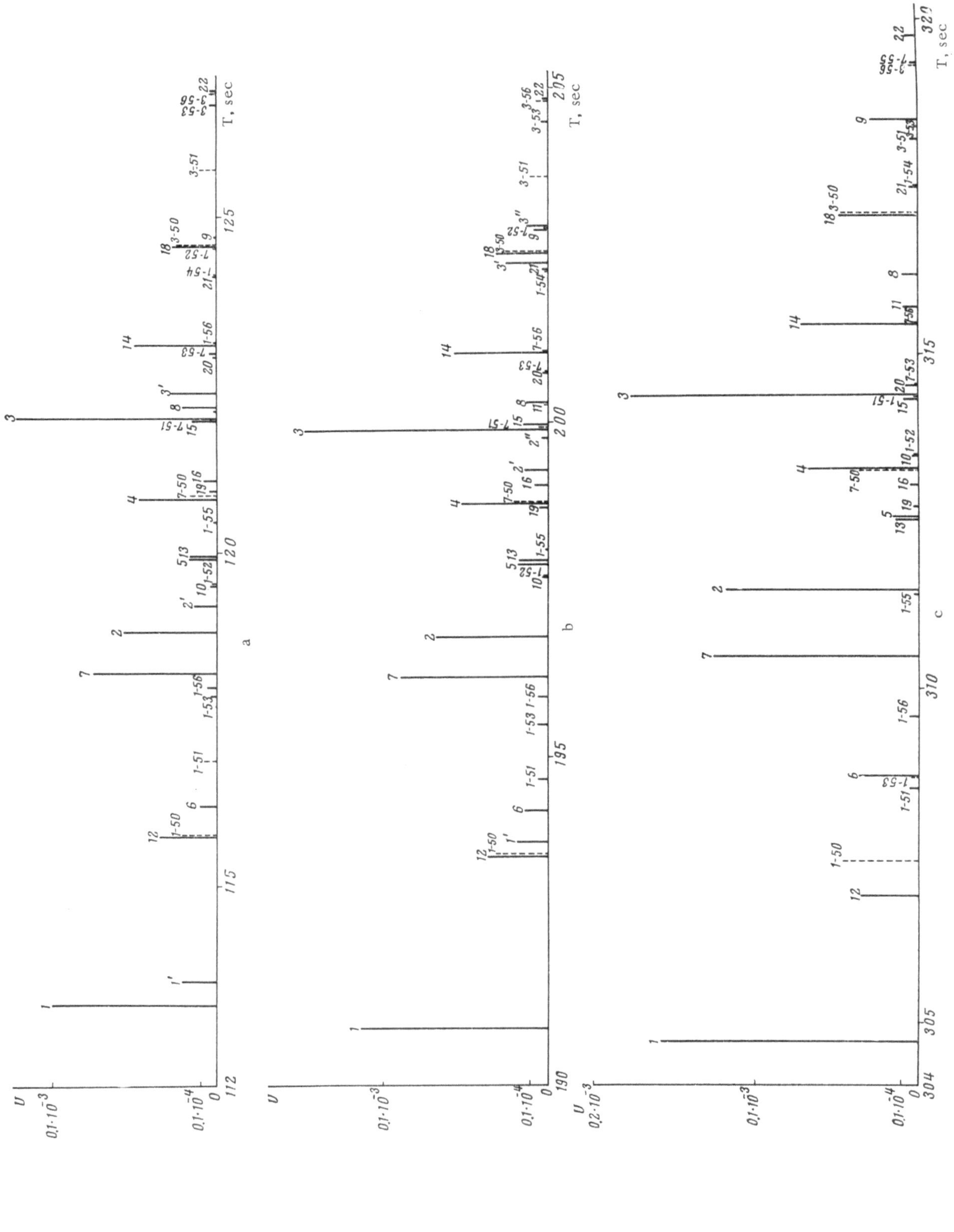

Fig. 3. Amplitudes and delays of waves for h = 20 km and a) Δ = 8°, b) Δ = 14°, and c) Δ = 24°.

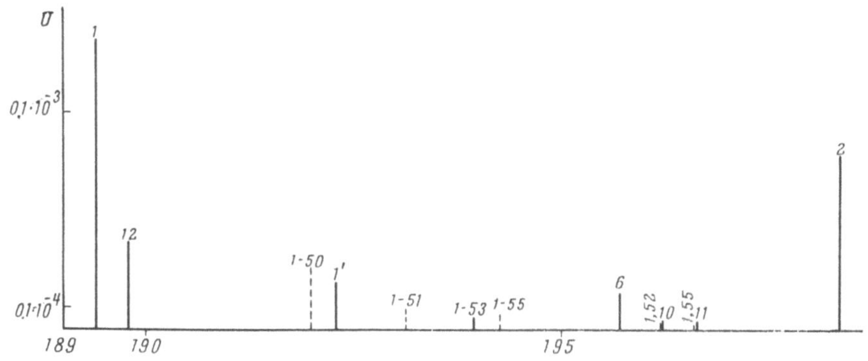

Fig. 4. Amplitudes and delays

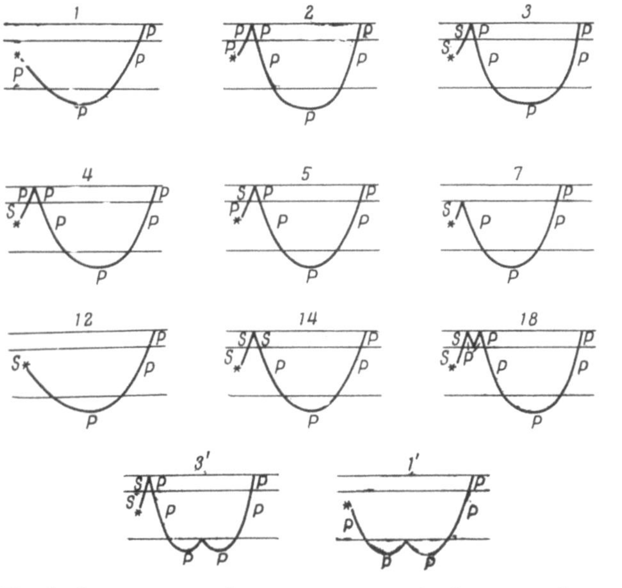

Fig. 5. Strongest waves for source in crust. The figures are the wave indices from Tables 1 and 2.

were unit impulses. The results of calculating $A_k(\Delta)$ and $\tau_k(\Delta)$ were represented as diagrams in which τ_k was plotted on the horizontal axis for each epicentral distance and A_k was plotted on the vertical axis (Figs. 2-4 and 6). This representation is convenient in that for any given waveform $f(t)$ we can draw qualitative conclusions about where to expect the arrivals of the individual phases on the recording and where the recording is substantially distorted by interference.

Figures 2, 3, and 4 show the arrivals of various types of waves and their relative amplitudes for a number of source depths in the crust and epicentral distances. Common to all cases is the fact that two groups of waves are distinguished in the given time interval from the beginning of the recording. The main (strongest) wave in the first group is P, which is the first arrival; in the second group, it is sP. For small source depths, these groups are practically inseparable: only when h becomes greater

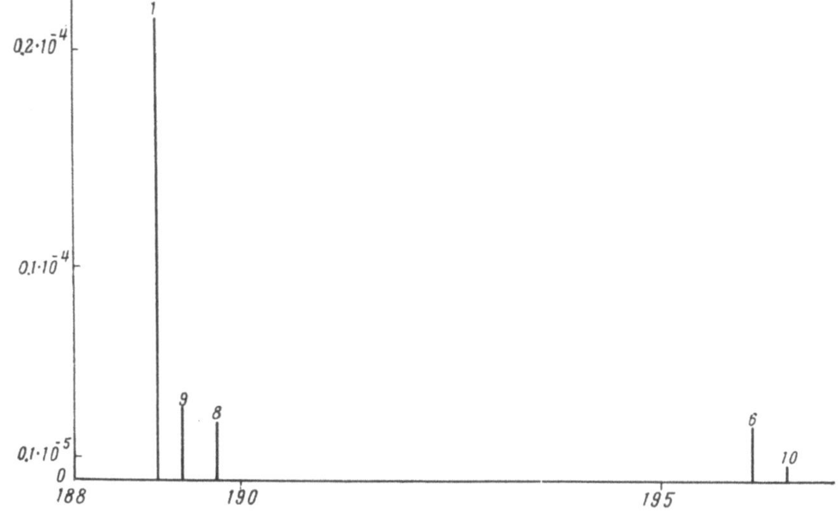

Fig. 6. Amplitudes and delays of

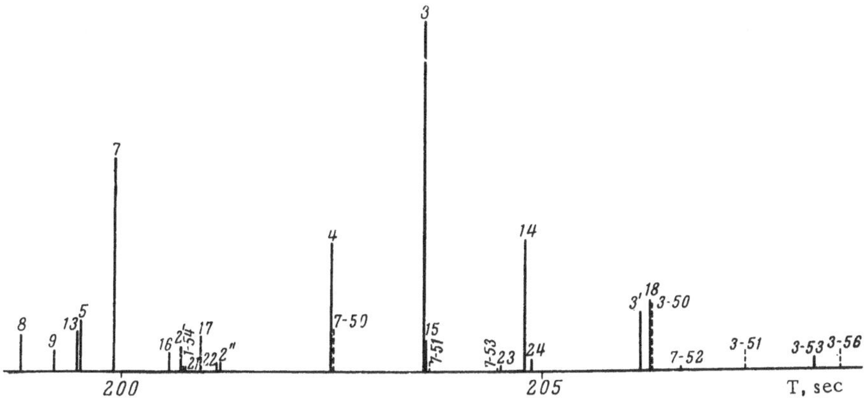

of waves for h = 37 km and Δ = 14°.

than 20 km does the second group lag behind the first by more than 5 sec. Moreover, as the source depth increases the groups become rearranged (waves from the first group move to the second), and the duration of the waves changes. For a fixed source depth, the relative arrival times and amplitudes of waves reflected and converted near the epicenter remain unchanged in the interval of epicentral distances 5-26°. This is because the angles of emergence of the rays from the source vary little for these epicentral distances.

The wave delays, which are chiefly determined by crustal structure, remain constant. Waves reflected downward from the Mohorovičić are an exception. The delay times of these waves increase with an increase in epicentral distance. At epicentral distances $\Delta > 20°$, they are practically outside of the given time interval, so that at these distances the recording of the first phases is no longer complicated by the superposition of multiple waves.

Waves 1, 2, 3, 4, 5, 6, 7, 12, 14, 18, 1', and 3' are the strongest of those considered for a source in the crystalline layer. They are shown schematically in Fig. 5.

Since it was assumed that transverse waves had five times greater radiation intensity than longitudinal, the waves that emerge from the source as transverse have the greatest intensity.

The first group is formed by a P wave and waves reflected from the nearest overlying interface. In our velocity profile, this is the boundary between the sedimentary and crystalline layers. As the source depth is increased, the delay of these waves relative to P is increased and they gradually move to the second group.

In the second group, the main waves are those reflected from the free surface (2 and 3). Intermediate converted waves are situated between them. As the source depth is increased, the duration of this group increases from 1.5 to 5 sec.

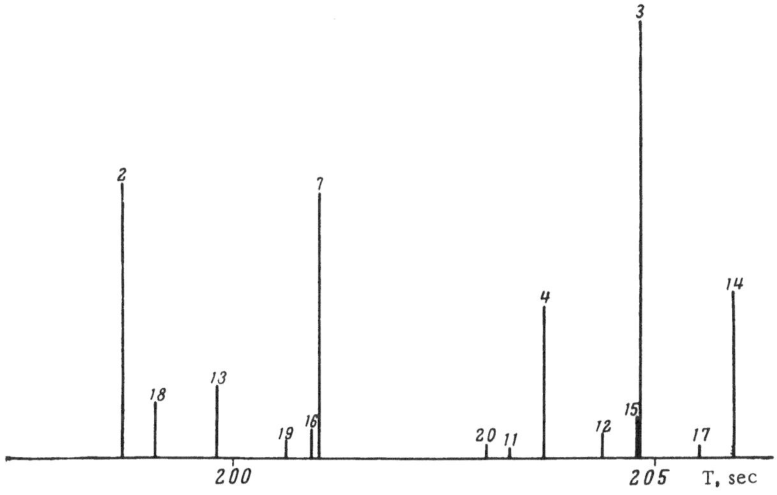

waves for h = 43 km and Δ = 14°.

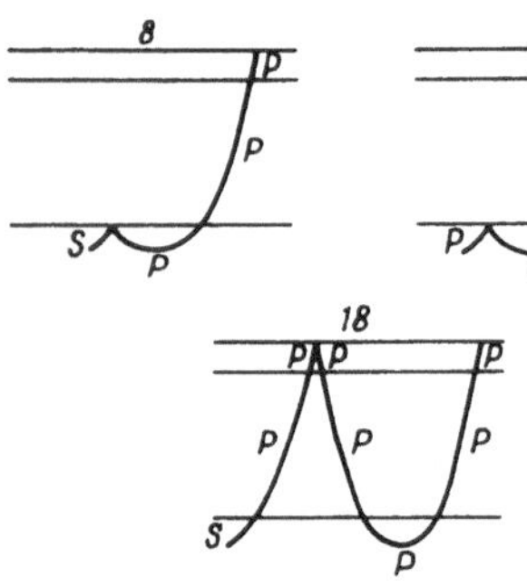

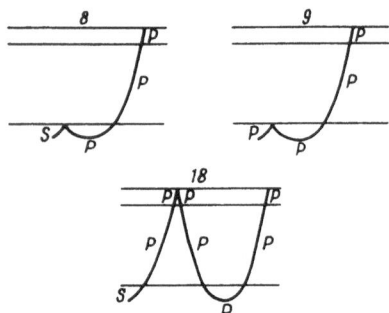

Fig. 7. Waves for source below Mohorovičić discontinuity. The figures are the wave indices from Tables 1 and 2.

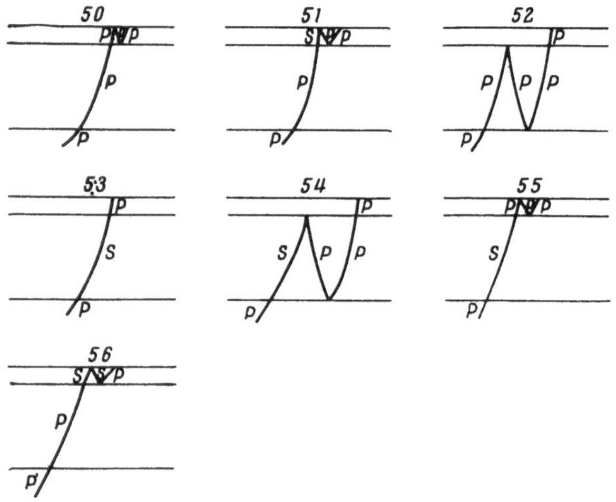

Fig. 8. Waves in area of station. The figures give the wave indices from Tables 1 and 2.

When the source is immediately below the Mohorovičić discontinuity the types of waves are changed, but the strongest waves remain the same and therefore the nature of the wave pattern is preserved.

For comparison with the preceding case (h = 37 km), the waves were calculated for the source depth h = 43 km (Fig. 6). The strongest of the new types of waves are 8, 18, and 9, which are shown in Fig. 7.

To clear up the problem of how waves reflected and converted in the vicinity of the recording station affect the wave pattern, we calculated A_k and τ_k for waves that were formed as follows. "Tails" of various waves that were converted and reflected near the station were added onto the strongest waves formed near the epicenter. The types of conversions and reflections are shown schematically in Fig. 8.

The amplitudes of these waves are shown by the broken lines in Figs. 2-4, where the first number indicates the initial wave and the second number the wave reflected near the station. It is apparent that except for waves of type n-50, all waves are so small that they can be ignored.

Now let us see how the foregoing results can be used to determine source depth. In seismology, source depth in the crust is determined from the difference in the arrivals of P and sP by a method developed by N. V. Kondorskaya [4]. When the phases are correctly identified, the depth is determined with an accuracy of ± 5 km.

Our calculations show that the sP wave is actually the strongest in the second group of arrivals. Table 3 shows the delay times of sP relative to the first arrival at various epicentral distances as functions of source depth for our crust velocity profile.

As the source depth increases the second group expands, and additional waves appear in the vicinity of sP, maximizing its arrival time. The wave (7) reflected downward from the bottom of the sedimentary layer can be especially deceptive. This wave is comparable in amplitude with sP and arrives earlier than it by 3.8 sec, on the average, for the given velocity profile. The frequencies of these waves must be identical, since they are waves of the same type (they emerge from the source as S and then undergo conversion). There is therefore a possibility of error in phase identification: the arrival of wave 7 can be mistaken for the arrival of sP.

As an example, Table 4 shows the differences between the arrival times of wave 7 and P.

TABLE 3

h = 8 km		h = 20 km		h = 37 km		h = 43 km	
Δ, deg	$(t_{sP} - t_P)$, sec	Δ, deg	$(t_{sP} - t_P)$, sec	Δ, deg	$(t_{sP} - t_P)$, sec	Δ, deg	$(t_{sP} - t_P)$, sec
5	4.75	5	8.80	5	13.8	5	15.5
17	4.90	17	9.10	17	14.5	17	17.0
24	5.00	24	9.70	24	15.7	24	17.5

TABLE 4

h = 20 km		h = 37 km	
Δ, deg	$(t_{(7)} - t_P)$, sec	Δ, deg	$(t_{(7)} - t_P)$, sec
5	5.1	5	10.1
17	5.3	17	10.7
24	5.8	24	11.8

If in determining source depth from the difference sP − P this error is made in phase identification, the source depth can be underestimated by the value $\Delta h = H \cdot (b_2 \cos i_2 / b_1 \cos i_1)$, where H is the thickness of the sedimentary layer and b_1, b_2, i_1, and i_2 are the velocities and angles of emergence of transverse waves in the sedimentary and crystalline layers, respectively. With the crustal structure used in our calculations the error of depth determination would be about 7 km.

The presence in the crust of intermediate boundaries (for example, between granitic and basaltic layers) does not change the basic conclusions of this work, since waves reflected and converted at these boundaries have low amplitudes. The presence of a boundary with a great velocity contrast, such as the boundary between sedimentary and crystalline layers, is significant. When the thickness of the sedimentary layer changes, therefore, the wave pattern must change appreciably.

LITERATURE CITED

1. Vol'vovskii, I. S., and B. S. Vol'vovskii (1962), "The nature of wave patterns in studies by the deep seismic sounding method in Uzbekistan," in: Deep Seismic Sounding of the Earth's Crust in the USSR, Leningrad, Gostopizdat, pp. 77-94.
2. Bulin, N. K. (1960), "Determination of the depth of a rock basement using PS earthquake waves," Izv. Akad. Nauk SSSR, Ser. Geofiz., No. 6, pp. 781-786.
3. Yanovskaya, T. B., and G. V. Golikova (1964), "Method and program for calculating P wave dynamics in the earth's mantle," in: Problems of the Dynamic Theory of Seismic Wave Propagation, No. 7, pp. 123-129.
4. Kondorskaya, N. V. (1956), "Discrimination of the sP wave in shallow earthquakes, and its use in determining focal depth," Tr. Geofiz. Inst. Akad. Nauk SSSR, No. 36, pp. 35-47.
5. Bessonova, E. N., O. D. Gotsadze, V. I. Keilis-Borok, I. V. Kirillova, S. D. Kogan, T. I. Kukhtikova, L. N. Malinovskaya, I. I. Pavlova, and A. A. Sorskii (1960), "Investigations of the mechanism of earthquakes," Tr. Geofiz. Inst. Akad. Nauk SSSR, No. 40, p. 166 [English translation: New York, Plenum Press (1960)].

APPROXIMATION OF VELOCITY DISTRIBUTIONS FOR CALCULATION OF P-WAVE TIMES AND AMPLITUDES†

I. Ya. Azbel' and T. B. Yanovskaya

An algorithm for calculating travel-time and amplitude curves of P waves by the ray theory is described in [1]. The earth model was a layered sphere. The dependence of velocity v upon depth r in each layer was represented by a cubical parabola. This approximation makes v(r) smooth enough to justify use of the ray theory (with the exception of regions where the ray theory is entirely inapplicable: caustics and shadow zones). The disadvantage of this approximation is that the travel-time curve (travel times $\tau(i_0)$ and epicentral distances $\Delta(i_0)$) is calculated by numerical integration, while the amplitudes are found by numerical differentiation of the travel-time curve.

This gives rise to two problems:

1. Finding a smooth approximation of the velocity distribution (without discontinuities in velocity gradient at the layer boundaries) that allows the integrals for $\tau(i_0)$ and $\Delta(i_0)$ to be calculated in closed form.

2. Determining the extent to which the amplitude–distance curve is distorted by an uneven approximation of the velocity–depth curve and by inaccuracy of the zeroth approximation of the ray theory for rays that come immediately below the boundaries of a gradient discontinuity.

1. Approximation of Profile

The velocity–depth curve is approximated in a depth interval by dividing this interval into a number of layers in each of which the dependence of velocity upon depth (or upon the distance r to the center of the earth) is defined by the same function but with different numerical parameters. In the simplest case only velocity continuity is ensured, and two parameters are sufficient. To ensure continuity of the velocity gradients as well three parameters are necessary. Thus the problem boils down to selecting a function v(r) that contains three parameters and permits integration in closed form of the expressions

$$\tau(i_0) = \int_{r_{q+1}}^{r_q} \frac{dr}{v(r)\sqrt{1-\frac{p_0^2 v^2}{r^2}}}; \tag{1}$$

$$\Delta(i_0) = p_0 \int_{r_{q+1}}^{r_q} \frac{v(r)\,dr}{r^2\sqrt{1-\frac{p_0^2 v^2}{r^2}}}. \tag{2}$$

The following function was proposed:

$$v(r) = \frac{r}{r_i}\left[(v_i + a_i)\left(\frac{r}{r_i}\right)^{-k_i} - a_i\right]. \tag{3}$$

Here we must determine the three parameters v_i, a_i, and k_i (it is obvious that r_i is not an independent parameter). The value v_i is the velocity at the boundary $r = r_i$, and the parameters k_i and a_i determine the nature of velocity variation in the layer $r_{i+1} < r < r_i$.

When $a_i = 0$, formula (3) becomes Bullen's law

$$v(r) = v_i\left(\frac{r}{r_i}\right)^{-k_i+1}, \tag{4}$$

and when $k_i = 1$, (3) becomes the linear law

$$v(r) = v_i + \frac{a_i}{r_i}(r_i - r). \tag{5}$$

† Translated from Vychislitel'naya Seismologiya, No. 2, Moscow, Nauka (1966), pp. 56-70.

It is easy to show that law (3) admits both an increase and a decrease in the velocity gradient in a layer. In fact, the second derivative of the velocity has the form

$$\frac{d^2v}{dr^2}=\frac{k_i(k_i-1)}{rr_i}(v_i+a_i)\left(\frac{r}{r_i}\right)^{-k_i}. \qquad (6)$$

Thus the sign of d^2v/dr^2 is determined by the product $k_i(k_i-1)(v_i+a_i)$. Since $(r/r_i)^{-k_i}>0$, the sign of d^2v/dr^2 in the layer is preserved, i.e., only monotonic variation of the velocity gradient is possible within one layer.

A disadvantage of the proposed approximation is that it does not permit a smooth transition from regions with a velocity increase to waveguides.

A waveguide is defined by the condition $(dv/dr)-(v/r)>0$. We shall show that $\xi(r)=(dv/dr)-(v/r)$ does not change sign within a layer.

It follows from (3) that

$$\xi(r)=-\frac{k_i}{r_i}(v_i+a_i)\left(\frac{r}{r_i}\right)^{-k_i}. \qquad (7)$$

Since $(r/r_i)^{-k_i}>0$, $\xi(r)$ will then have the same sign as $(k_i/r_i)(v_i+a_i)$.

Thus, if it is required that a velocity–depth curve with a waveguide be approximated by this law, a gradient discontinuity must be permitted at the boundaries of the layer, where $\xi(r)>0$. But this is not important for amplitude calculation by the ray method, since rays do not have their deepest points in regions where $\xi(r)>0$.

With this limitation, we can in principle smoothly approximate any velocity–depth curve. The following scheme is proposed for this.

The velocity distribution v(r) is assumed to be given at any point (for example, the velocities are given at successive discrete points and an interpolation method is specified for obtaining the velocities between these points). The velocity gradient $\beta(r)$ or a method for calculating it is also assumed to be given.

It is required to divide the depth interval into layers with boundaries $r_0, r_1, \ldots, r_N$ and to select v_i, k_i, and a_i such that at these layer boundaries r_i the velocities $v(r_i)$ and gradients $\beta(r_i)=dv/dr$ are continuous and differ from the given values by not more than ε_v and ε_β, respectively.

The parameters are determined successively from layer to layer by the same algorithm. The upper boundary of a layer is assumed to be given: in the first layer this is the free surface, and in the other layers it is defined as the lower boundary of the preceding layer. The velocity at the boundary $r=r_i$ is also known: it is given at the free surface, and at subsequent boundaries it is obtained from calculations in the preceding layer. Thus for each layer we must determine three values: k_i, a_i, and r_{i+1}, from which we then calculate v_{i+1}.

The lower-layer boundary r_{i+1} is sought in a depth interval such that the velocity gradient in the layer $r_{i+1}<r<r_i$ varies monotonically. To find k_i, a_i, and r_{i+1}, we use the continuity of the velocity gradient at the boundary $r=r_i$ and the condition that the velocity and gradient at the lower unknown boundary r_{i+1} be equal to the given values with some preassigned accuracy.

From the condition at the boundary $r=r_i$ we have

$$-\frac{k_i(v_i+a_i)}{r_i}+\frac{v_i}{r_i}=\beta(r_i), \qquad (8)$$

hence,

$$a_i=\frac{(1-k_i)v_i-\beta(r_i)r_i}{k_i}. \qquad (9)$$

The two conditions at the lower boundary are written as

$$\frac{r_{i+1}}{r_i}\left[(v_i+a_i)\left(\frac{r_{i+1}}{r_i}\right)^{-k_i}-a_i\right]=v(r_{i+1}); \qquad (10)$$

$$\frac{1}{r_i}\left[(1-k_i)(v_i+a_i)\left(\frac{r_{i+1}}{r_i}\right)^{-k_i}-a_i\right]=\beta(r_{i+1}). \qquad (11)$$

Substitution of (9) into (10) and (11) results in a system of two equations in k_i and r_{i+1}:

$$\frac{r_{i+1}}{r_i}\left[\frac{v_i-\beta(r_i)r_i}{k_i}\left(\frac{r_{i+1}}{r_i}\right)^{-k_i}-\frac{(1-k_i)v_i-\beta(r_i)r_i}{k_i}\right]=v(r_{i+1}), \qquad (12)$$

$$\frac{(1-k_i)}{r_i}\frac{[v_i-\beta(r_i)r_i]}{k_i}\left(\frac{r_{i+1}}{r_i}\right)^{-k_i}-\frac{(1-k_i)v_i-\beta(r_i)r_i}{r_ik_i}=\beta(r_{i+1}). \qquad (13)$$

These equations must be solved numerically; the left and right sides must be equal with preassigned accuracy.

2. Formulas

Integrals (1) and (2), which determine the increment to travel time and angular distance when a wave passes through a layer $r_{q+1}<r<r_q$, are calculated in closed form by substituting velocity functions (3) into them. The result of integration

is:

$$\tau^q(i_0) = \frac{r_q}{k_q a_q}[\Phi(i_{q+1}) - \Phi(i_q)]; \tag{14}$$

$$\Delta^q(i_0) = \frac{1}{k_q}[\Psi(i_{q+1}) - \Psi(i_q)], \tag{15}$$

where i_q and i_{q+1} are the angles of incidence of the ray at the upper and lower boundaries of the layer, respectively;

$$\Phi(i) = \ln \operatorname{tg} \frac{i}{2} + I(i); \tag{16}$$

$$\Psi(i) = i + \frac{p_0 a_q}{r_q} I(i), \tag{17}$$

and the expression I(i) has different forms, according to $p_0 a_q/r_q$, where p_0 is the ray parameter,

$$I(i) = \frac{1}{2\sin a} \ln \frac{1+\sin(\alpha+i)}{1-\sin(\alpha-i)}$$

$$\text{for } \left|\frac{p_0 a_q}{r_q}\right| < 1, \quad \frac{p_0 a_q}{r_q} = \cos\alpha; \tag{18}$$

$$I(i) = -\frac{1}{\operatorname{sh}\alpha} \operatorname{arctg} \frac{1 \pm \operatorname{ch}\alpha \sin i}{\operatorname{sh}\alpha \cos i} \quad \text{for } \left|\frac{p_0 a_q}{r_q}\right| = \operatorname{ch}\alpha > 1; \tag{19}$$

$$I(i) = \frac{\cos i}{1+\sin i} \quad \text{for } \left|\frac{p_0 a_q}{r_q}\right| = 1. \tag{20}$$

The sign in the numerator of (19) is the same as the sign of $p_0 a_q/r_q$.

If the deepest point of the ray is inside the layer, then $\pi/2$ must be substituted for i_{q+1} in (14) and (15). In this case there will be no singularities in expressions (16)-(18) and (20). In expression (19) we use $\pm\pi/2$ as $\operatorname{arctg}\frac{1 \pm \operatorname{ch}\alpha \sin i}{\operatorname{sh}\alpha \cos i}$; the sign is determined by the sign of the numerator. The $\tau(i_0)$ and $\Delta(i_0)$ thus calculated are the travel time and distance for half of the refracted ray; the values must be doubled to obtain the total travel time and distance of the refracted wave.

When $k_q = 0$ or $a_q = 0$, expressions (14) and (15) are indeterminancies of the form of 0/0. When k_q or a_q is close to zero, therefore, other expressions should be used.

When $k_q \simeq 0$

$$\tau^q(i_0) = \frac{p_0 \ln \frac{r_q}{r_{q+1}}}{\sin i_q \cos i_q}\left[1 - \frac{k}{2}\frac{\left(\sin i_q + \frac{p_0 a_q}{r_q}\right)\cos 2i_q}{\sin i_q \cos^2 i_q} \ln \frac{r_q}{r_{q+1}}\right]; \tag{21}$$

$$\Delta^q(i_0) = \operatorname{tg} i_q \ln \frac{r_q}{r_{q+1}}\left[1 + \frac{k}{2}\frac{\left(\sin i_q + \frac{p_0 a_q}{r_q}\right)}{\sin i_q \cos^2 i_q} \ln \frac{r_q}{r_{q+1}}\right]. \tag{22}$$

For small a_q

$$\tau^q(i_0) = \frac{p_0}{k_q}\left[(\operatorname{ctg} i_q - \operatorname{ctg} i_{q+1}) - \frac{\beta}{2}(A(i_{q+1}) - A(i_q))\right], \tag{23}$$

where

$$A(i) = \ln \operatorname{tg}\frac{i}{2} - \frac{\cos i}{\sin^2 i}, \tag{24}$$

and $\sin\beta = p_0 a_q/r_q$.

The expression for $\Delta^q(i_0)$ at small a_q does not have singularities.

It is easy to see that when $a_q = 0$ formulas (15) and (23) become the expressions obtained for Bullen's law of velocity variation.

The total travel time of the wave and the epicentral distance are obtained by summing the expressions for $\tau^q(i_0)$ and $\Delta^q(i_0)$ over all layers along the ray:

$$\tau = \sum_q \tau^q(i_0); \tag{25}$$

$$\Delta = \sum_q \Delta^q(i_0). \tag{26}$$

The intensity of the wave at the surface r = R is calculated by the formula

$$A(i_0) = \frac{\sqrt{\sin i_0 \sin i_R}\prod_i \varkappa_i}{v_0\sqrt{R p_0 \sin\Delta \left|\cos i_R \left(\frac{\partial\tau}{\partial i_0}\right)_{r=R}\right|}}, \tag{27}$$

where i_0 is the angle of emergence of the ray from the source, i_R the angle of emergence of the ray at the surface, and $\varkappa_i$ the refraction coefficients for all velocity discontinuities intersected by the ray.

An expression for $\partial\tau/\partial i_0$ is obtained by simple differentiation of (25). Thus, $d\tau/di_0$ is a sum similar to (25) and (26) over all layers whose terms $\partial\tau^q/\partial i_0$ are derivatives of expressions such as (14) or (21) and (23).

When k_q and a_q differ from zero

$$\frac{\partial\tau^q}{\partial i_0} = \frac{r_q \operatorname{ctg} i_0}{k_q a_q}\left\{\frac{1}{\cos i_{q+1}} + Y(i_{q+1}) - \frac{1}{\cos i_q} - Y(i_q)\right\}, \tag{28}$$

where

$$Y(i) = \operatorname{ctg}^2\alpha\, I(i) - \frac{\operatorname{ctg}\alpha \cos i + \operatorname{tg} i \sin\alpha}{\sin\alpha(\cos\alpha + \sin i)} \quad \text{for } \left|\frac{p_0 a_q}{r_q}\right| < 1; \tag{29}$$

$$Y(i) = -\operatorname{cth}^2\alpha I(i) - \frac{\operatorname{tg} i \operatorname{sh}\alpha \pm \operatorname{cth}\alpha \cos i}{\sin\alpha(\sin i \pm \operatorname{ch}\alpha)}$$

$$\text{for } \left|\frac{p_0 a_q}{r_q}\right| = \operatorname{ch}\alpha > 1; \tag{30}$$

$$Y(i) = -\frac{\operatorname{tg} i}{1+\sin i} - \frac{\cos i\,(2+\sin i)}{3\,(1+\sin i)^2} \quad \text{for} \quad \left|\frac{p_0 a_q}{r_q}\right| = 1. \tag{31}$$

For small k

$$\frac{\partial \tau^q}{\partial i_0} = \frac{p_0 \ln \frac{r_q}{r_{q+1}} \operatorname{tg} i_q}{\cos^2 i_q} \operatorname{ctg} i_0 \times$$

$$\times \left[1 + \frac{k}{2}\, \frac{\left(\sin i_q + \frac{p_0 a_q}{r_q}\right)(1+2\sin^2 i_q)}{\cos^2 i_q \sin i_q} \ln \frac{r_q}{r_{q+1}}\right]. \tag{32}$$

For small a

$$\frac{\partial \tau^q}{\partial i_0} = \frac{p_0 \operatorname{ctg} i_0}{k_q}\Bigg[\operatorname{tg} i_{q+1} - \operatorname{tg} i_q -$$

$$- \beta\left(A(i_{q+1}) - A(i_q) + \frac{1}{\cos i_{q+1} \sin^2 i_{q+1}} - \frac{1}{\cos i_q \sin^2 i_q}\right)\Bigg]. \tag{33}$$

3. Algorithm

The final result of the calculation is the first-arrival travel-time curve $t(\Delta)$ and the amplitude-distance curve $A_m(\Delta)$ (the maximum amplitude in a group of first arrivals as a function of epicentral distance), since it is precisely these data that are obtained by observation.

The travel time curve $\tau(\Delta)$ and amplitude-distance curve $A(\Delta)$ are given in parametric form: τ, Δ, and A are represented as functions of the angle of emergence i_0 (formulas (25), (26), and (27)) But to obtain the travel times τ and amplitudes A as functions of the epicentral distance Δ we must interpolate these values at points corresponding to the successive values of Δ, i.e., Δ_0, $\Delta_0 + h_\Delta$, $\Delta_0 + 2h_\Delta \ldots \Delta_0 + Nh_\Delta$.

In some cases, the functions $\tau(\Delta)$ and $A(\Delta)$ can be multiple-valued, forming a loop or angle on the travel-time curve. Cases when $\tau(i_0)$ and $\Delta(i_0)$ are discontinuous functions are possible; in this case, the curves $\tau(\Delta)$ and $A(\Delta)$ do not exist in a certain range of epicentral distances, and there is a shadow zone. Therefore, after interpolation from the obtained functions $\tau(\Delta)$ and $A(\Delta)$ it is necessary to construct single-valued functions $t(\Delta)$ and $A_m(\Delta)$ that are defined over the entire range of epicentral distances.

Interpolation is accomplished in the program as follows. The function $\Delta(i_0)$ is calculated by formula (26) with a given constant step h_Δ. As soon as $\Delta(i_0)$ is greater than one of the values $\Delta_k = \Delta_0 + kh_\Delta$ (or, in the case of a downward branch of the travel-time curve, less than Δ_k), the angle i_{0k}, which must correspond to Δ_k, is calculated by linear interpolation formulas. We calculate $\Delta(i_{0k})$ for the obtained angle. If $\Delta(i_{0k})$ does not agree with Δ_k with given accuracy, i.e., the condition

$$|\Delta(i_{0k}) - \Delta_k| < \delta, \tag{34}$$

is not satisfied, the step for i_0 is reduced by a factor of 10, $\Delta(i_0)$ is calculated starting with i_0^{n-1} with the new step, and the interpolation procedure is repeated. If in this case $\Delta(i_{0k})$ still does not agree with Δ_k, the step is again reduced. These successive subdivisions of the step are made not more than four times, since at this point the step is smaller than the initial one by a factor of 10^4.

If condition (34) is satisfied in any of the interpolation steps, then $\tau(i_{0k})$ and $A(i_{0k})$ are calculated for the obtained angle i_{0k}. These values are assumed to correspond to the epicentral distance Δ_k.

But if after four subdivisions of the step condition (34) is not satisfied, it is assumed that the point Δ_k is in the shadow zone. In a shadow zone the travel-time is associated with a diffracted wave and is calculated by linear extrapolation of the travel-time curve; the amplitude is assumed to be zero. The travel-time curve is extended into the shadow zone up to the point at which the refracted wave emerges.

It was assumed that A_m was the maximum of A at a point in question if more than one wave arrived at this point.

Thus, $t(\Delta)$ and $A_m(\Delta)$ are selected in the program by the following rules:

1. If one wave arrives at the point Δ_k, then its travel time $\tau(\Delta_k)$ and amplitude $A(\Delta_k)$ are taken as $t(\Delta_k)$ and $A_m(\Delta_k)$.

2. If several waves with nonzero amplitudes arrive at the point Δ_k (the travel-time curve is multiple-valued), then we take as $t(\Delta_k)$ the minimum of the arrival times of these waves, and as $A_m(\Delta_k)$ we take the maximum amplitude of the waves that arrive within a given interval δt after the first arrival.

3. If several waves arrive at the point in question but the first arrival is a wave with zero amplitude (angle on travel-time curve after shadow zone), the preceding rule applies only to subsequent arrivals, i.e., to waves with nonzero amplitudes. Thus in this case the first arrival $t(\Delta_k)$ will no longer pertain to a diffracted wave but will pertain to a refracted wave.

4. Effect of Approximation of the Velocity–Depth Curve on the Calculated Amplitude–Distance Curve

Approximation of a velocity profile by functions (3) according to the scheme described in Section 1 is rather complicated and is unsuitable for bulk calculations (for example, in solving the inverse problem by sorting many variants). This gives rise to the question of whether or not a velocity–depth curve can be represented by simpler functions, which will, however, give velocity gradient discontinuities at the layer boundaries.

The following questions, related to Problem 2 at the beginning of this paper, will be considered here:

1. How do we approximate a velocity distribution (i.e., divide an interval of depth into layers and specify the velocity laws in the layers) so that the gradient discontinuities at the layer boundaries do not substantially distort the amplitude–distance curve?

2. Can an amplitude curve calculated for a velocity approximation with velocity gradient discontinuities be smoothed so that the smoothed amplitude–distribution curve is with sufficient accuracy equivalent to the $A(\Delta)$ for the smoothed $v(r)$?

These questions were considered by analyzing calculations for various approximations of the same velocity distribution in the upper mantle ($5700 < r < 6335$). The initial distribution to be approximated was the smoothed curve represented by functions (3). Bullen's law (4) was used as the approximating function for the gradient discontinuity at the layer boundaries. The calculations were made for one example of velocity distribution but this does not limit the generality of the results, since this distribution represented all the main characteristics that are subject to determination in seismic research: regions with increasing, decreasing, and approximately constant velocity gradients. The initial velocity–depth curve and the corresponding velocity gradient are shown in Fig. 1.

The initial velocity–depth curve was approximated in several ways. The number of given layers, boundary depths, and velocity gradients were varied.

When the approximation of the velocity–depth curve is not smooth, the ray method is unsuitable for calculating amplitude in certain ranges of emergence angle and epicentral distance. These ranges correspond to the emergence distances of rays whose deepest points are immediately below a gradient discontinuity boundary. The greater the discontinuity, the wider the range. It must be expected, therefore, that when the velocity profile is represented by a small number of layers (consequently, with great gradient discontinuities at the layer boundaries), the amplitude–distance curve will have rather extended areas that do not agree with the amplitude–distance curve for the smoothed profile. With a greater number of layers, these areas become smaller, but their number increases and they can overlap.

In order to find the best division of the profile into layers, calculations were made for an approximation of the original profile with 5, 16, and 64 layers. The velocity gradient and amplitude curves for these cases are shown in Fig. 2, a and b.

The first version of the approximation, with the smallest number of layers, was set up so that only one layer was associated with each extremum on the velocity gradient curve $\beta(r) = -dv/dr$. Since each extremum on the curve $\beta(r)$ corresponds to an extremum of the same type (maximum or minimum on curve $A(\Delta)$), the main characteristics of the amplitude–distance curve (maxima and minima at particular epicentral distances) are preserved in this version. But the levels of the amplitude–distance curve at certain Δ differ very greatly (up to 0.3 in $\log A$). This is particularly true of $A(\Delta)$ regions corresponding to boundaries with decreasing $\beta(r)$.

When the number of layers is increased, the agreement between the amplitude–distance is at first not improved (version 2). On the contrary, two maxima appear at a distance of about 15° instead of one maximum. This is explained by the fact that when $\beta(r)$ increases at the successive boundaries, the sequence of rises in the level of $A(\Delta)$ that correspond to these boundaries can be different from the sequence of boundaries. In par-

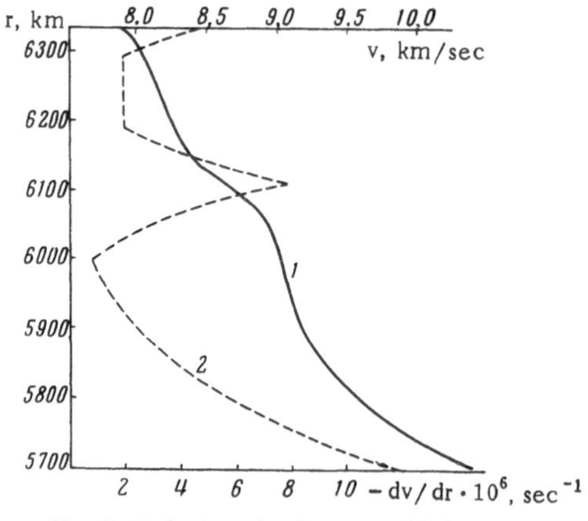

Fig. 1. Velocity–depth curve v(r) (curve 1) and corresponding velocity gradient dv/dr as function of r (curve 2).

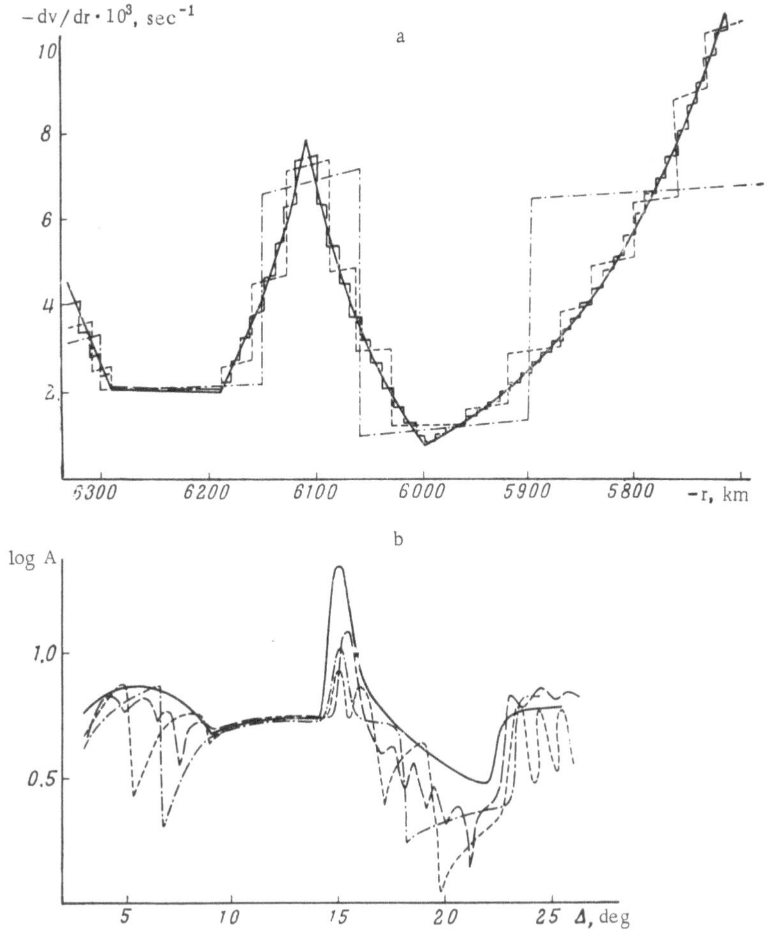

Fig. 2. Approximations of velocity distribution different numbers of layers: a) velocity gradients; b) amplitude–distance curves. The solid curve is the initial velocity–depth curve.

ticular, in version 2 the first maximum (at $\Delta = 15°$) corresponds to the boundary r = 6130 km, while the next maximum (at $\Delta = 16°$) corresponds to the overlying boundary r = 6160 km. This inversion of the increments of $A(\Delta)$, which results in splitting of one maximum into several, depends upon the relationship between the gradient discontinuities at the successive boundaries and upon the layer thicknesses. With step approximation of a rapidly increasing gradient (in this case, 6190 > r > 6100) it is difficult to avoid splitting the maximum on the curve $A(\Delta)$. Thus it is better to approximate this case by one boundary with a gradient discontinuity.

The introduction of additional layers in regions with a smoothly decreasing $\beta(r)$ results in additional steps on the curve $A(\Delta)$. The general tendency of $A(\Delta)$ to increase in the corresponding region is preserved, and the difference between the average levels of the amplitude–distance curves is reduced.

The third approximation version was set up so that the layer thicknesses were 10 km, so that there was a total of 64 layers. The drop in $\beta(r)$ at a layer boundary did not exceed 0.001 sec^{-1}.

In this case, the amplitude–distance is a wavy line which on the average follows all variations of the initial amplitude–distance curve. The oscillatory nature of $A(\Delta)$ is a result of the presence of velocity gradient discontinuities in the approximated profile. These oscillations can be eliminated by smoothing the amplitude–distance curve by, for example, the operator

$$\bar{A}^* = \frac{A^*_{k-1} + A^*_k + A^*_{k+1}}{3}, \tag{35}$$

where $A^* = \log A$ (the successive points of $A(\Delta)$ are calculated with a step size of 0.5°). Figure 3 shows the result of smoothing by this formula for version 3. The smoothed curve is similar in shape to the initial amplitude–distance curve, but its level in

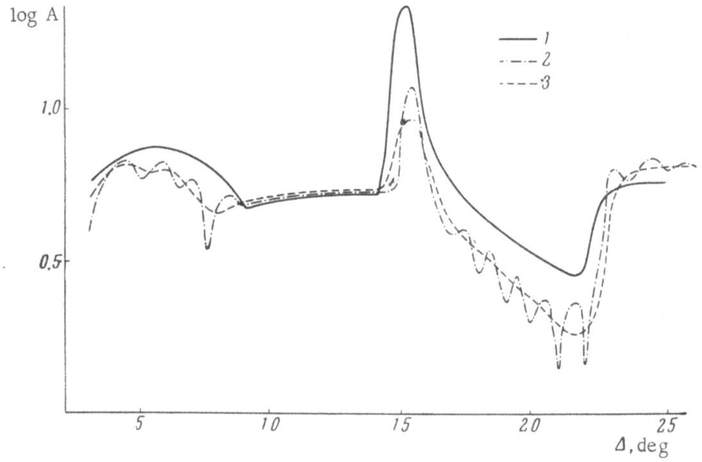

Fig. 3. Example of smoothing the amplitude–distance curve: 1) A*(Δ) for the initial velocity distribution; 2) A*(Δ) for version 3 (Fig. 2); 3) smoothed amplitude–distance curve.

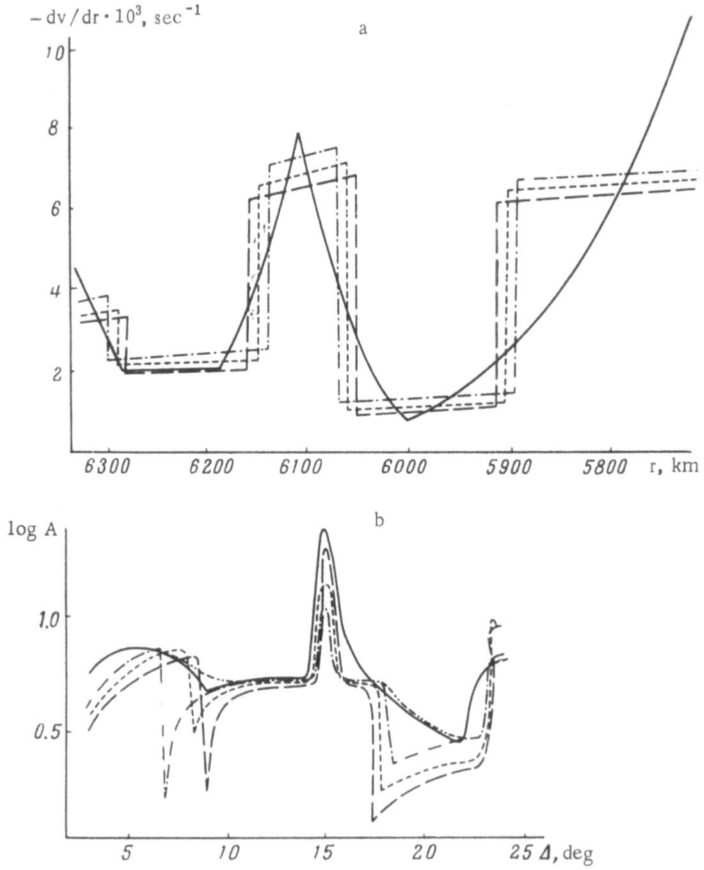

Fig 4. Versions of velocity gradient approximation with the same number of layers: a) velocity gradient curves; b) corresponding amplitude–distance curves. The curve with two dots and two dashes smooths the minima on the calculated (dot-and-dash) curve (see below). The solid curve is A(Δ) for the initial velocity distribution.

intervals corresponding to a decrease in velocity gradient ($5° < \Delta < 10°$ and $16° < \Delta < 22°$) is lower, and in intervals corresponding to a velocity gradient increase ($\Delta > 23°$) is higher than the level of $A(\Delta)$ for the smoothed v(r). This is evidently a result of the fact that when the profile is divided into such a great number of layers, the rays always have their deepest point in regions where formula (27) is invalid.

Thus although the most detailed approximation with this method of smoothing $A(\Delta)$ gives the best agreement between the amplitude curves, it has two disadvantages. The average level of the calculated $A(\Delta)$ is either overestimated or underestimated in various intervals of Δ; the method also requires a great many calculations for $A(\Delta)$ as well as for specifying the velocity approximation.

In practice, however, it is sufficient to obtain on the calculated $A(\Delta)$ only the fundamental characteristics (maxima and minima), and for this we can make a rough approximation of the initial velocity–depth curve, as was done in the example of version 1. Here we must investigate beforehand the extent to which the approximation is stable, i.e., the degree to which the position and nature of the extrema on $A(\Delta)$ are preserved when the boundaries and, accordingly the velocity gradients, in the layers vary.

Figures 4a and 4b show three different methods of approximating $\beta(r)$ and the amplitude–distance curves $A(\Delta)$ that correspond to these approximations. The layer boundaries were varied within a range of 20 km. As can be seen from Fig. 4b, the overall nature of the amplitude–distance curves is retained for all three cases. The extrema are shifted within very small limits. As might be expected, there is an appreciable change in the level of $A(\Delta)$ only in regions of amplitude–distance curve instability [2] – in regions with small velocity gradients and a sharp gradient rise.

It is interesting to note that the curves $A(\Delta)$ seem to follow all basic changes in the curves $\beta(r)$: the maxima on $\beta(r)$ correspond to the maxima on $A(\Delta)$ and the minima on $\beta(r)$ correspond to the minima on $A(\Delta)$.

We can propose the following method for smoothing the minimum: if after the minimum, $A(\Delta)$ reaches an approximately constant level, the curve should be drawn smoothly from a point on $A(\Delta)$ before the minimum to a part of $A(\Delta)$ of the constant level. Such a curve should be close to $A(\Delta)$ for the smoothed velocity distribution (see Fig. 4).

Note that even with a rough approximation of the profile, the deviation of the travel time curves does not exceed 0.2 sec, i.e., this method of velocity approximation is quite suitable for calculating travel-time curves.

The main conclusions are the following:

1. To obtain the main characteristics (extrema) on the curve $A(\Delta)$ it is sufficient to approximate the velocity distribution v(r) with the number of layers that corresponds to the number of principal velocity gradient extrema. If v(r) is parametrized on the basis of the shape of the amplitude curve $A(\Delta)$, it is sufficient to make the number of layers in this profile equal to the number of extrema on $A(\Delta)$.

2. To make the amplitude–distance curve calculated for such an approximated v(r) sufficiently close to the curve for the smooth profile, the sharp minima on the calculated curve should be replaced by smooth transitions from one level of $A(\Delta)$ to another.

3. With detailed approximation of velocity distribution and subsequent smoothing of the calculated amplitude–distance curve, the obtained $A(\Delta)$ is similar in nature to the $A(\Delta)$ for the smoothed profile. Regions where the absolute levels of these curves differ are possible, however.

LITERATURE CITED

1. Yanovskaya, T. B., and G. V. Golikova (1964), "Method and program for calculating P wave dynamics in the earth's mantle," in: Problems of the Dynamic Theory of Seismic Wave Propagation, No. 7, pp. 123-129.
2. Yanovskaya, T. B., G. V. Golikova, and Yu. A. Surkhov (1964), "Amplitude curves of P waves," in: Problems of Dynamic Theory of Seismic Wave Propagation, No. 7, pp. 104-114.

AMPLITUDES OF FUNDAMENTAL SEISMIC WAVES IN THE MANTLE†

L. M. Kilinchuk and T. B. Yanovskaya

Introduction

The interpretation of seismological observations at all distances, except the very nearest (out to 10°), is based on the basic compendium of seismic-wave travel times composed by Jeffreys and Bullen in 1940 [1] and later refined and supplemented by a number of authors [10].

The tables [1] contain travel-time curves for 24 body wave phases. They are based in part on observations and in part on calculations. The travel-time curves of P, S, PKP, and SKS for normal focal depth are entirely based on observations. These waves are easier to observe, since they arrive earliest in the corresponding wave groups.

The velocity distribution in the earth's mantle was determined from the travel-time curves of P and S waves; the travel-time curves of waves reflected from the surface of the core (PcP and ScS) were theoretically calculated from the distribution. These calculations are partially based on observations, since the radius of the core in the calculations was made such that the calculated travel-time curves best matched the observed travel times. But the travel-time curves of all other waves and for other focal depths were calculated on the stationary-time principle from the travel-time curves of the above-mentioned waves [2].

Such calculations have meaning, however, only when the comparative amplitude of each wave is estimated.

In this case, some travel-time curves can be "superfluous," in the sense that the corresponding waves can have such little amplitude that for most earthquakes they would be practically impossible to detect against the background of noise and of other waves. On the other hand some fairly strong waves may not be included in the tables (since in [1], naturally only a small portion of the geometrically possible phases are tabulated).

It is interesting, therefore, to calculate theoretically the amplitude curves (dependence of amplitude upon distance and focal depth) for all seismic waves whose travel-time curves are given in [1] and which are widely used for treatment of seismological data.

This paper gives the results of calculations of amplitude–distance curves for waves whose rays lie almost entirely in the mantle: they intersect the crust almost vertically and do not penetrate into the core. We considered the case of a surface source in greater detail (Figs. 1-5). Amplitude–distance curves for focal depths of 100, 200, and 700 km are also given (Figs. 6-17).

It should be borne in mind that the amplitude of seismic waves depends upon a great number of factors that are unimportant in a study of travel times: absorption, source radiation pattern, discontinuities of the vertical velocity gradient, etc.

As a rule, these factors, are known only roughly, if at all. Naturally, they have been ignored or taken into account very approximately in the calculations. Therefore only the basic rough characteristics of the calculated amplitude curves can be compared with observations.

Our calculations were made by the method described in [3, 4] using the program described in [5].

† Translated from Vychislitel'naya Seismologiya, No. 4, Moscow, Nauka (1968), pp. 226-251.

TABLE 1

r, km	v_P, km/sec	v_S, km/sec	r, km	v_P, km/sec	v_S, km/sec	r, km	v_P, km/sec	v_S, km/sec
6371	5.56	3.36	5775	10.17	5.64	4800	12.22	6.71
6338	7.00	3.90	5750	10.31	5.71	4700	12.34	6.76
6338	7.75	4.35	5725	10.44	5.78	4600	12.48	6.81
6300	7.83	4.39	5700	10.55	5.84	4500	12.62	6.86
6250	7.97	4.47	5675	10.65	5.91	4400	12.77	6.91
6200	8.12	4.55	5650	10.74	5.97	4300	12.87	6.96
6150	8.28	4.63	5625	10.82	6.02	4200	12.99	7.01
6100	8.44	4.71	5600	10.90	6.07	4100	13.10	7.05
6075	8.52	4.75	5575	10.97	6.11	4000	13.22	7.10
6050	8.60	4.79	5550	11.04	6.15	3900	13.34	7.15
6025	8.68	4.83	5500	11.16	6.22	3850	13.40	7.17
6000	8.77	4.88	5450	11.27	6.28	3800	13.46	7.19
5975	8.86	4.95	5400	11.36	6.33	3750	13.51	7.22
5950	8.97	5.04	5350	11.44	6.37	3700	13.57	7.23
5925	9.12	5.13	5300	11.53	6.41	3650	13.60	7.24
5900	9.30	5.22	5200	11.67	6.47	3600	13.62	7.26
5875	9.48	5.31	5100	11.82	6.54	3550	13.64	7.27
5850	9.66	5.40	5000	11.96	6.60	3500	13.64	7.29
5825	9.85	5.48	4900	12.10	6.65	3471	13.64	7.30
5800	10.02	5.56						

1. Data Used for Computations

The following data are required for calculating the amplitude curves: the longitudinal and transverse wave velocities v_P and v_S; the density discontinuities at the interfaces; the absorption coefficients for longitudinal and transverse waves; and the source radiation pattern.

We shall describe how these data were specified in this work.

The parameters of the medium were assumed to be functions only of the distance r from the center of the earth.

The velocity–depth curves $v_P(r)$ and $v_S(r)$ were constructed on the basis of the Jeffreys model [6], which corresponds to his travel-time curves. The cross-sections were divided into layers, in each of which the velocity was approximated by Bullen's law:

$$v(r) = v_i \left(\frac{r}{r_i}\right)^{-k_i}, \quad r_i \geqslant r \geqslant r_{i+1}, \; i = 1, 2, \ldots . \tag{1}$$

The values $v = v_i$ at the layer boundaries $r = r_i$ are given in Table 1. These values were obtained by interpolation of Jeffreys' data with smooth variation from layer to layer of the mean velocity gradient $(v_{i+1} - v_i)/(r_{i+1} - r_i)$.

Approximation of the velocity–depth curve in the form of (1) is convenient for calculations but gives at each boundary $r = r_i$ gradient discontinuities that do not exist in the Jeffreys model. To obtain this model, our velocity–depth curves have to be smoothed. The velocity gradient discontinuities result in peaks on the amplitude–distance curve, so that the directly calculated curves had a sawtooth shape. They were then smoothed by the method described in [4], which is equivalent to the necessary smoothing of the velocity–depth curve.

The densities above and below the boundaries $\rho_i^+ = \rho(r_i + 0)$ and $\rho_i^- = \rho(r_i - 0)$ are given in Table 2.

We ignored a number of possible profile characteristics. In particular, we ignored the low-velocity zone in the upper mantle (its effect can be considerable at distances to 20° for P and S waves and, accordingly, at greater distances for waves of greater multiplicity) and the boundaries in the crust, which were allowed for by the travel-time curves.

Absorption was allowed for by introducing into the amplitude formulas the factor

$$\exp\left(-\frac{\pi}{T}\sum_j \frac{t_j}{Q_j}\right), \tag{2}$$

where T is the predominant wave period, t_j the travel time of the wave through the j-th layer, and Q_j the anelastic absorption factor in the layer.

The profile Q(r) for the mantle was constructed on the basis of [7, 8]. The Q_j data we used are given in Table 3. A wave period T was assumed to be 5 sec.

TABLE 2

r, km	ρ^+, g/cm³	ρ^-, g/cm²
6371	—	2.30
6338	2.30	3.34
3471	5.66	9.95

TABLE 3

Layer boundaries	Q_P	Q_S	Layer boundaries	Q_P	Q_S
6371—6338	1012.5	450	6025—5975	900	400
6338—6300	155	70	5975—5550	1350	600
6300—6250	225	100	5550—5350	2000	900
6250—6150	337.5	150	5350—4800	2900	1400
6150—6025	450	200	4800—3471	3400	1600

These estimates of Q(r) were obtained from observations of surface waves, and it is unclear to what extent they can be carried over to the case of body waves. Therefore, we shall give calculations for a medium without absorption (which corresponds to T = ∞). From these calculations we can obtain the amplitude curves A* (Δ, T) for any other period. According to (2),

$$A^*(\Delta, T) = A^*(\Delta, \infty) - \frac{5}{T}\left[A^*(\Delta, \infty) - A^*(\Delta, 5)\right],$$

where Δ is the epicentral distance and A* is the logarithm of the amplitude.

Similarly, we can recalculate A* for any other law $\tilde{Q}(r)$ that is geometrically similar to the one chose, i.e., for $\tilde{Q}(r) = kQ(r)$.

Radiation Pattern. It was assumed in the calculations that the source radiated longitudinal and transverse waves uniformly in all directions. Actually, of course, radiation is nonuniform: in all focus models considered, the wave intensity is a function of the angles between the axes of the source and the take-off angle of the ray at the focus. But since the axes are oriented differently for different foci, the radiation pattern can be considered uniform, on the average, especially since for the waves we consider here, rays emerge at the surface that leave the focus not in all possible directions but only within an aperture angle of about 80-90° (the other rays undergo total internal reflection at the base of the crust). A more detailed examination of the effect of the radiation pattern is irrelevant for the purposes of this study.

It was assumed that the radiation intensity of S waves at the source was thrice that of P waves, i.e., that their intensities were inversely proportional to the square of their velocities. This assumption is not essential: to take another ratio it is sufficient to shift up or down the curves that emerge from the source as S waves.

2. Calculated Amplitude Curves

1. Besides direct refracted P and S waves, we calculated the waves reflected from the free surface once (PP, SS, PS, SP) and twice (PPP, SSS, PSS, PPS, PSP, SPP, SPS, SSP), the waves reflected from the core boundary (PcP, ScS, PcS, ScP), and also mixed waves (ScSP, PScS). Figures 1a and 1b, 2a and 2b, show the amplitude–distance curves A*(Δ) for the vertical component Z and horizontal component H of the displacement on the earth's surface caused by these waves, as a function of epicentral distance. The curves in Figs. 1 and 2 were calculated without allowance for absorption in the mantle. The results when absorption was taken into account are shown in Figs. 3a and 3b, and 4a and 4b. In these figures the curves are not shown for all of the waves represented in Figs. 1 and 2 (the waves doubly reflected from the free surface and ScSP and PScS are absent).

For brevity, we shall henceforth denote the logarithms of the various waves by Latin letters: the first indicates the component and the others indicate the wave type (for example, HS means horizontal displacement in an S wave).

2. First, let us consider the effect of absorption on wave intensity. Taking absorption into account, with T = 5 sec, the amplitudes of HS (Fig. 4b) are approximately equal to the amplitudes of ZP (Fig. 3a). Actually, HS is usually appreciably greater than ZP. We can introduce a correction for the fact that the periods of S waves are greater than those of P. This correction is introduced under the assumption that T = 10 sec. The curves for T = 10 sec are also shown in Figs. 3a, 3b and 4a, 4b. Here, HS for T = 10 sec becomes greater by a factor of 2-3 than ZP for T = 5 sec, which is in better agreement with observations.

Comparison of the amplitudes of the other waves, however, indicates that the absorption of transverse waves was overestimated. In reality, when absorption is taken into account, the calculated amplitudes of HSS is smaller by at least one order of magnitude than that of HS; if so, SS and, all the more, SSS are usually not observed.

With longitudinal waves, allowance for absorption has practically no effect on the relationships between the intensities of the individual waves, and these relationships for all T (including T = ∞) remain close to the observed values (the exception

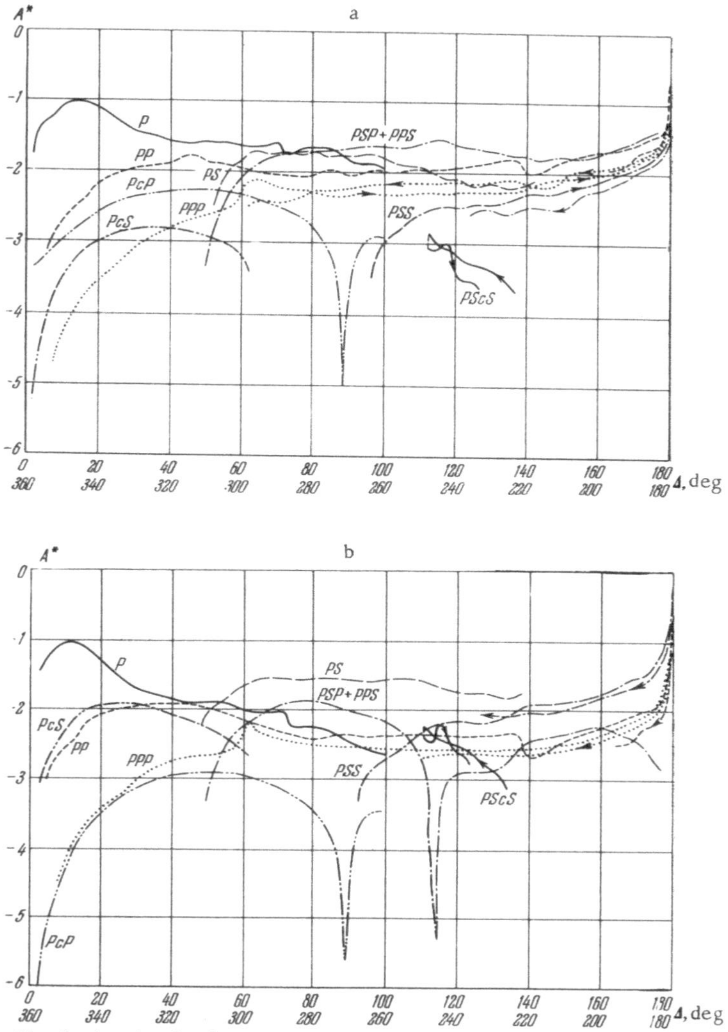

Fig. 1. Amplitude-distance curves of waves emerging from the source as longitudinal, calculated without allowance for absorption in mantle: a) vertical and b) horizontal displacement components.

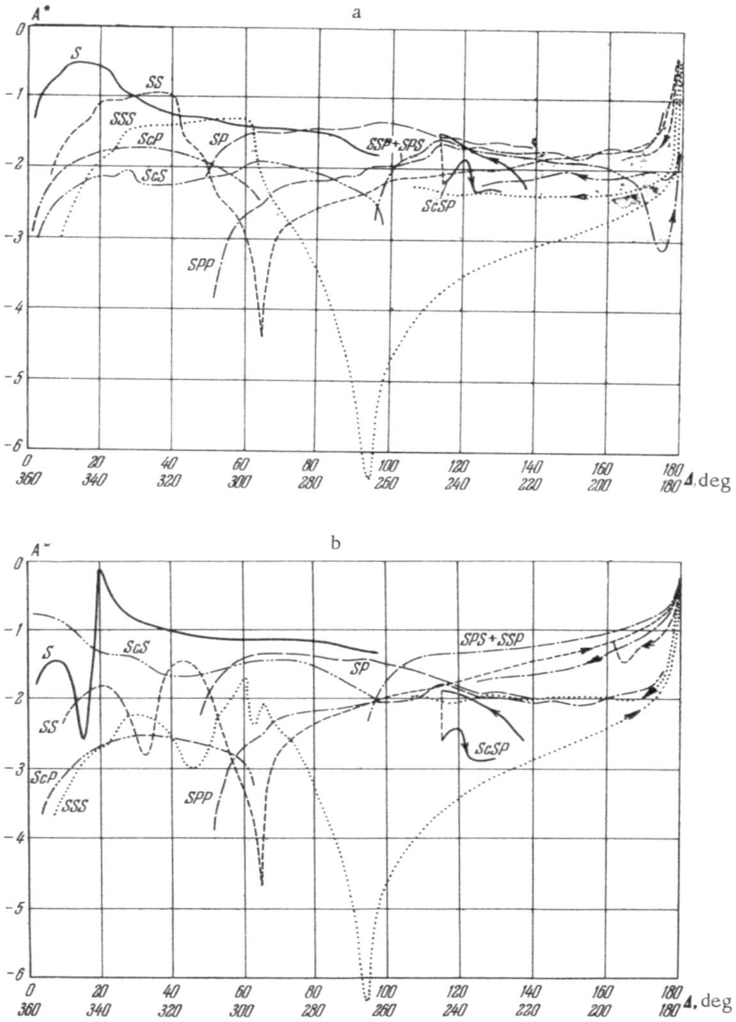

Fig. 2. Amplitude-distance curves of waves emerging from the source as transverse, calculated without allowance for absorption: a) vertical and horizontal displacement components.

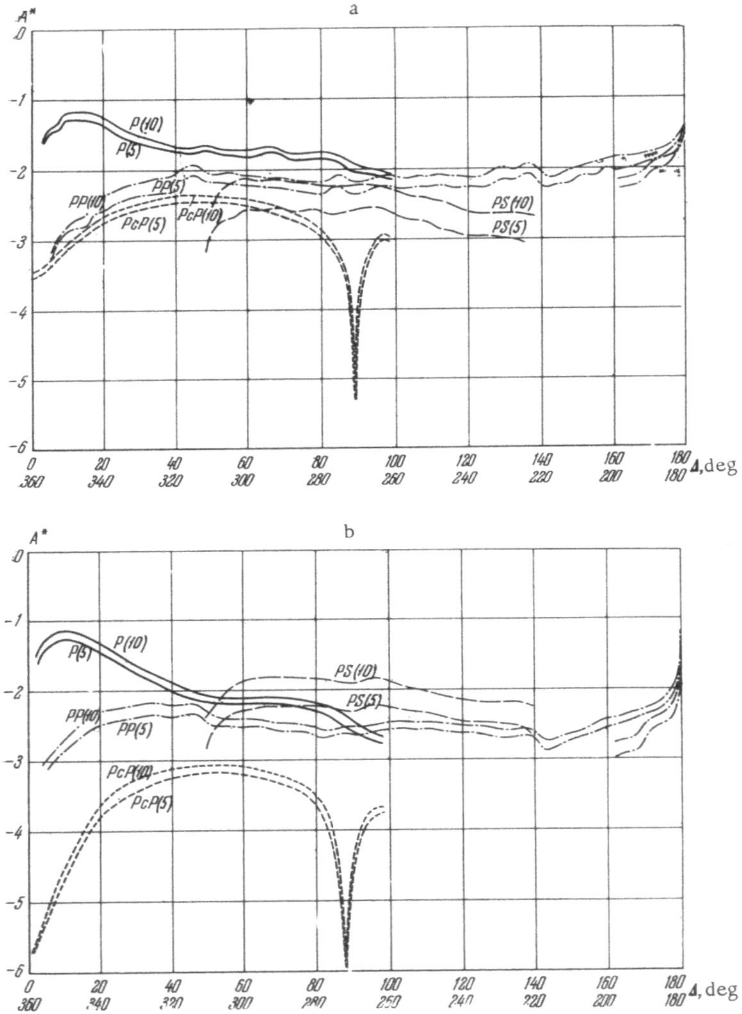

Fig. 3. Amplitude–distance curves of waves emerging from the source as longitudinal, calculated with allowance for absorption in mantle: a) vertical and b) horizontal displacement components. Numbers in parentheses are periods (sec) of corresponding waves in seconds.

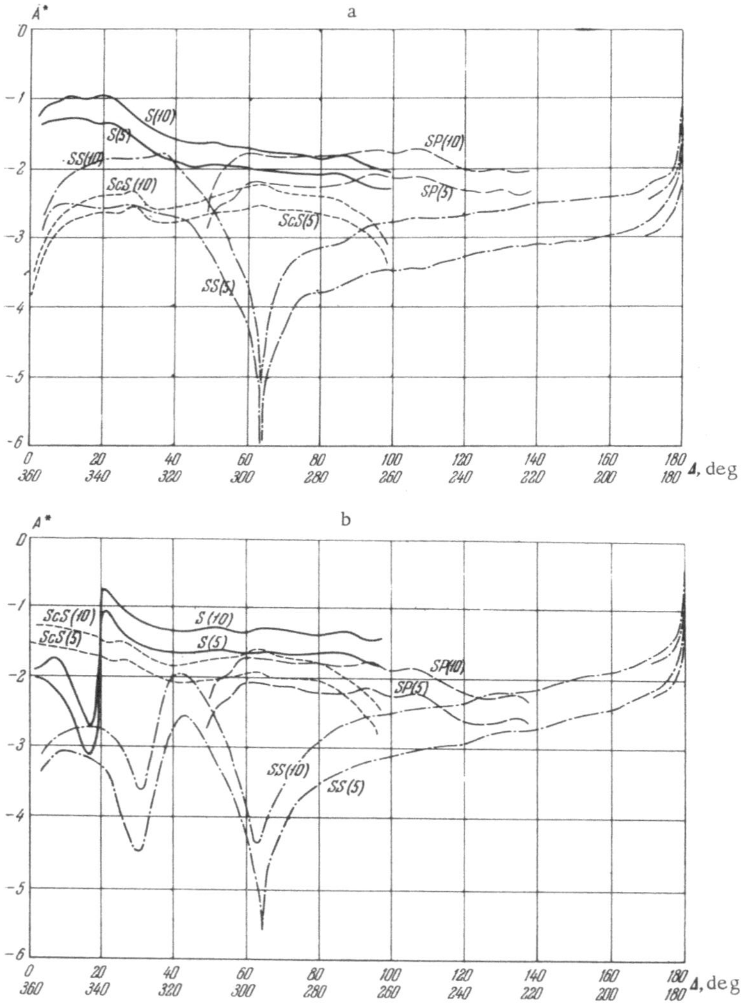

Fig. 4. Amplitude–distance curves of waves emerging from the source as transverse. Numbers in parentheses are periods (sec) of corresponding waves in seconds.

is PcP, the underestimated amplitude of which is evidently explained by incorrect parameters for the core boundary). Thus, the correctness of the Q(r) values assumed for the P waves cannot be judged; we can only conclude that the $Q_P(r)$ values from Table 3 can be used as the lower limit of possible Q for longitudinal waves, and $Q = \infty$ must be used for the upper boundary.

Since the most likely amplitude relationships for individual waves are obtained for transverse waves when absorption is ignored, and since absorption has practically no influence on the amplitude relationships for longitudinal waves, we shall analyze amplitude curves that have been calculated without allowance for absorption, and take the possible effect of absorption into account qualitatively.

3. The very low calculated amplitude of SS in the interval of 50-90° and of SSS in the interval of 70-170° is noteworthy. These waves could not have been observed at this amplitude. In actuality, however, they are among the most intense. To be sure, we calculated only the radial displacement of SV, but we cannot assert that the SS and SSS waves observed at these distances are always type SH, especially as they often have a clearly nonzero Z component.

The decrease in the calculated intensity of SS and SSS in these distance intervals stems from the fact that the coefficient of reflection from the earth's surface is close to zero at the corresponding emergence angles. It is possible that S'S and S'S'S reflected downward from the bottom of the crust were mistaken for SS and SSS. As can be seen from Fig. 5, the intensity of S'S in this interval is high enough to mask the entires of SS. (A similar effect occurs for SSS.) The possibility, finally, that the observed SS and SSS were interference waves formed by the superposition of a number of waves reflected from the crustal boundaries is not ruled out. Then the reflection would have to be taken into account by considering the crust as a thin reflecting layer. The reflection coefficient would be a function of the period in this case.

Distortion of the spectrum of an SS wave due to its passage through the thin crustal layer can cause the apparent arrival time of this wave to deviate from that predicted by the Jeffreys–Bullen tables. In fact, a great spread of travel times is observed for all waves reflected from the earth's surface [2]. In this connection it would be interesting to study the relationship between the observed spectra of SS and S and to make the corresponding theoretical calculations.

4. The amplitudes of the converted waves SP and PS are of the same order; this also applies in some distance intervals to PSP, PPS, and SPP, and to a lesser degree to SPS, PSS, and SSP. With a surface source these waves interfere; depending upon the source radiation pattern at the source they can either cancel or intensify each other. Thus the observed amplitude of these waves with a surface source appears to vary at random.

5. The theoretical amplitude of PcP is too small: it is smaller by a factor of 5-10 than that of P. Absorption does not change this relationship (Fig. 3). Evidently the density ratio at the core boundary used in the calculations does not correspond to reality.

6. The conclusions are approximately the same for deeper foci (Figs. 6-17).

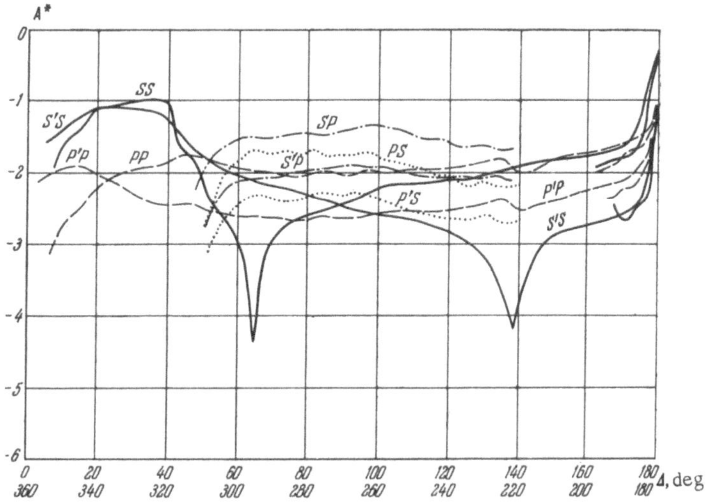

Fig. 5. Amplitude–distance curves (vertical displacement component) of waves reflected from the free surface (PP, SS, PS, SP) and waves reflected downward from bottom of crust (P'P, S'S, P'S, S'P).

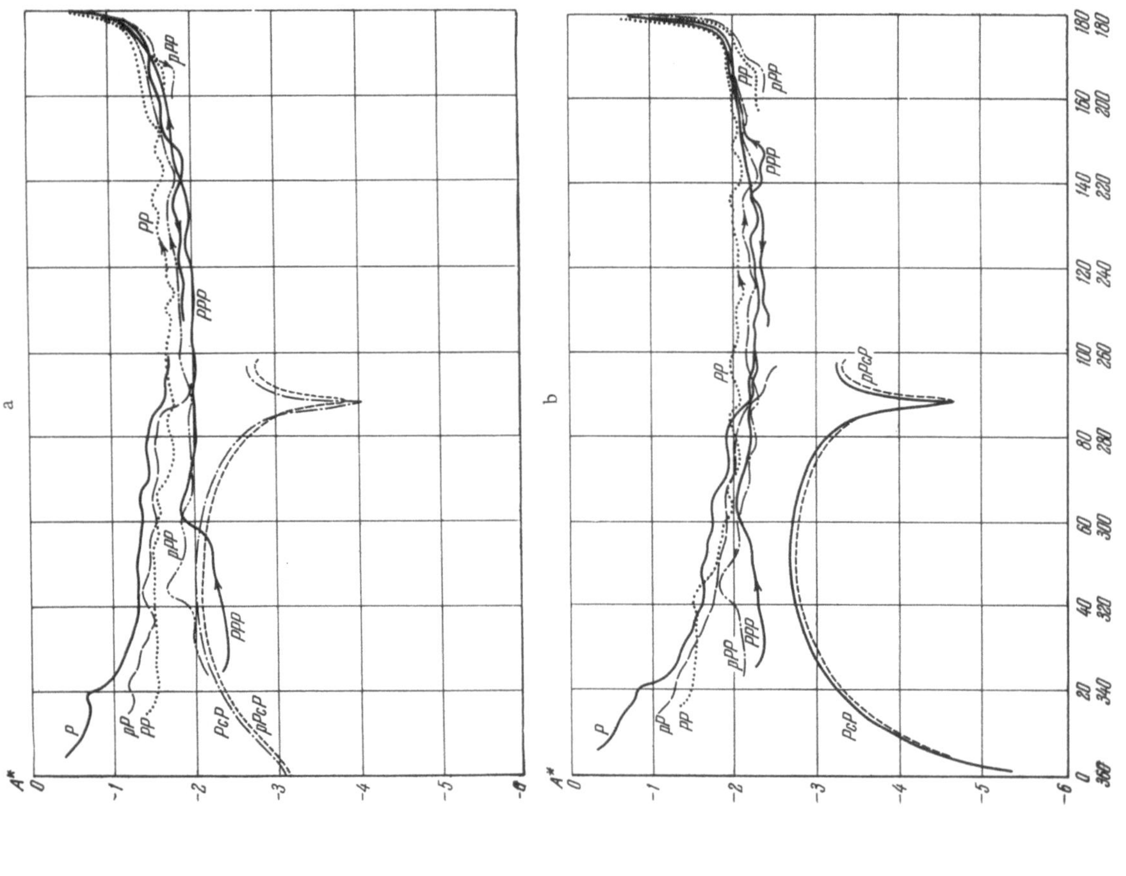

Fig. 6. Amplitude-distance curves of waves propagated only as longitudinal calculated without allowance for absorption, for focal depth of 100 km.

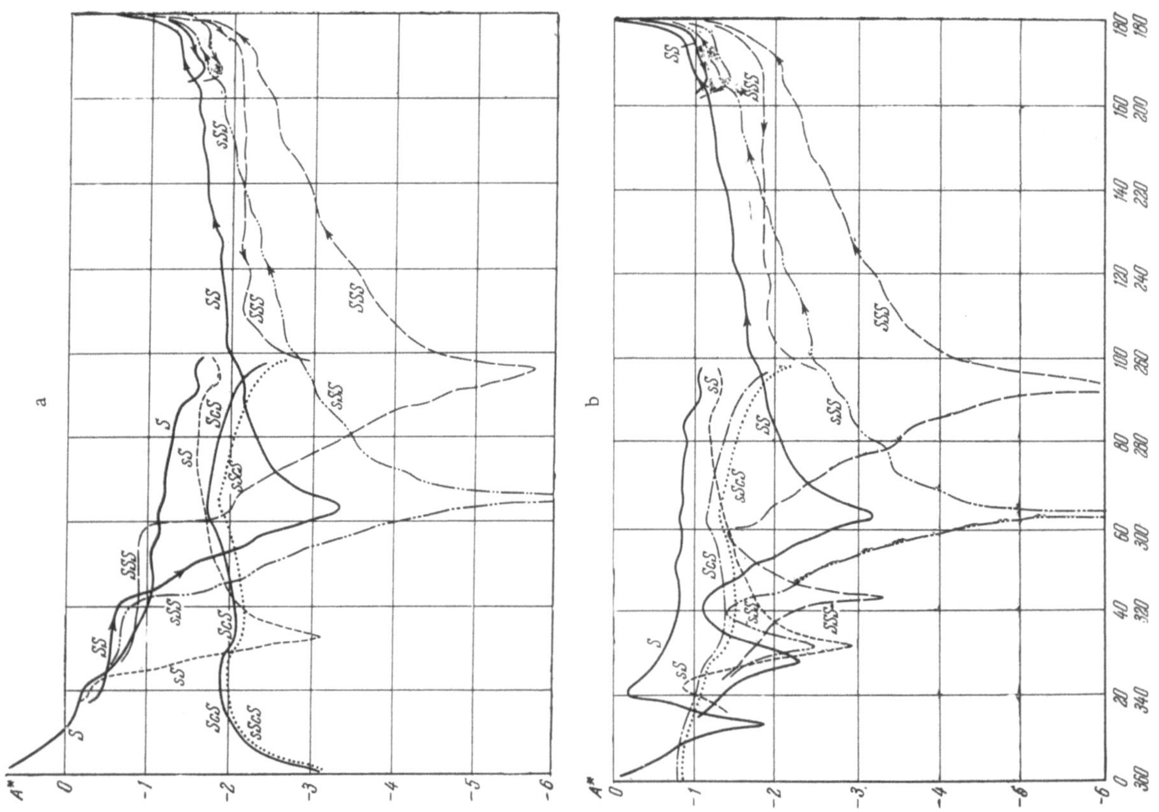

Fig. 7. Amplitude-distance curves of waves propagated only as transverse calculated without allowance for absorption, for focal depth of 100 km.

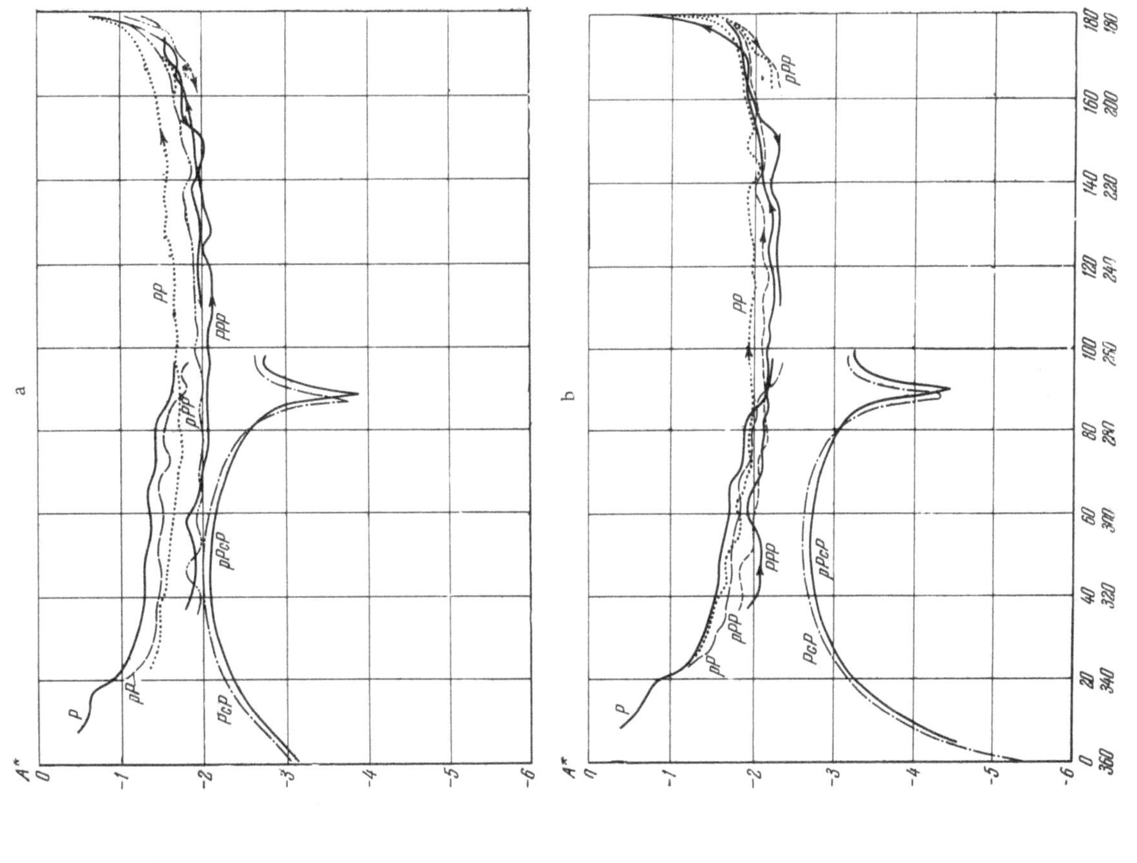

Fig. 8. Amplitude–distance curves of waves propagated only as longitudinal calculated without allowance for absorption, for focal depth of 200 km.

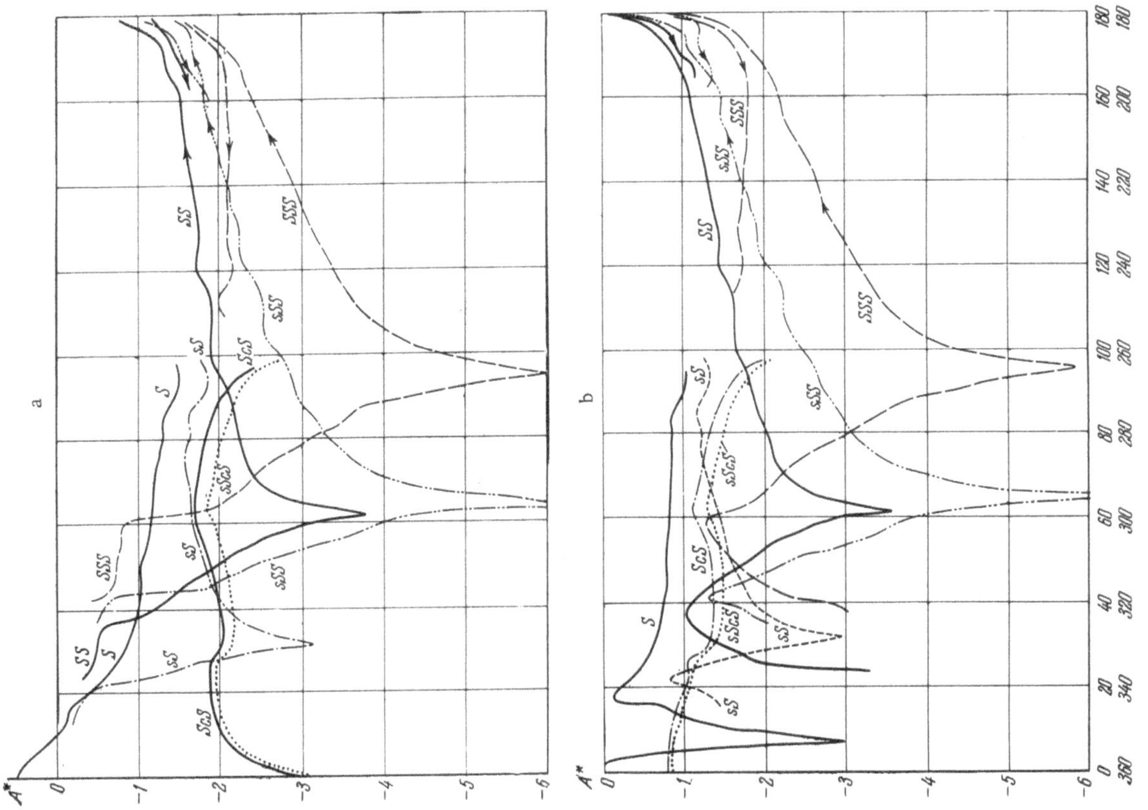

Fig. 9. Amplitude–distance curves of waves propagated only as transverse calculated without allowance for absorption, for focal depth of 200 km.

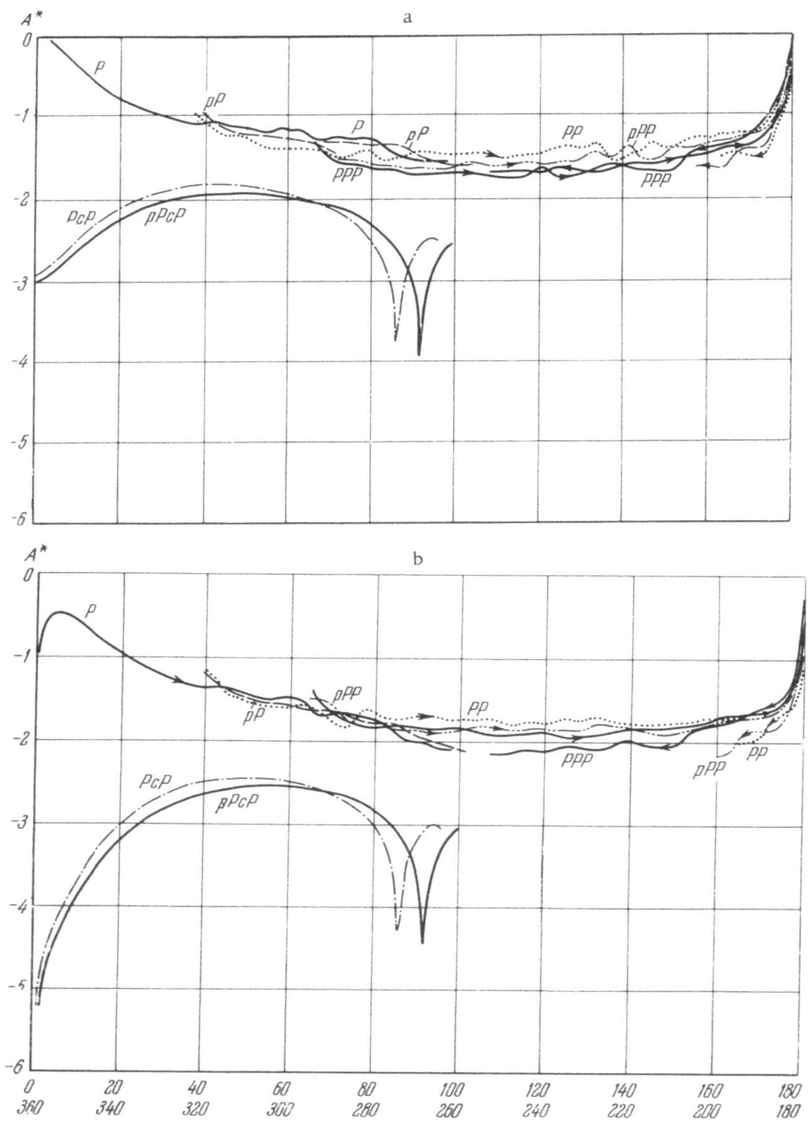

Fig. 10. Amplitude–distance curves of waves propagated only as longitudinal calculated without allowance for absorption, for focal depth of 700 km.

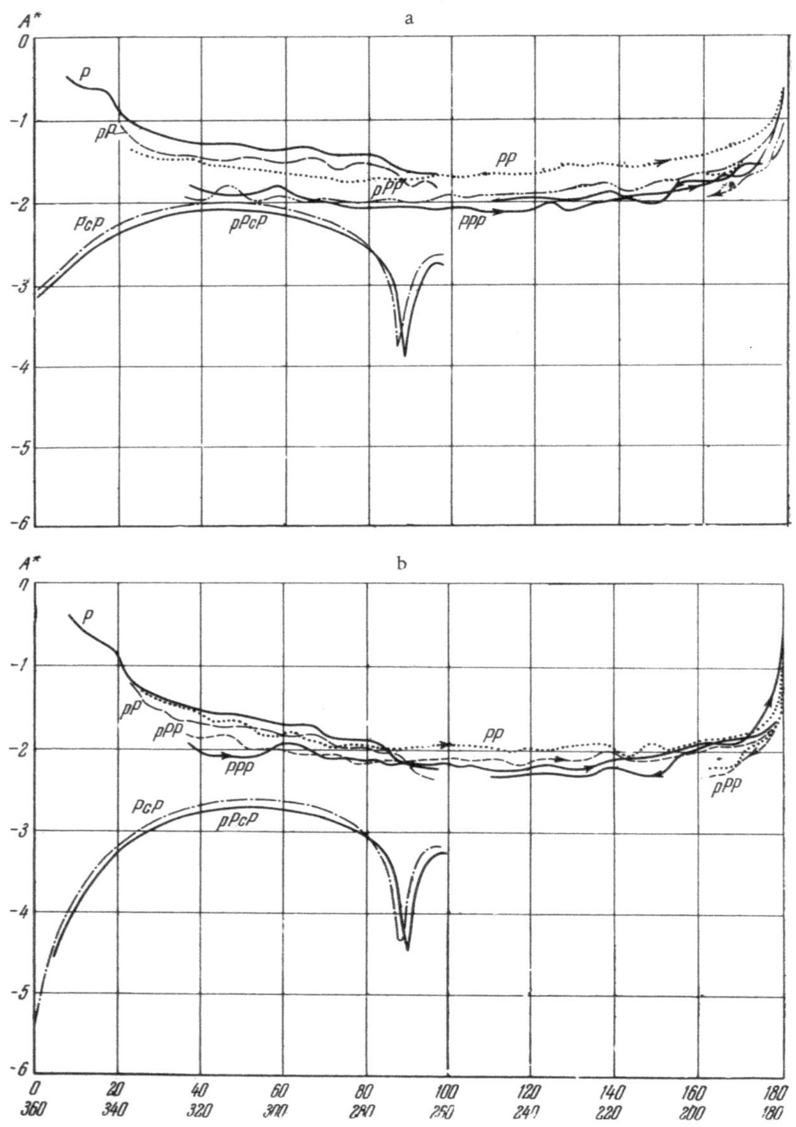

Fig. 11. Amplitude–distance curves of waves propagated only as transverse calculated without allowance for absorption, for focal depth of 700 km.

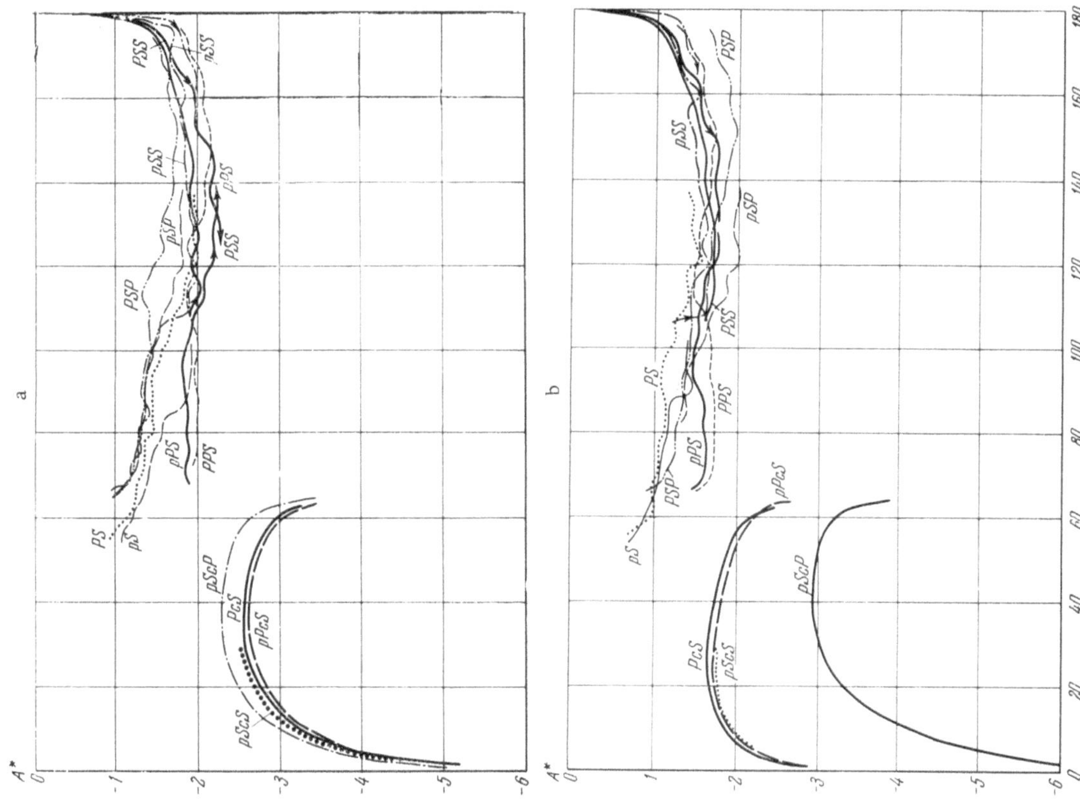

Fig. 13. Amplitude–distance curves for converted waves emerging from source as transverse, calculated without allowance for absorption; for source depth of 100 km.

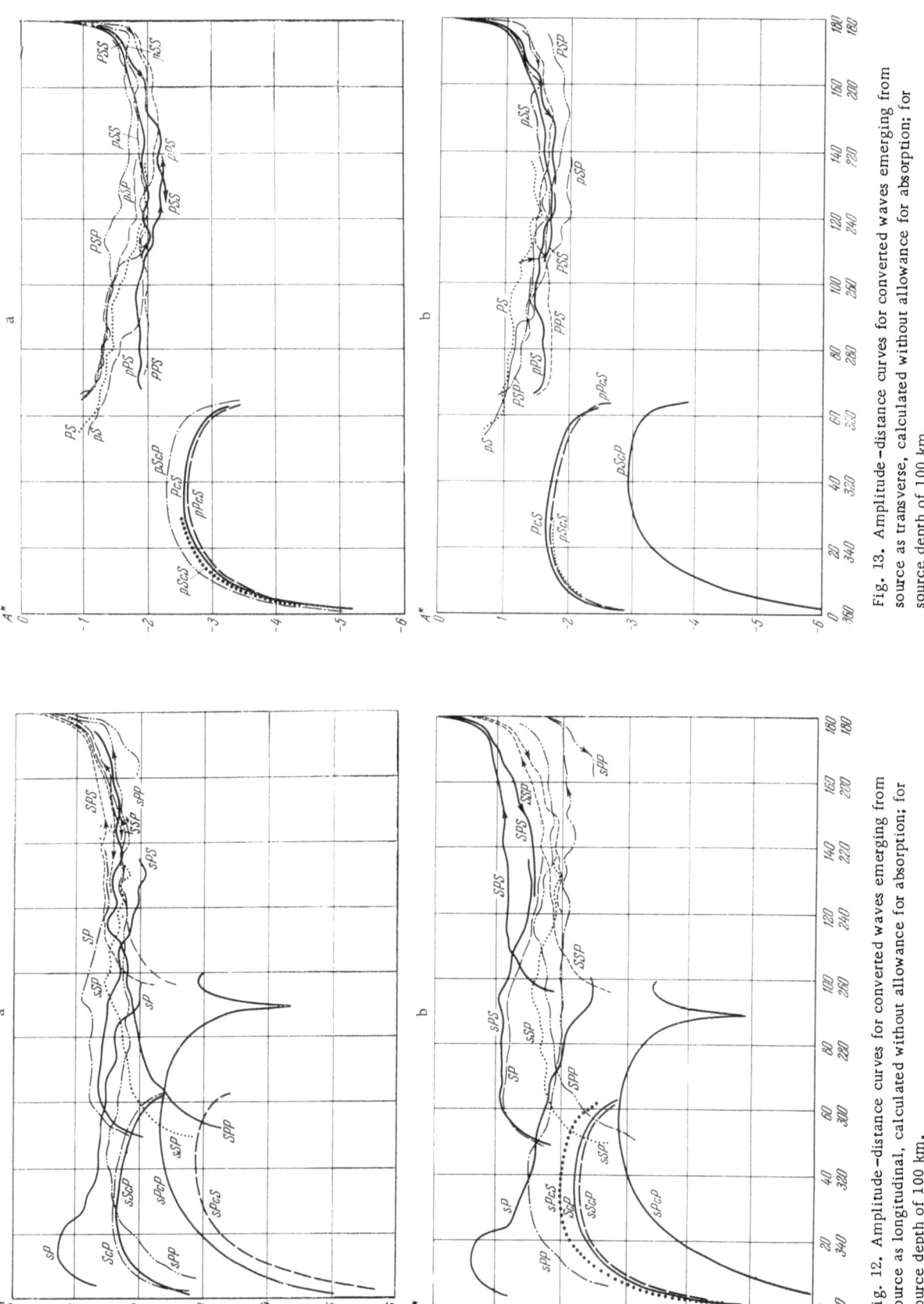

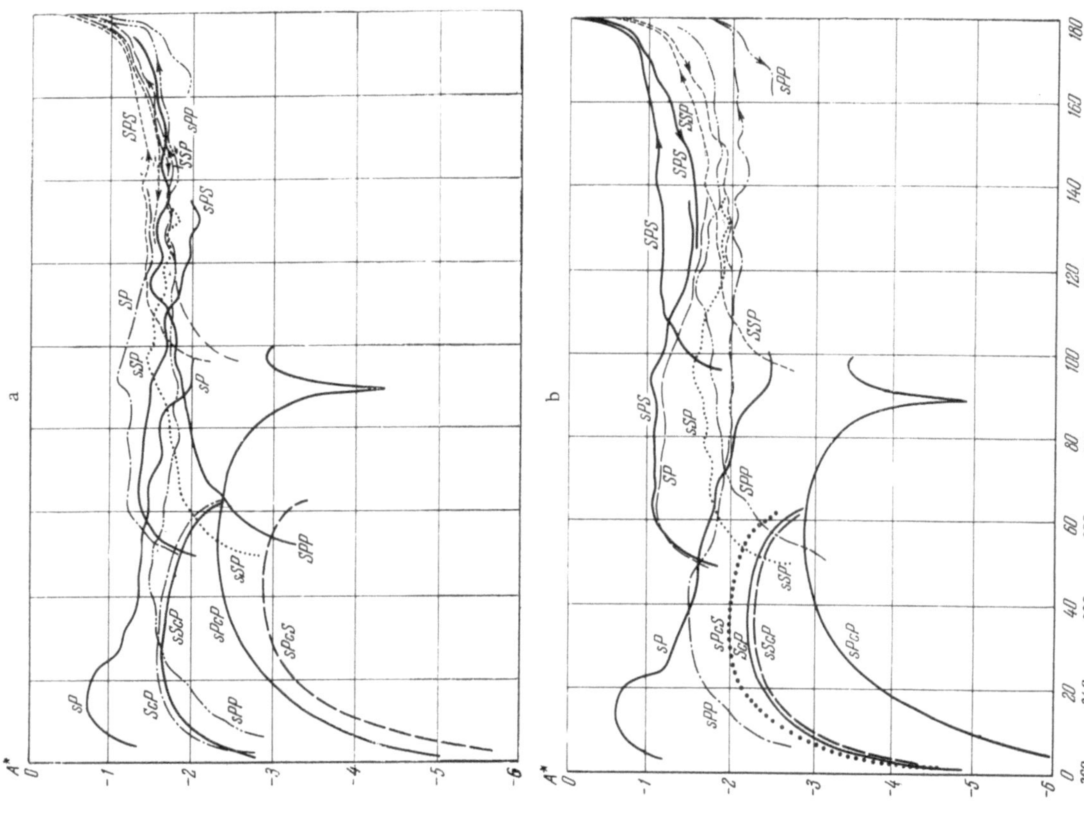

Fig. 12. Amplitude–distance curves for converted waves emerging from source as longitudinal, calculated without allowance for absorption; for source depth of 100 km.

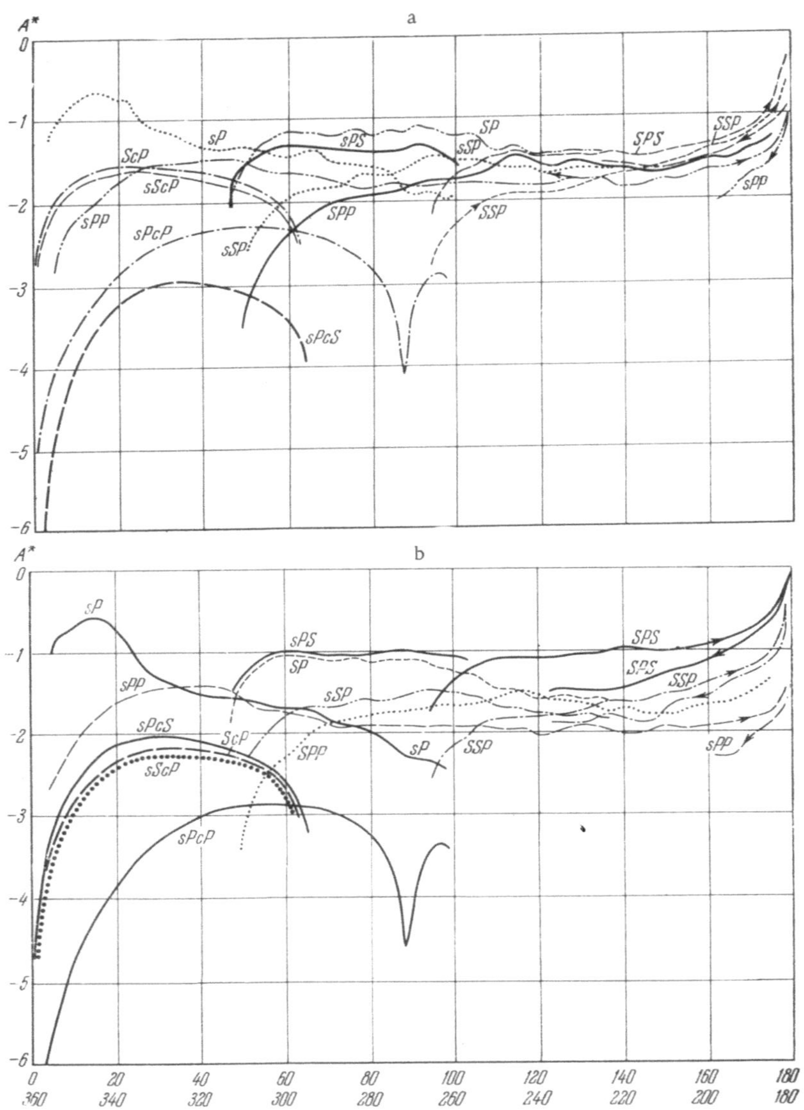

Fig. 14. Amplitude–distance curves for converted waves emerging from source as longitudinal, calculated without allowance for absorption; for source depth of 200 km.

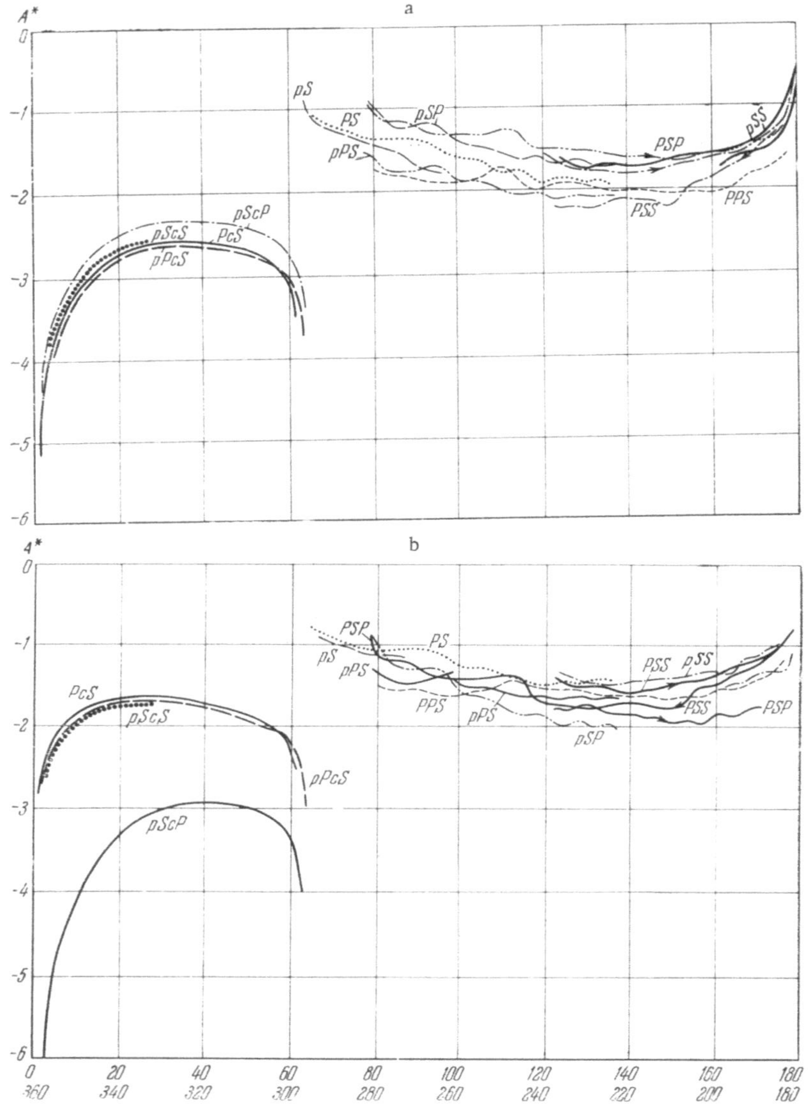

Fig. 15. Amplitude–distance curves for converted waves emerging from source as transverse, calculated without allowance for absorption; for source depth of 200 km.

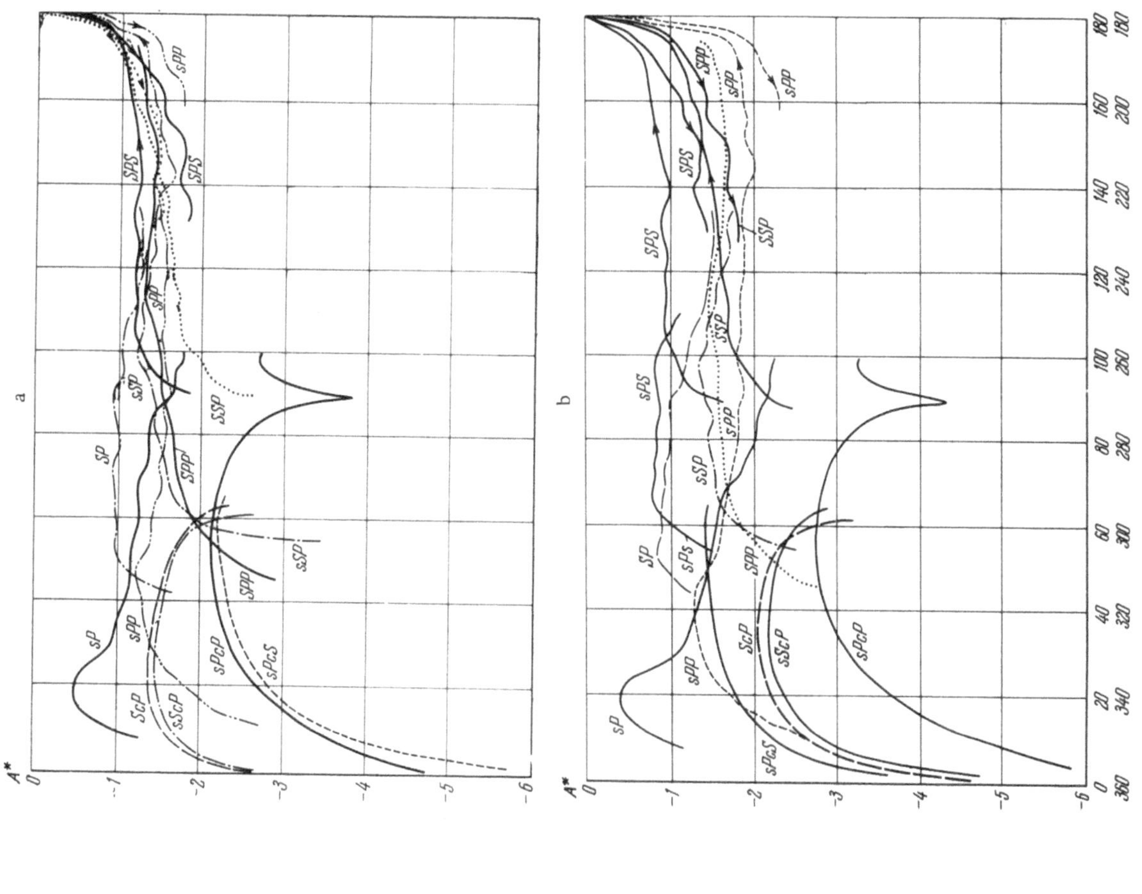

Fig. 16. Amplitude–distance curves for converted waves emerging from source as longitudinal, calculated without allowance for absorption; for source depth of 700 km.

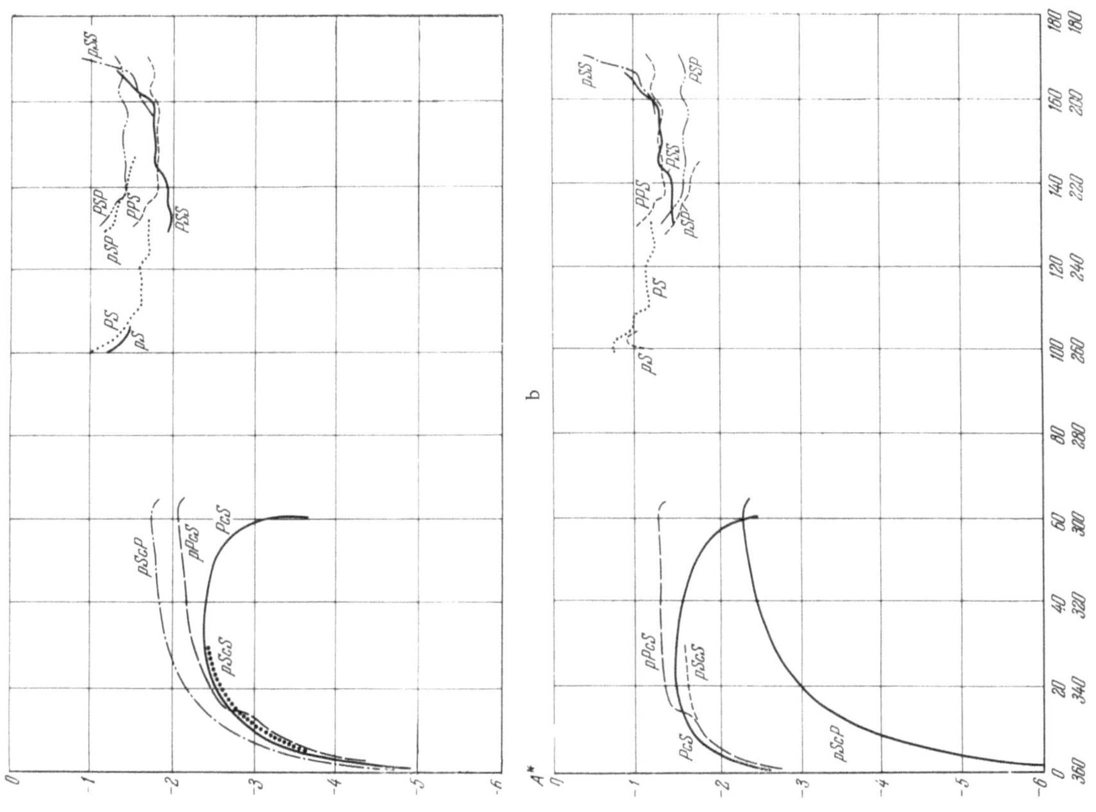

Fig. 17. Amplitude–distance curves for converted waves emerging from source as transverse, calculated without allowance for absorption; for source depth of 700 km.

LITERATURE CITED

1. Jeffreys, H., and K. E. Bullen (1940), Seismological Tables, London, British Association for the Advancement of Science, Gray-Milne Trust.
2. Jeffreys, H. (1952), The Earth: Its Origin, History, and Physical Constitution, Cambridge, Cambridge University Press.
3. Alekseev, A. S., and B. Ya. Gel'chinskii (1959), "On the ray method for calculating wave fields for heterogeneous media with curvilinear interfaces," in: Problems of the Dynamic Theory of Seismic Wave Propagation, No. 3, pp. 107-160.
4. Azbel', I. Ya., and T. B. Yanovskaya, "Approximation of velocity distributions for calculation of P-wave times and amplitudes," this volume, p. 62.
5. Yanovskaya, T. B. (1970), "Program for calculating travel-time and amplitude curves of body waves in layered media," in: Problems of the Dynamic Theory of Seismic Wave Propagation, No. 8, pp. 87-92.
6. Jeffreys, H. (1939), "The times of P, S, and SKS and the velocities of P and S," in: Mon. Not. Roy. Astr. Soc., Geophys. Suppl., 4:498-533.
6a. Bullen, K. E. (1938), "Theoretical travel times of the phases PP, PPP, PS, PPS, SS, PSS, and SSS," in: Mon. Not. Roy. Astr. Soc., Geophys. Suppl., 4:583-593.
7. Anderson, D. L., and C. B. Archambeau (1964), "The anelasticity of the earth," J. Geophys. Res., 69:2071-2084.
8. Anderson, D. L., A. Ben-Menahem, and C. B. Archambeau (1965), "Attenuation of seismic energy in the upper mantle," J. Geophys. Res., 70:1441-1448.
9. Kilinchuk, L. M., and T. B. Yanovskaya (1966), "The amplitude ratio PP/P," in: Computational Seismology, No. 2, pp. 71-82, Moscow, Nauka.

INTERFERENCE HEAD WAVES†

É. N. Bessonova and G. G. Mikhota

In a number of problems of seismic prospecting and earthquake seismology, one must study how head waves are influenced by the structure of the medium near the boundary at which they are formed. This influence is often due to interference phenomena in the transitional zone adjacent to the boundary. The effect of the interference of a small number of waves is easily computed by direct summation of the individual waves [1, 2, 5].

The present paper discusses a method for direct calculation of the spectrum of the sum interference wave in a medium consisting of an arbitrary number of horizontally homogeneous layers lying in a half-space or between the half-space. An interference head wave is defined as the sum of all waves that are propagated as head waves along the boundary of the lower half-space but undergo reflections and refractions on the upper boundaries. The spectra of the waves propagated along the boundary as longitudinal and transverse are calculated separately; if necessary, the waves that approach the observation point or have emerged from the source as longitudinal and transverse can be separated.

1. Formulation of Problem and Initial Formulas

Given an ideally elastic medium consisting of plane parallel layers between two half-spaces; each layer and half-space will be characterized by the longitudinal and transverse wave velocities a_i and b_i, the shear modulus μ_i, and layer thickness H_i, $i = 1, 2, \ldots n$. We shall assign to the upper half-space index $i = 0$ and to the lower half-space $i = (n+1)$. A rectangular coordinate system is introduced such that $z = 0$ coincides with the boundary of the lower half-space, and the z axis is directed downward (Fig. 1).

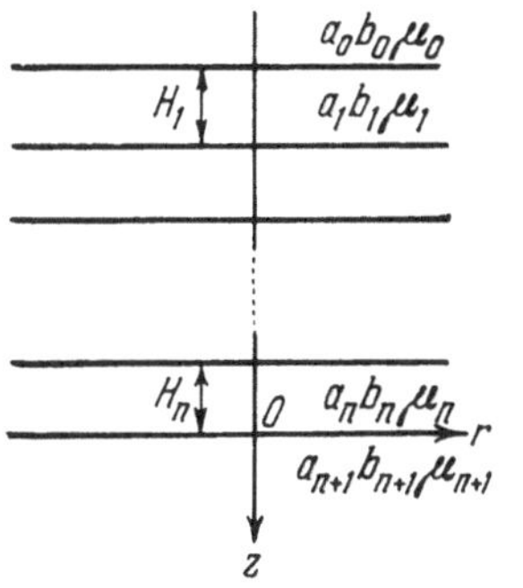

Fig. 1

Let a point source characterized by the potential

$$\varphi = \varphi(r, t) = \varphi_1(r) \cdot \varphi_2(t) \tag{1}$$

be applied in the upper half-space at the point $z = -z_0$. We must write out the displacement at an arbitrary point in the upper half-space caused by the given source and corresponding to a wave propagating along the boundary $z = 0$ with velocity a_{n+1} or b_{n+1}.

The total displacement at the point $(-z)$ is given by the formula [3, 4, 7]

$$\left.\begin{aligned} q &= \int_{-\infty}^{\infty} \theta(p) e^{ipt} \int_0^{\infty} g_1 F(\xi z) J_1(\xi r)\, \xi d\xi\, dp, \\ w &= \int_{-\infty}^{\infty} \theta(p) e^{ipt} \int_0^{\infty} g_2 F(\xi z) J_0(\xi r)\, \xi d\xi\, dp, \end{aligned}\right\} \tag{2}$$

where

$$\theta(p) = k \int_{-\infty}^{\infty} \varphi_1(t) e^{-ipt}\, dt, \qquad F(\xi z) = k_1 \int_0^{\infty} \varphi_2(r) J_0(\xi r)\, r dr,$$

† Translated from Vychislitel'naya Seismologiya, No. 4, Moscow, Nauka (1968), pp. 263-274.

$$\left.\begin{aligned} g_1 &= \xi A_0 e^{-\alpha_0(H-h)} - i\beta_0 C_0 e^{-\beta_0(H-h)}, \\ g_2 &= \alpha_0 A_0 e^{-\alpha_0(H-h)} + i\xi C_0 e^{-\beta_0(H-h)}, \\ \alpha_0 &= \sqrt{\xi^2 - \frac{p^2}{a_0^2}}, \quad \beta_0 = \sqrt{\xi^2 - \frac{p^2}{b^2}}, \end{aligned}\right\} \tag{3}$$

H is the vertical distance from the boundary of the upper half-space to the point at which the displacement is sought; h is the corresponding distance to the source; r is the horizontal distance from the source to the observation point; and q and w are the horizontal and vertical displacement components. A_0 and C_0 satisfy the matrix equation

$$\xi \begin{pmatrix} 1 & -i|\bar\beta_0| \\ -i\bar\alpha_0 & 1 \\ -\xi\mu_0\bar\gamma_0 & 2i\xi|\bar\beta_0|\mu_0 \\ 2i\xi\bar\alpha_0\mu_0 & -\xi\mu_0\bar\gamma_0 \end{pmatrix} \begin{pmatrix} A_0 e^{-\alpha_0 h} \\ C_0 e^{-\beta_0 h} \end{pmatrix} + \begin{pmatrix} i \\ i\bar\alpha_0 \\ i\bar\gamma_0\mu_0\xi \\ -2i\mu_0\xi\bar\alpha_0 \end{pmatrix} \xi^2 = \xi \prod_{q=1}^{n} M_q \cdot \begin{pmatrix} i & i|\bar\beta_{n+1}| \\ i\bar\alpha_{n+1} & 1 \\ -i\xi\mu_{n+1}\bar\gamma_{n+1} & -2i|\bar\beta_{n+1}|\mu_{n+1}\xi \\ -2i\xi\mu_{n+1}\bar\alpha_{n+1} & -\mu_{n+1}\xi\bar\gamma_{n+1} \end{pmatrix} \begin{pmatrix} A_{n+1} \\ C_{n+1} \end{pmatrix}, \tag{4}$$

$$M_q = \begin{pmatrix} -2\xi^2 \operatorname{ch}\alpha_q H_q + \gamma_q \operatorname{ch}\beta_q H_q & \xi\left(-\frac{\gamma_q}{\alpha_q}\operatorname{sh}\alpha_q H_q + 2\beta_q \operatorname{sh}\beta_q H_q\right) & \xi/\mu_q\,(-\operatorname{ch}\alpha_q H_q + \operatorname{ch}\beta_q H_q) & \frac{1}{\mu_q}\left(\frac{\xi^2}{\alpha_q}\operatorname{sh}\alpha_q H_q + \beta_q \operatorname{sh}\beta_q H_q\right) \\ -\xi\left(2\alpha_q \operatorname{sh}\alpha_q H_q - \frac{\gamma_q}{\beta_q}\operatorname{sh}\alpha_q H_q\right) & \gamma_q \operatorname{ch}\alpha_q H_q - 2\xi^2 \operatorname{ch}\beta_q H_q & \frac{1}{\mu_q}\left(\alpha_q \operatorname{sh}\alpha_q H_q - \frac{\xi^2}{\beta_q}\operatorname{sh}\beta_q H_q\right) & \xi/\mu_q\,(\operatorname{ch}\alpha_q H_q - \operatorname{ch}\beta_q H_q) \\ 2\xi\gamma_q\mu_q(\operatorname{ch}\alpha_q H_q - \operatorname{ch}\beta_q H_q) & \mu_q\left(\frac{\gamma_q^2}{\alpha_q}\operatorname{sh}\alpha_q H_q - 4\xi^2\beta_q \operatorname{sh}\beta_q H_q\right) & \gamma_q \operatorname{ch}\alpha_q H_q - 2\xi \operatorname{ch}\beta_q H_q & \xi\left(\frac{\gamma_q}{\alpha_q}\operatorname{sh}\alpha_q H_q - 2\beta_q \operatorname{sh}\beta_q H_q\right) \\ \mu_q\left(-4\xi^2\alpha_q \operatorname{sh}\alpha_q H_q + \frac{\gamma_q^2}{\beta_q}\operatorname{sh}\beta_q H_q\right) & 2\xi\gamma_q\mu_q(-\operatorname{ch}\alpha_q H_q + \operatorname{ch}\beta_q H_q) & \xi\left(-2\alpha_q \operatorname{sh}\alpha_q H_q + \frac{\gamma_q}{\beta_q}\operatorname{sh}\beta_q H_q\right) & -2\xi^2 \operatorname{ch}\alpha_q H_q + \gamma_q \operatorname{ch}\beta_q H_q \end{pmatrix} \tag{5}$$

where

$$\alpha_q = \sqrt{\xi^2 - \frac{p^2}{a_q^2}}; \quad \beta_q = \sqrt{\xi^2 - \frac{p^2}{b_q^2}}; \quad \gamma_q = 2\xi^2 - \frac{p^2}{b_q^2};$$

the overscored values are found from the expressions

$$\alpha_q^2 = -\xi^2\bar\alpha_q^2; \quad J_q = \xi^2 \bar J_q; \quad \beta_q^2 = -\xi^2\bar\beta_q^2.$$

At great distances r from the source, most of the displacements consist of surface waves, and the solution of our problem is given by the second term of an asymptotic expansion of the displacements, which decreases with distance as r^{-2}:

$$q = \frac{k_1}{r^2}\int_{-\infty}^{\infty} Q(p)\,e^{ipt}\left(e^{-ip\frac{r}{a_{n+1}}}\frac{\partial}{\partial\alpha_{n+1}}(\xi g_1)\big|_{\xi=h_{n+1}} + e^{-ip\frac{r}{b_{n+1}}}\frac{\partial}{\partial\beta_{n+1}}(\xi g_2)\big|_{\xi=k_{n+1}}\right)dp,$$

$$w = \frac{k_1}{r^2}\int_{-\infty}^{\infty} Q(p)\,e^{ipt}\left(e^{-ip\frac{r}{a_{n+1}}}\frac{\partial}{\partial\alpha_{n+1}}(\xi g_2)\big|_{\xi=h_{n+1}} + e^{-ip\frac{r}{b_{n+1}}}\frac{\partial}{\partial\beta_{n+1}}(\xi g_2)\big|_{\xi=k_{n+1}}\right)dp, \tag{6}$$

where

$$h_{n+1} = \frac{p}{a_{n+1}} \text{ and } k_{n+1} = \frac{p}{b_{n+1}}.$$

Formulas (6) give the total field of all waves that propagate along the boundary z = 0 as P and S head waves with apparent velocities a_{n+1} and b_{n+1}, respectively. To separate the waves that emerge from the source as longitudinal or as transverse, it is sufficient to give the source as a center of dilatation or of rotation. To categorize the waves that approach the receiver as either longitudinal or transverse, we must consider the values related to α_0 and β_0 individually in formulas (3) for g_1 and g_2, i.e., in formulas (6) instead of g_1 we must take $g_{11} = \xi A_0 e^{-\alpha_0(H-h)}$ for a longitudinal wave and $g_{12} = -i\beta_0 C_0 e^{-\beta_0(H-h)}$ for a transverse wave. The same holds for g_2. A program was written to calculate the functions

$$F(h_{n+1}z)\frac{\partial}{\partial\alpha_{n+1}}(\xi g_{ij})\big|_{\xi=h_{n+1}}, \tag{7}$$

$$i = 1,2, \quad j = 1,2$$

for F corresponding to a center of dilatation:

$$F=\frac{e^{-\alpha_0|z+z_0|}}{\alpha_0}. \qquad (8)$$

These functions are the amplitude–frequency response of the medium for the interference head wave when a longitudinal wave emerges from the source. To obtain the wave spectrum, it is sufficient to multiply expression (7) by Q(p), which is the spectral function of the source.

2. Some Calculations of the Effect of a Transitional Layer on Waves Recorded in Deep Seismic Sounding

In deep seismic sounding on land (the Caucasus) and at sea (Sea of Okhotsk), it was established [6] that the spectrum of the wave P^0, which travels along the surface of the consolidated crust (bottom of sediments), varies significantly along the profiles. For most of the profile, the spectrum of the P^0 wave has a simple form (one maximum), and frequencies of 3-6 Hz at sea and 3-8 Hz on land predominate in the spectrum. In other parts of the profiles, when the sediments are thin at the observation point, two maxima are found in the spectrum of P^0. In observations at sea, the first maximum occurs in the frequency range of 3-6 Hz and the second around 11-14 Hz; on land, the first maximum occurs at frequencies of 3-8 Hz and the other at 9-14 Hz. The hypothesis that this variation in the spectrum is related to the presence of thin layers near the bottom of the sediments was considered.

To check, we calculated the functions described by expression (7). The source was taken as a center of dilatation

$$\varphi(r,t)=\frac{1}{R}\varphi\left(t-\frac{R}{a_0}\right),$$

where $R=\sqrt{(z-z_0)^2+r^2}$.

Expressions (7) were calculated separately for the vertical (w) and horizontal (q) displacement components for longitudinal (PPP) and converted (PPS) waves.

According to Section 1, they are the amplitude–frequency response of the medium for the corresponding types of waves. Below, they are called simply the amplitude response.

In order to obtain a physically clear picture, we restrict ourselves to the simplest model of a medium: one thin layer is placed above the boundary with which the formation of P^0 is associated. This approximation follows from experimental data. In studying the basement in many regions, two waves are recorded whose apparent velocities differ by 0.2-0.5 km/sec. The first wave corresponds to the top of the weathering; the second, to the bottom of the weathering, i.e., to the basement itself. When the belt of weathered zone is thin, both waves merge into one interference oscillation. Such compound oscillations are usually recorded in deep seismic sounding over extended parts of profiles.

Calculations were made for two models (Figs. 2a and 3a). The first model characterizes the tran-

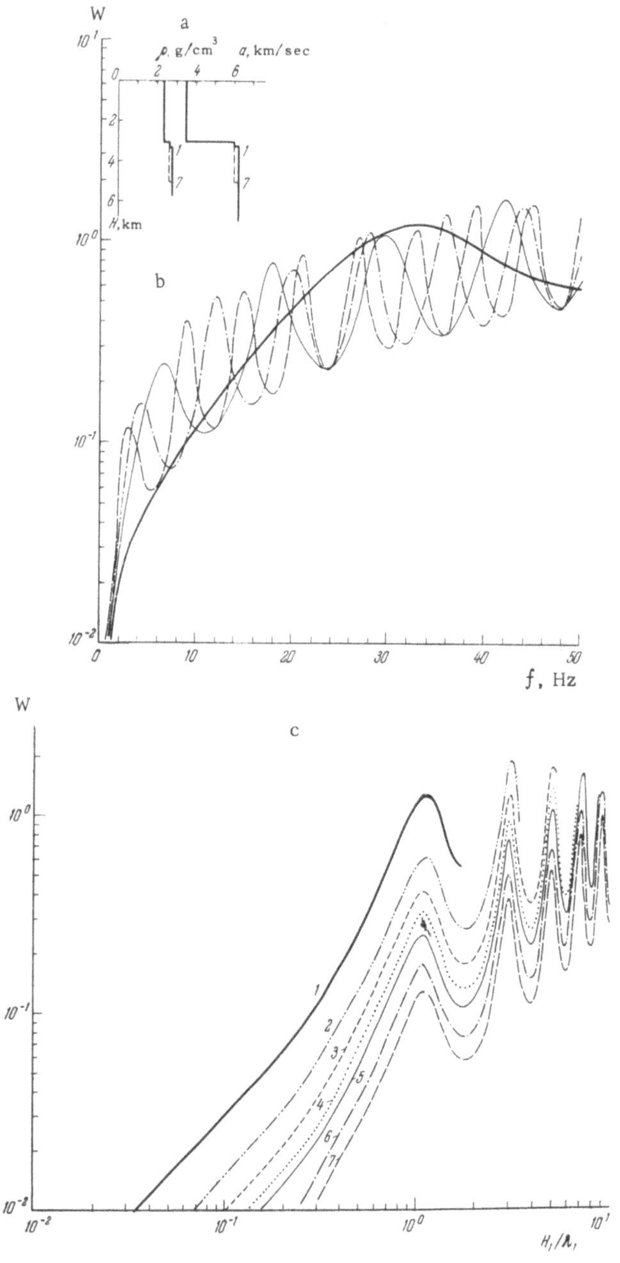

Fig. 2. Spectral characteristics of the medium for model 1: a) velocity and density profiles; b, c) amplitude characteristics of thin layers of variable thickness. The parameter is the layer thickness H_1 in km: 1) 0.2; 2) 0.4; 3) 0.6; 4) 0.8; 5) 1.0; 6) 1.5; 7) 2.0.

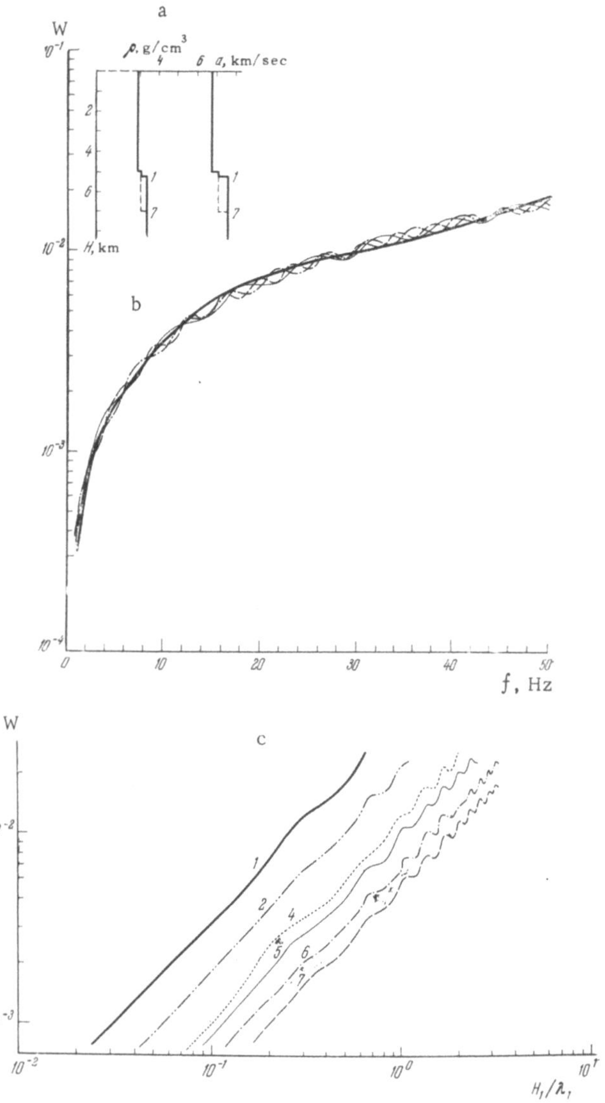

Fig. 3. Spectral characteristics of the medium for model 2. Symbols as in Fig. 2.

sition from sediments to the basement, for continental crust and the second, to the transition from the crust to the mantle in oceanic regions.

The upper part of the medium is replaced by a half-space to eliminate waves reflected from the free surface.

In the first model, the sediment velocities were assigned to the upper half-space and the basement velocities were assigned to the lower half-space. The velocities in the layer were lower by 0.2-0.5 km/sec than those below it.

In the second model, the velocities of the "basalt" layer were assigned to the upper half-space, and the velocities determined from observations for the Mohorovičić discontinuity were assigned to the lower half-space. The velocities in the layer correspond to values obtained for the lower part of the oceanic crust [6]. It was assumed that $a_i/b_i = \sqrt{3}$.

The thickness of the upper half-space was 3 km in the first and 5 km in the second model. The assumed values of the layer thicknesses used were: 0.2, 0.4, 0.6, 0.8, 1.0, 1.5, and 2.0 km.

Calculations were made for different ratios of velocities, at the upper boundary of the layer of 0.33-0.70 (continent) and 0.90-0.96 (ocean); at the bottom of the layer the velocity ratio was assumed to be 0.87-0.97 (continent) and 0.88-0.92 (ocean). In all variants of these models, the velocity increased discontinuously with depth.

Below we shall consider only the amplitude response of the vertical component of the P wave.

Figures 2b and 2c, show the calculation results for model 1, and Figs. 3b and 3c, show the results for model 2. The spectral characteristics are given as functions of frequency $\omega = 2\pi f$ (Figs. 2b and 3b) and as functions of the ratio of layer thickness H_1 to wavelength in the layer λ_1 ($\lambda_1 = 2\pi a_1/\omega$) (Figs. 2c and 3c).

It can be seen that the amplitude response has extrema at frequencies which depend on layer thickness. The thinner the layer, the smoother the response. As layer thickness increases, the number of extrema increases, but each extremum falls within a narrower frequency range. This has a simple physical explanation: the thicker the layer, the greater the difference between the paths of the individual components of the interference wave.

The relative amplitude of the individual extrema is a function of the ratio of the velocity at the top of the layer to the velocity at the bottom. If this ratio is close to unity, the response is so smooth that it is difficult to detect the presence of the thin layer above the boundary. The amplitudes of the extrema increase when the velocity does not change much between the top and the bottom of the layer (Figs. 2b and 2c).

At the working frequencies of deep seismic sounding (5-20 Hz), the ratio of the amplitudes of the maxima to the amplitudes of the minima is 4-5 for a layer thickness greater than 0.4 km, i.e., in this case the thin layer can in principle be distinguished by analyzing the spectral characteristics of the waves.

The spectral characteristics have a similar form in the coordinate system w, H_1/λ_1 (Figs. 2c and 3c). The principal extrema are confined to the same values of H_1/λ_1, but as can be seen from Figs. 2c and 3c, when H_1/λ_1 is constant the level

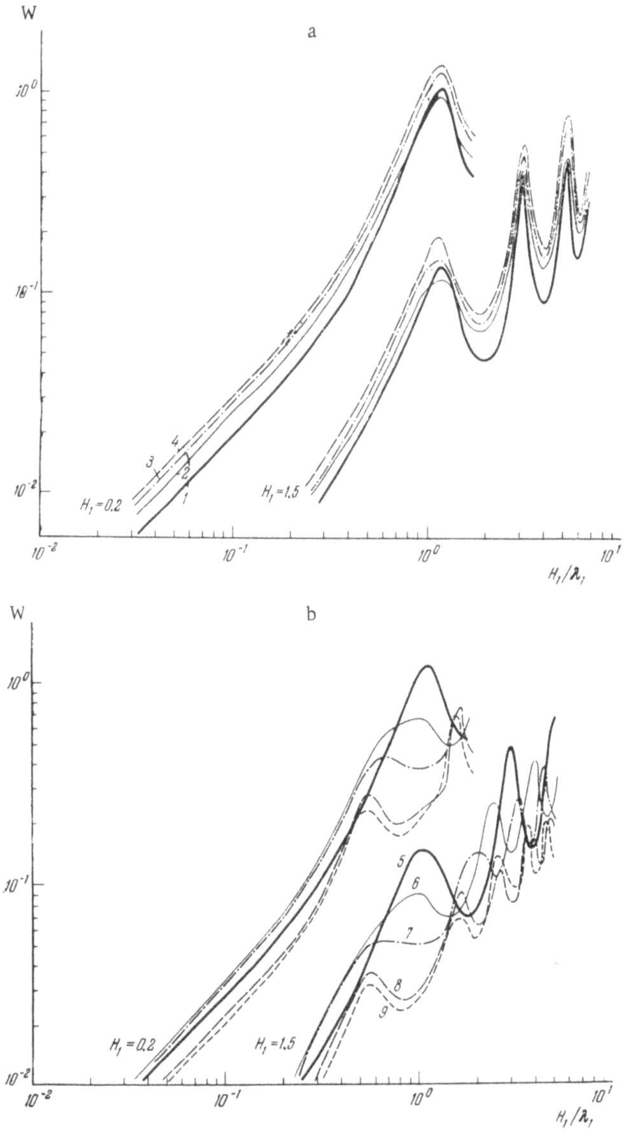

Fig. 4. Spectral characteristics of thin layer for models with the parameters a_1/a_2 = 0.97 and a_0/a_1 = 0.33-0.58 (a) and a_0/a_1 = 0.545 and a_1/a_2 = 0.87-0.97 (b). The spectral characteristics are given for two layer thicknesses (H_1 = 0.2 km and H_1 = 1.5 km); a) curve parameters a_0/a_1: 1) 0.33; 2) 0.42; 3) 0.50; 4) 0.58; b) curve parameters a_1/a_2: 5) 0.97; 6) 0.95; 7) 0.917; 8) 0.89; 9) 0.87.

of the characteristic decreases as the layer thickness increases.

Figures 4a and 4b show the amplitude response of a thin layer for several models in which only the velocity ratio either at the top a_0/a_1 (Fig. 4a) or at the bottom a_1/a_2 (Fig. 4b) varied. It is apparent that for models with a_1/a_2 = const and variable a_0/a_1, the maxima and minima of the spectral characteristics are in the same ranges of H_1/λ_1. The velocity ratio a_0/a_1 affects only the absolute level of the components of the spectral characteristic (Fig. 4a).

For the models with a_0/a_1 = const and variable a_1/a_2, the position of the extrema on the frequency axis changes: as a_1/a_2 decreases, the maxima and minima move in the direction of smaller H_1/λ_1, i.e., toward lower frequencies (Fig. 4b).

The position of the extrema on the frequency axis is therefore determined by the velocity ratio a_1/a_2 at the bottom of the layer.

Our main conclusion is that the presence of transitional layers near the boundary in certain media greatly affects the spectra of the head waves.

We can also study multilayered transition zones using this program. By varying the model we can also study the effect on the head wave of other factors such as low-velocity zones, thin layers above the discontinuity, etc.

LITERATURE CITED

1. Berzon, I. S., A. M. Epinat'eva, G. N. Pariiskaya, and S. P. Starodubrovskaya (1962), Dynamic Characteristics of Seismic Waves in Real Media, Moscow, Izd. Akad. Nauk SSSR.
2. Berzon, I. S., L. I. Ratnikova, and M. I. Rats-Khizgiya (1966), Seismic Converted Reflected Waves, Moscow, Nauka.
3. Naimark, M. A. (1948), "Oscillations of a thin elastic layer lying on an elastic half-space," Tr. Seism. Inst. Akad. Nauk SSSR, No. 119; No. 127.
4. Keilis-Borok, V. I. (1960), Interference Surface Waves, Moscow, Izd. Akad. Nauk SSSR.
5. Petrashen', G. I. (1957-1959), Materials for a Quantitative Study of Seismic Wave Dynamics, Leningrad, Izd. Leningrad Gos. Univ.
6. Gal'perin, E. I., and I. P. Kosminskaya (eds.) (1964), Crustal Structure in the Transitional Region from the Asiatic Continent to the Pacific Ocean, Moscow, Nauka.
7. Haskell, N. A. (1953), "The dispersion of surface waves on multilayered media," Bull. Seism. Soc. Amer., 43:17-34.

METHOD FOR CALCULATING SURFACE RAYLEIGH WAVES IN A VERTICALLY INHOMOGENEOUS HALF-SPACE†

M. G. Neigauz and G. V. Shkadinskaya

1. Statement of Problem

As is well-known [1], displacements in Rayleigh waves are expressed in terms of the eigenvalues and eigenfunctions of the following boundary value problem.

We consider the vector equation

$$\frac{d}{dz}\left(A\frac{d\mathbf{w}}{dz}+\xi B\mathbf{w}\right)-\xi B^*\frac{d\mathbf{w}}{dz}+(p^2\rho E-\xi^2 C)\mathbf{w}=0 \quad (1)$$

in the segment $(0, +\infty)$ with the boundary conditions

$$A\frac{d\mathbf{w}}{dz}+\xi B\mathbf{w}=0 \quad \text{for} \quad z=0;$$
$$|\mathbf{w}(z)|\to 0 \quad \text{for} \quad z\to+\infty, \quad (2)$$

where p and ξ are real parameters; p is the angular frequency; ξ is the wave number; p/ξ is the phase velocity; $w = (w_1, w_2)$; A, B, C, and E are square matrices of order 2:

$$A=\begin{pmatrix}\mu(z) & 0\\ 0 & \lambda(z)+2\mu(z)\end{pmatrix};$$

$$B=\begin{pmatrix}0 & \mu(z)\\ -\lambda(z) & 0\end{pmatrix};$$

$$C=\begin{pmatrix}\lambda(z)+2\mu(z) & 0\\ 0 & \mu(z)\end{pmatrix};$$

E is the unit matrix:

$$E=\begin{pmatrix}1 & 0\\ 0 & 1\end{pmatrix}$$

and matrix B* is the transpose of B, i.e.,

$$B^*=\begin{pmatrix}0 & -\lambda(z)\\ \mu(z) & 0\end{pmatrix}.$$

$$\mu(z)=b^2(z)\rho(z);\quad \lambda(z)=(a^2(z)-2b^2(z))\rho(z),$$

and $b(z)$, $a(z)$, and $\rho(z)$ are given piecewise continuous positive functions of the parameter Z which are constant when $z \geq Z$, and we assume $a^2(z) > 2b^2(z)$, $b(z) \leq b(Z)$, $a(z) \leq a(Z)$, and $\rho(z) \leq \rho(Z)$.

The physical meaning of the last five functions is as follows: $\rho(z)$ is the density of the medium, $b(z)$ and $a(z)$ are respectively the velocities of transverse and longitudinal waves in the medium and $\mu(z)$ and $\lambda(z)$ are Lamé parameters.

The vector functions w(z) and $Aw' + \xi Bw$ are assumed to be continuous for all z.

The vector w(z) gives the horizontal and vertical displacement components in the medium and the vector $Aw' + \xi Bw$ gives the normal and tangential stress components in the medium.

The value $p_k^2(\xi)$ of the parameter p^2 for which a nonzero solution $w_k(z)$ of Eq. (1) exists and satisfies boundary conditions (2) is called the eigenvalue of the boundary value problem (1) and (2), and $w_k(z)$ is the corresponding eigenfunction of this problem. The number k of the eigenvalue $p_k^2(\xi)$, (k = 0, 1, 2 ... ∞) is exactly the number of a normal mode of the Rayleigh wave.

Below we shall discuss a method for calculating the eigenvalues $p_k^2(\xi)$ and associated eigenfunctions $w_k(z, \xi)$ with any degree of accuracy for any k.

2. Factorization Method

First let us study the problem for fixed ξ.

It can be shown that the operator given by Eq. (1) and boundary conditions (2) is self-adjoint.

† Translated from Vychislitel'naya Seismologiya, No. 2, Moscow, Nauka (1966), pp. 121-129.

We shall solve the boundary value problem (1) and (2) by the factorization method proposed by I. M. Gel'fand and O. V. Lokutsievskii [2]. We shall use the version of the method proposed by V. B. Lidskii and M. G. Neigauz [3].

We introduce the new unknown complex matrix U(z) by the formula

$$AW' + \xi BW = im(E-U)(E+U)^{-1}W, \qquad (3)$$

where W is a second-order matrix whose columns are two linearly independent solutions of Eq. (1); m is a normalization factor.

It can be shown that matrix U(z) is unitary, i.e., $U(z)U^*(z) = E$, where $U^*(z)$ denotes the adjoint of U(z).

If we differentiate Eq. (3) with respect to z and substitute into Eq. (1), we obtain a differential equation for the matrix U(z):

$$U'(z) = iS(U(z))\cdot U(z), \qquad (4)$$

where

$$S(U) = -\frac{1}{2}\Big\{m(E-U)A^{-1}(E-U^*) - i\xi(E-U)A^{-1}B(E+U^*) + i\xi(E+U)B^*A^{-1}\times (E-U^*) + \frac{1}{m}(E+U)M(E+U^*)\Big\},$$

and

$$M = \xi^2 B^*A^{-1}B + p^2\rho E - \xi^2 C.$$

It is easy to show that S(U) is Hermitian, i.e., $S^*(U) = S(U)$.

From (3) we can derive a formula that defines matrix U(z) in terms of W(z) – the matrix of the two linearly independent solutions of Eq. (1)

$$U(z) = -(V - imE)(V + imE)^{-1}, \qquad (5)$$

where

$$V(z) = (AW^1 + \xi BW)\cdot W^{-1}.$$

Let $w_1(z)$ and $w_2(z)$ be the two linearly independent solutions of Eq. (1), and let $w_3(z) = C_1w_1(z) + C_2w_2(z)$ and $w_4(z) = C_3w_1(z) + C_4w_2(z)$, where $w_3(z)$ and $w_4(z)$ are linearly independent. It is not difficult to show that the matrix U(z) defined by the pair of solutions $w_1(z)$ and $w_2(z)$ coincides with the matrix U(z) defined by the pair of solutions $w_3(z)$ and $w_4(z)$.

It is obvious that the matrix U(z) is uniquely determined by Eq. (4) when its initial value at some point is given.

Theorem 1. The eigenvalue p_k^2 of problem (1) and (2) is a value of the parameter p^2 such that the matrix U(z) satisfies Eq. (4) and the boundary conditions: when z = Z, matrix U(z) is defined by equation (5), when $z \geq Z$, W(z) is a matrix composed of the two exponentially decreasing solutions of Eq. (1); when z = 0, det [E − U(0)] = 0.

The theorem is proved without difficulty. The boundary conditions for matrix U(z) are obtained from conditions (2) for the vector w.

It follows from theorem 1 that to find any eigenvalue p_k^2 of the initial problem (1) and (2), it is sufficient to be able to solve the boundary value problem for the matrix U(z).

Let $\tilde{p}^2$ be an eigenvalue of problem (1) and (2). If we know matrix U(z) only for $p^2 = \tilde{p}^2$, how do we determine the number of eigenvalue $\tilde{p}^2$? To answer this question, let us examine the eigenvalues of matrix U(z) as functions of the parameters z and p^2 (we shall call them the characteristic roots of matrix U(z)).

A unitary second-order matrix has two characteristic roots of the form $\exp[i\varphi_1(z,p^2)]$ and $\exp[i\varphi_2(z,p^2)]$, where the functions $\varphi_i(z,p^2)$ (i=1, 2) are real. Matrix $U(z,p^2)$, which is a continuous matrix, is a function of its arguments; therefore its characteristic roots, and also the functions $\varphi_i(z,p^2)$, can be defined as continuous functions of the parameters z and p^2. The functions $\varphi_i(z,p^2)$ can always be defined so that $\varphi_1(z,p^2) \geq \varphi_2(z,p^2)$.

Let us define the functions $\varphi_i(z,p^2)$ at the point z = Z so that $-\pi < \varphi_i(z,p^2) < \pi$.

We shall give without proof two lemmas on the behavior of the functions $\varphi_i(z,p^2)$, i = 1, 2.

Lemma 1. The functions $\varphi_i(z,p^2)$ are monotonically increasing functions of the parameter p^2 when the parameter z is fixed.

Lemma 2. $\partial\varphi_i(z_0,p^2)/\partial z < 0$, if $\varphi_i(z_0,p^2) = \pi + 2m\pi$ (m = 0, ±1, ...).

Using these lemmas, we shall prove the following theorem.

Theorem 2. If $\varphi_1(0,\tilde{p}_k^2) = 2m\pi$, then $\tilde{p}_k^2$ is an eigenvalue of problem (1) and (2), and its number is k = 2m − 1 (m = 1, 2, ...). If $\varphi_2(0,\tilde{p}_k^2) = 2m\pi$, then the number of the eigenvalue is k = 2m (m = 0, 1, ...).

We shall not give a complete proof of this theorem. Note only that either $\exp[i\varphi_1(0,\tilde{p}_k^2)] = 1$ or $\exp[i\varphi_2(0,\tilde{p}_k^2)] = 1$ follows from the conditions of the theorem. This means that $\det[E - U(0,\tilde{p}_k^2)] = 0$. Hence from theorem 1 we find that $\tilde{p}_k^2$ is an eigenvalue of problem (1) and (2).

3. Integration of Equations

The characteristic vector functions w(z) cannot be calculated from Eq. (3), since the right side of this equation has a singularity at the point z_0 when $\det[E + U(z_0)] = 0$.

Similarly, as in [3], we can introduce a new unknown vector function $\eta(z)$ by the formula

$$\eta(z) = \frac{i}{2}(E - U)(Aw' + \xi Bw) + \frac{m}{2}(E + U)w. \quad (6)$$

If we differentiate (6) with respect to z and substitute it into Eq. (1), we obtain an equation for $\eta(z)$:

$$\eta' = \frac{1}{2}\Big\{-imA^{-1}(E - U^{*}) - \xi A^{-1}B(E + U^{*}) + \\ + \xi B^{*}A^{-1}(E - U^{*}) - \frac{i}{m}M(E + U^{*})\Big\}\eta. \quad (7)$$

It is easy to obtain formulas that express w and w' in terms of $\eta(z)$:

$$w = \frac{1}{2m}(E + U^{*})\eta; \; w' = -A^{-1}\Big\{\xi Bw + \frac{i}{2}(E - U^{*})\eta\Big\}. \quad (8)$$

To find the eigenvalues and eigenfunctions of problem (1) and (2), we must solve initial value problems for Eqs. (4) and (7).

We integrate Eqs. (4) and (7) from right to left, i.e., from infinity to 0.

Equation (4) is integrated stably in both directions only if the difference scheme is properly chosen. If matrix U(z) is calculated by integrating Eq. (4) from left to right, i.e., from 0 to $+\infty$, the family of solutions of (1) corresponding to it does not have an exponentially decreasing solution when $z \to +\infty$, due to roundoff errors in determining the eigenvalue. Therefore a linear combination of these solutions cannot give a good approximation of the eigenfunction.

The initial value for matrix U(z), as follows from theorem 1, could have been taken at the point $z = Z$. To reduce calculation time, it is advisable that the point for beginning the integration be $\tilde{z} \le Z$. We shall do this as it was done in [1].

First we select the reversal point $\bar{z}$, namely such that when $z \ge \bar{z}$ neither of the exponentially decreasing solutions of Eq. (1) vanishes, i.e., $w(z) \ne 0$ when $z \ge \bar{z}$. As this point we take the greatest root of the equation $b(z) = p/\xi$.

From the point $z = \bar{z}$ we integrate in the direction of increasing z the equation

$$\frac{dV}{dz} = -\sqrt{\xi^2 - \frac{p^2}{b^2(z)}}\,V \quad (9)$$

with the initial condition $V(z) = 1$. Generally speaking, the function V(z) maximizes the absolute value of any exponentially decreasing solution of Eq. (1).

We integrate Eq. (9) to the first point $\tilde{z}$ where $V(\tilde{z}) < 1/N$. The number N is chosen according to the designated accuracy of solution of the problem. The point $\tilde{z}$ is selected as the point for beginning the integration of Eqs. (4) and (7). When $z \ge \tilde{z}$, we assume that the functions $b(z)$, $a(z)$, and $\rho(z)$ are constants.

4. Calculation of Eigenvalues

The eigenvalue p_k^2 is calculated as a zero of the function

$$f(p^2) = \begin{cases} \varphi_1(0, p^2) - 2m\pi & \text{for } k = 2m - 1\ (m = 1, \ldots), \\ \varphi_2(0, p^2) - 2m\pi & \text{for } k = 2m\ (m = 0, 1, \ldots). \end{cases}$$

We shall use a combination of the Newton–Raphson method and the binary section method to find the zero of the function $f(p^2)$. We calculate the functions $\varphi_1(0, p^2)$ and $\varphi_2(0, p^2)$ by solving the initial value problem for matrix U(z). We set the initial conditions for matrix U(z) at the point $z = \tilde{z}$ as in theorem 1. We select a difference approximation of Eq. (4) such that matrix U(z) – the solution of the difference equations – is unitary at each point

$$\frac{U_{n+1} - U_n}{h_n} = i\,\frac{S_{n+\frac{1}{2}}\,(U_n + U_{n+1})}{2}, \quad (10)$$

where

$$S_{n+1} = S[U_n + \frac{ih_n}{2}S(U_n)U_n].$$

It is easy to verify that if the matrix U_n is unitary, then the matrix U_{n+1} given by Eq. (10) is also unitary. The arguments of the roots of matrix U_n are calculated at each point: $\varphi_1^{(n)}$ and $\varphi_2^{(n)}$.

System (10) is solved with a variable step h_n. Step h_n is chosen such that, in the first place the given approximation error is not exceeded and, secondly the functions $\varphi_1(z)$ and $\varphi_2(z)$ in one step vary by not more than π/l, where the integer l is selected in accordance with the given solution accuracy ($l \ge 8$).

5. Calculation of Eigenfunctions

Let the eigenvalue p_k^2 be found. We calculate the characteristic eigenfunction as follows.

Solving Eq. (4) and Eq. (7) jointly from the point $z = \tilde{z}$ to zero, we immediately calculate two linearly independent solutions of Eq. (1), which are exponentially decreasing when $z \to +\infty$. The desired eigenfunction is a linear combination of these two solutions of Eq. (1): $w(z) = C_1 w_I(z) + C_2 w_{II}(z)$. The coefficients C_1 and C_2 are such that the vector function w(z) satisfies boundary condition (2) when z = 0. Thus to find the eigenfunction we must be able to calculate two linearly independent solutions that are exponentially decreasing at ∞. It was necessary to use the factorization method to solve Eq. (7). Otherwise, two solutions that were linearly independent at the point $\tilde{z}$ would soon become linearly dependent for all practical purposes.

We shall use an extension of the method of orthogonal factorization proposed by S. K. Godunov [4].

Let w_I and w_{II} be two solutions of Eq. (1) orthogonal at $\tilde{z}$, i.e.,

$$(w_I(\tilde{z}), w_{II}(\tilde{z})) = w_I^{(1)} \cdot w_{II}^{(1)} + w_I^{(2)} \cdot w_{II}^{(2)} = 0.$$

We integrate Eq. (7) to the first such point z_1

$$\frac{(w_I(z_1), w_{II}(z_1))}{\sqrt{(w_I(z_1), w_I(z_1)) \cdot (w_{II}(z_1), w_{II}(z_1))}} > \alpha,$$

where the number α $(0 < \alpha < 1)$ is selected in accordance with the given solution accuracy. At the point z_1, we orthogonalize the solutions w_I and w_{II}, correcting w_I by the formula

$$w_I^{new}(z_1) = w_I^{old}(z_1) + d_1 w_{II}(z_1),$$

where

$$d_1 = -\frac{(w_I^{old}(z_1), w_{II}(z_1))}{(w_{II}(z_1), w_{II}(z_1))}.$$

From the point z_1 we seek two linearly independent solutions of Eq. (7) with new initial conditions. The point z_2 and the coefficient d_2 are found similarly, etc. All the coefficients d_k are stored.

In each of the segments $[z_1, \tilde{z}]$, $[z_2, z_1]$, etc., the eigenfunction w(z) is some linear combination of the two linearly independent solutions: $w(z) = C_1^{(k)} w_1(z) + C_2^{(k)} w_2(z)$ when $z_{k+1} \geq z \geq z_k$. Let n be the number of segments. The coefficients $C_1^{(k)}$ and $C_2^{(k)}$ are calculated by the formula

$$C_1^{(k)} = C_1; \quad C_2^{(k)} = C_2 - \sum_{i=k}^{n-1} d_i; \quad C_2^{(n)} = C_2.$$

6. Calculation of Phase and Group Velocities

The integrals

$$\int_0^\infty (Aw', w')\,dz; \quad \int_0^\infty (Bw, w')\,dz;$$

$$\int_0^\infty (Cw, w)\,dz; \quad \int_0^\infty \rho(w, w)\,dz$$

are computed along with the eigenfunction and its derivative. Then the eigenvalue p_k^2 is refined by a formula from the calculus of variations:

$$p_k^2 = \frac{\int_0^\infty (Aw', w')\,dz + 2\xi \int_0^\infty (Bw, w')\,dz + \xi^2 \int_0^\infty (Cw, w)\,dz}{\int_0^\infty \rho(w, w)\,dz}.$$

This formula is obtained by scalar multiplication of Eq. (1) by w(z), integration of both sides from 0 to ∞, and taking $\int_0^\infty \left(\frac{d}{dz}(Aw' + \xi Bw), w\right) dz$ by parts, taking the boundary conditions into account.

Then we calculate the phase velocity $v_k = p_k/\xi$ and the group velocity $C_k = dp_k/d\xi$, which are found by using a formula from perturbation theory:

$$\frac{dp_k}{d\xi} = \frac{\xi \int_0^\infty (Cw, w)\,dz + \int_0^\infty (Bw, w')\,dz}{p_k \int_0^\infty \rho(w, w)\,dz}.$$

To derive this formula, Eq. (1) is differentiated with respect to the parameter ξ, dotted into w(z), and integrated from 0 to ∞.

The eigenfunctions and their derivatives are normalized by the factor

$$\frac{\sqrt{\xi}}{2p_k \frac{dp_k}{d\xi} \int_0^\infty \rho(w, w)\,dz} \cdot w_2(0),$$

where $w_2(0)$ is the second component of the eigenfunction w(z) at the point z = 0.

The first variations of the phase and group velocities with respect to the coefficients $a(z)$, $b(z)$, and $\rho(z)$ and their derivatives are also computed. The functions $a(z)$, $b(z)$, $\rho(z)$ are assumed to be piecewise linear. With these variations we can calculate the phase and group velocities for slightly different cross sections and approximately calculate the absorption of Rayleigh waves given the absorption of body waves.

Summary

The method described here has the following virtues:

1. Universality: the eigenvalues and eigenfunctions of problem (1) and (2) are calculated auto-

matically for any coefficients $a(z)$, $b(z)$, and $\rho(z)$, value of ξ, and mode number k.

2. Simultaneous calculation of the phase and group velocities and the derivatives of the phase velocity, and also the normalized intensities.

LITERATURE CITED

1. Andrianova, Z. S., V. I. Keilis-Borok, A. L. Levshin, and M. G. Neigauz (1965), Surface Love Waves, Moscow, Nauka [English translation: Seismic Love Waves, New York, Consultants Bureau, 1967].
2. Gel'fand, I. M., and O. V. Lokutsievskii (1962), "The pursuit method for solving difference equations," in: (S. K. Godunov and V. S. Ryaben'kii, eds.), Introduction to the Theory of Difference Schemes, Moscow, Fizmatgiz.
3. Lidskii, V. B., and M. G. Neigauz (1962), "Pursuit method in the case of a second-order self-adjoint system," Zh. vychisl. matem. i matem. fiz., 2(1):161-165.
4. Godunov, S. K., and V. S. Ryaben'kii (1962), Introduction to the Theory of Difference Schemes, Moscow, Fizmatgiz.
5. Keilis-Borok, V. I., M. G. Neigauz, and G. V. Shkadinskaya (1965), "Application of the theory of eigenfunctions to the calculation of surface wave velocities," Revs. Geophys., 3:105-110.

HIGHER MODES OF RAYLEIGH WAVES AND UPPER MANTLE STRUCTURE†

V. I. Frantsuzova, A. L. Levshin, and G. V. Shkadinskaya

The effect of the seismic velocity distribution of the upper mantle on the dispersion and amplitude of Love waves was studied in [1, 4, 7] by numerical experimentation. The main result of these investigations was the discovery of anomalous properties of the higher modes in models of the mantle with a low-velocity zone. The conclusions of [1, 4, 7] for a vertically inhomogeneous flat earth were later confirmed by calculations for a radially inhomogeneous spherical earth [3] and explained by an asymptotic theory in [2]. Similar results for dispersion curves were obtained in [9]. In the present paper, the same studies are continued for Rayleigh waves: the fundamental mode (the Rayleigh wave proper) and the higher modes (shear modes). The dispersion and amplitudes of the first three higher Rayleigh modes are calculated for a number of vertically inhomogeneous models of the earth. Combined theoretical seismograms for various models of the upper mantle and possibilities for experimental detection of waveguides from recordings of higher modes will be considered later.

Brief Description of Program

The calculation method, which is based on the spectral theory of differential operators and one version of the factorization method, is described in detail in [5, 8]. We shall therefore give only certain characteristics of the program.

Input Data

1. The velocity and density profiles of the earth in the form of tables of depths z_i ($i = 0, \ldots, m < 64$; $z_0 = 0$, $z_m = Z$) and the corresponding longitudinal wave velocities a_{i-}, a_{i+}, transverse wave velocities b_{i-}, b_{i+}, and densities ρ_{i-}, ρ_{i+}. The minus and plus signs in the subscripts refer to the values of the parameters immediately above and below z_i; the parameters coincide when z_i does not correspond to an interface. The interpolation of the parameters between points z_{i+}, z_{i+1-} is linear.

2. A table of the source depths h_j ($0 < h_j < Z$) for which the intensities of free-surface displacement are calculated.

3. A table of the wave numbers ξ_j for which the calculation is made.

4. The first k_f and last k_l mode numbers ($k = 1, 2, 3, \ldots$) for which a calculation is made.

Output Data

For each k in the interval $k_f \le k_j \le k_l$ and a given sequence of ξ_j, the following are calculated and printed out: the angular frequency p_k; period T_k; phase velocity v_k; group velocity C_k; the value $\sqrt{|dC_k/d\xi|}$, which is used in calculations by the stationary-phase method; the ratio ν_k of the amplitudes of the horizontal and vertical components of free-surface displacement; and also

$$\overline{W}_{kQ}(0, h_j), \quad \frac{d\overline{W}_{kQ}}{dh}(0, h_j)$$

for all h_j.

Here $\overline{W}_{kQ}$ is proportional to the amplitude of the vertical component of free-surface displace-

† Translated from Vychislitel'naya Seismologiya, No. 4, Moscow, Nauka (1968), pp. 183-196.

TABLE 1

i	z_i, km	Bullen A model	Jeffreys model		Gutenberg model	
		ρ_i, g/cm^2	b_i, km/sec	a_i, km/sec	b_i, km/sec	a_i, km/sec
0	0	2.74	3.55	6.14	3.55	6.14
1—	19	2.74	3.55	6.14	3.55	6.14
1+	19	3.0	3.80	6.58	3.80	6.58
2—	38	3.0	3.80	6.58	3.80	6.58
2+	38	3.32	4.37	7.75	4.65	8.20
3	60	3.35	4.40	7.82	4.58	8.15
4	80	3.36	4.44	7.88	4.48	8.06
5	100	3.39	4.46	7.95	4.38	8.00
6	140	3.41	4.50	8.07	4.35	7.86
7	170	3.44	4.55	8.18	4.36	7.92
8	200	3.47	4.60	8.26	4.40	8.04
9	250	3.51	4.68	8.42	4.50	8.22
10	300	3.55	4.76	8.58	4.62	8.55
11	400	3.65	4.94	8.93	4.93	8.96
12	500	3.92	5.32	9.66	5.32	9.60
13	600	4.12	5.66	10.24	5.62	10.12
14	800	4.48	6.13	11.01	6.14	10.88
15	1000	4.68	6.36	11.43	6.37	11.44
16	1200	4.80	6.50	11.71	6.50	11.76
17	1400	4.90	6.62	11.99	6.62	12.08
18	1600	5.00	6.73	12.26	6.75	12.28

ment in the k-th mode when a concentrated pulsed force acts at depth h_j (Q = T is a horizontal force and Q = Z is a vertical force). The value $d\overline{W}_{kQ}/dh$ is used in calculations of displacement from multipoles.

If necessary, the partial derivatives of the phase and group velocities with respect to the layer parameters (a_{i+}, b_{i+}, and ρ_{i+} at the roof of the layer and their gradients in the layer) are calculated for each ξ_j and each layer between the depths z_i and z_{i+1} ($0 < i < m-1$) and the half-space ($z > z_m$).

Rayleigh Waves in the Gutenberg and Jeffreys Models

Using the described program, we calculated the surface waves in two models of the upper mantle with Gutenberg and Jeffreys velocity profiles. The densities corresponded to the Bullard model. The data on the models are given in Table 1 (below, we shall call the models G–B and J–B).

The dispersion of the first four modes in these models is shown in Fig. 1, a-d. The differences in dispersion for k = 1 are comparatively small, reaching a maximum at T = 150-200 sec. However, the high observation accuracy at these periods allows such differences to be detected experimentally and preference to be given to the G–B model or other models with a waveguide in the mantle [6, 11].

The differences in the dispersion of the higher modes are more pronounced, especially for k = 3 and k = 4, where flat sections in the phase velocity curves and sharp additional group velocity minima are noted for the G–B model. For k = 3, such a minimum is at T ≈ 10 sec with a velocity differential of 0.3 km/sec; for k = 4, at T ≈ 4 sec with a velocity drop of 0.8 km/sec. Let us compare the frequency characteristics of the medium $\overline{W}_{kZ}$ for the various modes when a vertical force acts at the surface and at a depth of 200 km. For the higher modes, they are given in Fig. 2. For the fundamental mode, they are given in Fig. 3 of the following paper. It can be seen that the differences are not great for the fundamental mode, while for the higher modes they increase with the mode number. For the G–B model, some regions of the spectra are relatively sharply attenuated and some are very intense. The former coincide with the flat sections of the v_k curves and stationary C_k; the latter coincide with sharp inflections of $v_k(T)$ (minima of C_k).

The ellipticities of the fundamental and higher modes in both models are qualitatively similar (Fig. 3): for k = 1, the horizontal component is 0.7-0.8 of the vertical; in the higher modes, v_k = 0.6-0.3 for waves in the crust and v_k = 0-0.6 for waves in the mantle. The particle motion for periods T < 50 sec and k ≤ 4 is retrograde. The anomalies in the behavior of the higher modes in the G–B model we attribute (just as for Love waves [1, 4]) to the

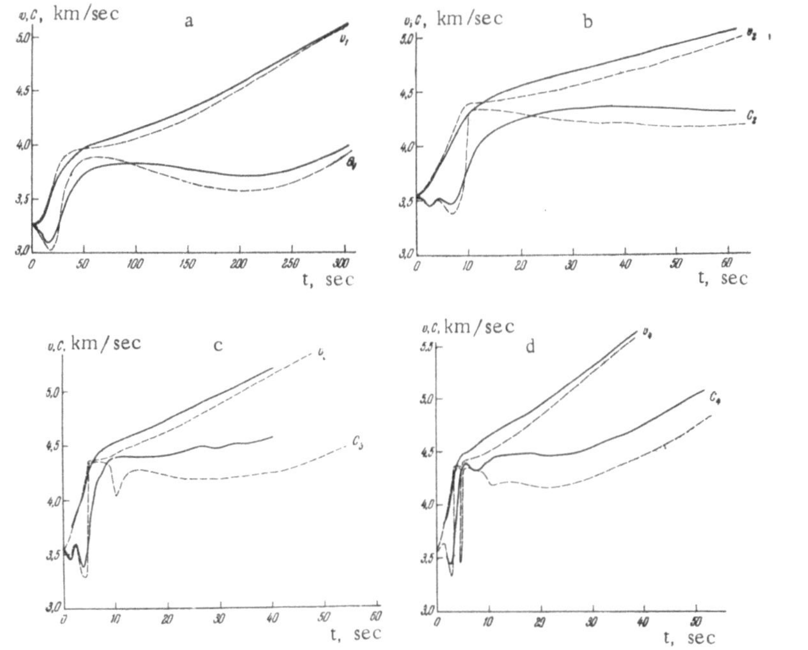

Fig. 1. Dispersion of phase and group velocities of Rayleigh waves in Jeffreys (solid lines) and Gutenberg (broken lines) models: a) fundamental mode; b-d) higher modes.

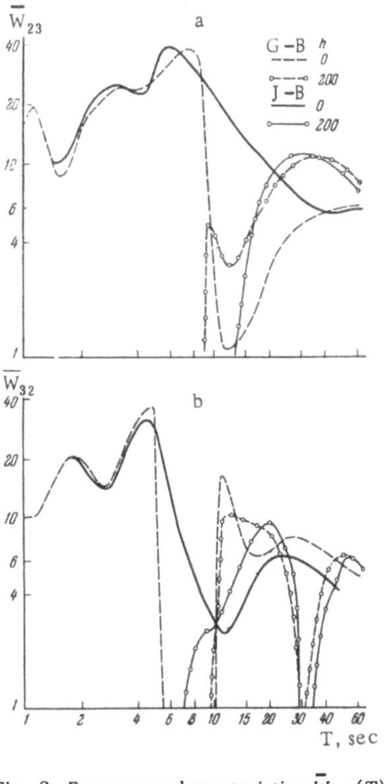

Fig. 2. Frequency characteristics $\bar{W}_{kZ}(T)$ of higher modes of Rayleigh waves in J-B (solid lines) and G-B (broken lines). Models for source on surface and at depth of 200 km.

existence of an internal low-velocity zone. Special calculations were made for a detailed study of the nature of the waves in various regions of the higher-mode spectrum. We studied the waves in a simplified model of the mantle without a surface layer M and in the same model with an additional homogeneous surface layer with thicknesses H = 5, 10, 40, and 60 km, simulating the earth's crust. The parameters of model M and the layer are given in Table 2.

Rayleigh Waves and Higher Modes in the Simplified Model

First let us consider the propagation of surface and channel waves in model M, where there is a unique minimum of velocities $b(z)$ and $a(z)$ at depth $\tilde{z}$ from the surface.

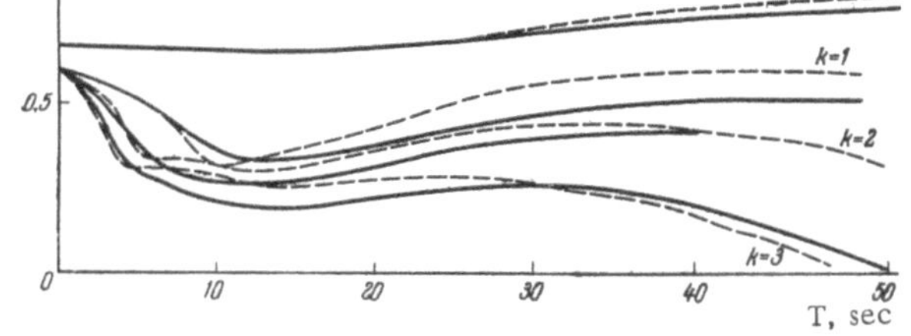

Fig. 3. Ellipticities of fundamental and higher modes in J-B (solid lines) and G-B (broken lines) models.

TABLE 2

i	z_i, H, km	ρ_i, g/cm^3	a_i, km/sec	b_i, km/sec
1	0	3.32	8.05	4.65
2	30	3.35	7.95	4.60
3	70	3.38	7.60	4.40
4	120	3.425	7.50	4.35
5	170	3.47	7.60	4.40
6	270	3.55	7.95	4.60
7	370	3.63	8.55	4.95
8	470	3.83	9.15	5.30
9	570	4.13	9.70	5.60
Layer	—	2.75	6.15	3.55

Fundamental Mode. As is known, the fundamental mode of a Rayleigh wave retains the properties of the surface wave in inhomogeneous models. The presence of an internal velocity minimum should not create anomalies in the behavior of the wave if the Rayleigh velocity at the surface $\beta(z_0) < \min b(z)$.† At very small T, the wave is concentrated in the immediate vicinity of the surface, having phase velocities that are close to the Rayleigh velocity $\beta(z_0)$ (Fig. 4). As T increases, the wave goes deeper into the medium. The oscillations also cover the low-velocity zones in this case. There is, however, no wave concentration in the minimum-velocity zone; at first, the phase velocity v_1 decreases with an increase in T, and then it begins to increase again. And, as can be seen from calculations, the minimum value of $v_1(T)$ at T = 60 sec is greater by 0.12 km/sec than the minimum value of $\beta(z)$ in the medium.

The graphs of $\partial v_1/\partial b_i$ are dome-shaped, and as i increases the tip of the dome moves smoothly to higher T (Fig. 5a). The graphs of $\partial v_1/\partial a_i$ and $\partial v_1/\partial \rho_i$ behave similarly, and the ratios

$$\max\left|\frac{\partial v_1}{\partial a_i}\right| \bigg/ \max\left|\frac{\partial v_1}{\partial b_i}\right|$$

are little dependent upon i and are close to 0.08-0.12.

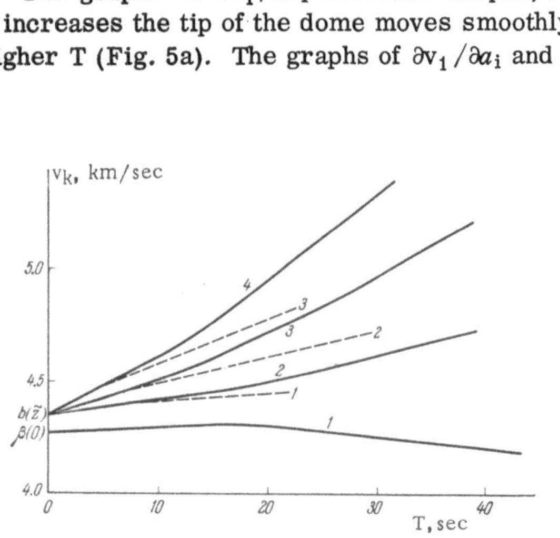

Fig. 4. Dispersion of phase velocities of Rayleigh waves in model M. The broken lines show Love-wave dispersion.

The amplitude data also attest to the absence of wave concentration in a waveguide. The only maximum of $\overline{W}_{1Z}(h)$ is near the surface, and it sinks monotonically into the medium as T increases. The maximum of $\overline{W}_{1T}(h)$ is at the surface, and the minimum of $\overline{W}_{1T}(h)$ sinks monotonically as T increases (Fig. 6a). The maxima of the frequency characteristics of the medium $\overline{W}_{1Z}(T)$ are shifted regularly in the direction of greater T with an increase in h. Thus, anomalous phenomena are absent. The ratio $\nu_1(T)$ approaches an asymptotic value when T → 0 in a homogeneous model with parameters as at the free surface, passes through a range of minimum values at T = 100-200 sec, and then increases smoothly as T increases.

Higher Modes. As has already been noted [10], interferential SV waves that have been trapped by surface or interior waveguides make the principal contribution to the higher modes of Rayleigh waves. Since the ray geometries are the same for SV and SH, we should expect coincidence of the properties of the high-frequency interference field of SV and SH and, as a consequence, coincidence of the dispersion curves of Love waves and the higher modes of Rayleigh waves for phase velocities that are close to the velocity $b(\tilde{z})$ in a waveguide.

Calculations for model M show that the dispersions of the first higher mode of Rayleigh waves

† Here and below, the Rayleigh velocity $\beta(z)$ is the velocity of a Rayleigh wave in a homogeneous half-space with the parameters $a(z)$, $b(z)$, and $\rho(z)$.

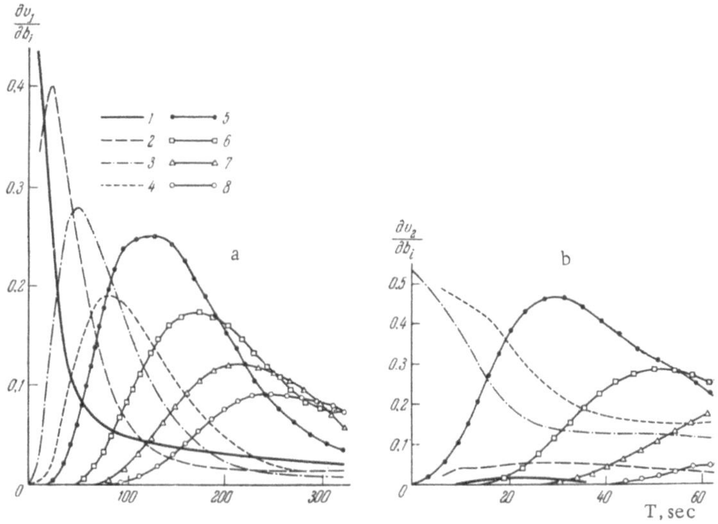

Fig. 5. Derivatives of phase velocity with respect to profile parameters: velocities b_i in model M: a) k = 1; b) k = 2.

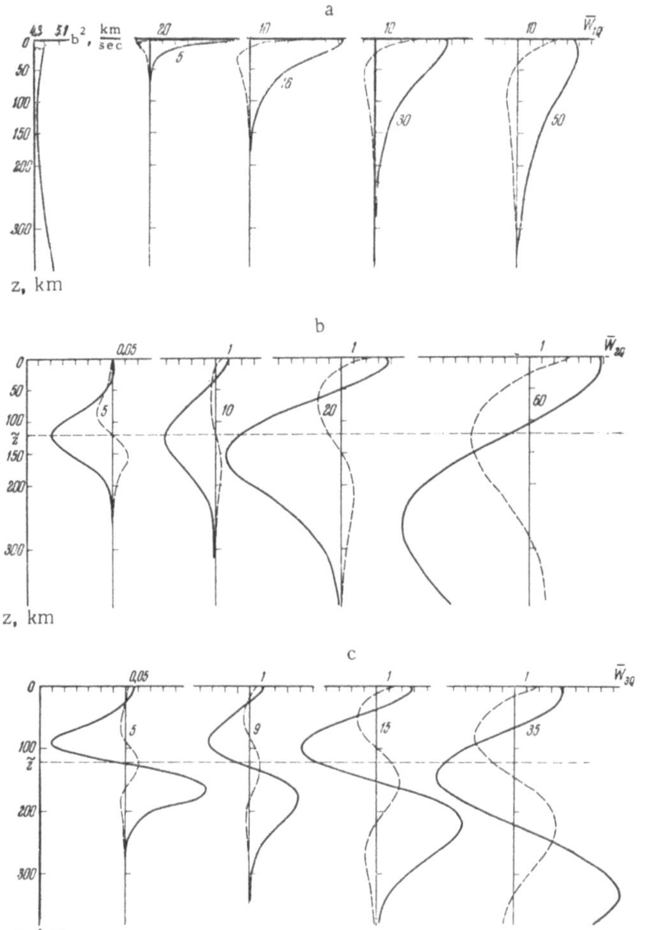

Fig. 6. Graphs of $\overline{W}_{kQ}(h)$ for various periods in model M: a) k = 1; b) k = 2; c) k = 3. Solid lines for $\overline{W}_{kZ}$; broken lines for $\overline{W}_{kT}$; numbers near curves are periods T in seconds.

and of the fundamental mode of Love waves practically coincide in the interval T = 0-10 sec, the dispersions of the second higher mode of Rayleigh waves and of the second mode of Love waves practically coincide at T = 0-8 sec, and the dispersions of the third higher mode of Rayleigh waves and of the third mode of Love waves practically coincide at T = 0-6 sec (Fig. 4).

The graphs of the derivatives $\partial v_k/\partial b_i$ behave as follows: for short periods the derivatives reach their maximum values for the internal low-velocity layers that form a waveguide (i = 3, 4); as T increases, the role of the layers above and below the waveguide gradually increases (Fig. 5b). The parameters a_i and ρ_i have practically no effect on the dispersion of the higher modes in the intervals where it coincides with the dispersion of Love waves.

In these intervals, the oscillation energy is concentrated in the waveguide. The amplitude distribution is symmetric relative to the waveguide axis. A vertical force within a low-velocity layer excites (for the same source power) a considerably stronger field than does a horizontal force. This is apparently due to the predominantly vertical polarization of interference SV waves (Figs. 6b and c). As T increases, the concentration of Rayleigh waves decreases more rapidly than for Love waves (which is possibly due to the formation of converted reflected waves at the free surface).

Waves that are concentrated in a low-velocity layer are naturally called channel waves. The maximum cut-off periods of channel waves decrease as k increases. Thus, the short-period parts of the higher-mode spectra form channel waves. The fundamental mode and the long-period parts of the higher-mode spectra form surface waves (in the generally accepted sense).

Effect of Surface Layer. An additional surface layer with relatively low velocities (lower than in the interior waveguide) makes substantial changes in the propagation conditions of Rayleigh waves.

The fundamental mode retains the properties of the boundary wave, just as in model M. At small T, the oscillation energy is concentrated in the upper layer near the free surface. As T increases the oscillations begin to feel the medium below the surface layer and gradually penetrate into the interior waveguide and the part of the model underlying it. The oscillations are not concentrated in the interior waveguide or near the bottom of the upper layer at any T.†

The behavior of the higher modes is considerably more complex. At sufficiently high k, the

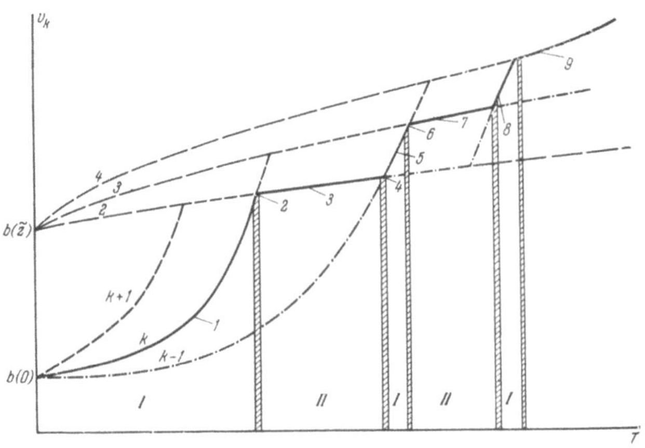

Fig. 7. Schematic representation of dispersion curves of higher modes in models with two waveguides.

spectrum of the k-th mode can be divided into alternating regions of type I and II (Fig. 7) in which the oscillations are concentrated either in the surface layer (crustal waves) or in the internal waveguide (channel waves). Regions of one type differ from one another in the number of extrema in the zone of concentration. The regions have phase velocity curves $v_k(T)$ with different shapes: a rapid increase with period for crustal waves and a slow increase with period for channel waves. Thus the $v_k(T)$ curves are ladder-shaped (Fig. 7).

The various types of regions are separated by narrow transition zones. In each such zone, oscillations in both waveguides can be excited by a source located in either of them. Discontinuities of $dv_k(T)/dT$ and C_k and pairwise convergence of adjacent $v_k(T)$ and $v_{k+1}(T)$ or $v_{k-1}(T)$ and $v_k(T)$ curves occur in the transition zones. The graphs of $\overline{W}_{kQ}(h)$ for two adjacent modes in the zone of convergence of the $v_k(T)$ curves almost coincide in absolute value and differ in sign in the lower waveguide (Fig. 8). As k increases and T decreases, the transition zones narrow, and the phase velocity curves $v_k(T)$ in these zones converge even more closely.

The situation is different with long periods that exceed the cutoff period of the k-th channel wave in model M. The $\overline{W}_{kQ}(h)$ curves are periodic over the entire low-velocity zone, including both waveguides and the intervening structure between them.

Comparison of the amplitude and phase spectra of adjacent modes, and also comparison with the results of [1, 2, 4], allows the following conclusions to be drawn:

† The contact conditions at the crust-mantle boundary are such that a Stoneley wave cannot be formed there.

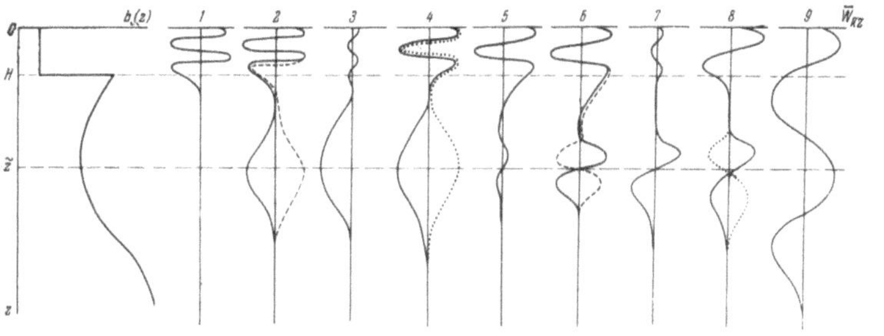

Fig. 8. Schematic graphs of $\overline{W}_{lZ}(h)$ in models with two waveguides: $l = k - 1$ (broken lines); $l = k$ (solid lines); $l = k + 1$ (dotted lines). The numbers above the curves correspond to those in Fig. 7.

a) In a model with two waveguides, the higher modes cannot be considered separately: their total field must be taken into account.

b) Regions I together form a field of surface waves in the crust that is similar to that which exists in models without a lower waveguide. This field is excited only by sources in the crust or immediately below it.

c) Regions II together form a channel-wave field similar to that in model M, which has no surface waveguide. This field is excited by sources within or near the channel and cannot be observed on the free surface.

d) An interior waveguide can be detected on the surface by the interference phenomena related to the interaction (resonance) of the fields in the two waveguides in the transition zones of the spectra. We shall consider these phenomena in greater detail.

Interference in Transition Zones of Spectra. For simplicity we shall assume that some transition zone of the k-th mode is represented by the period $\tilde{T}$, the phase velocity $v_k(\tilde{T})$, and the amplitude $\overline{W}_{kQ}(h, \tilde{T})$. Let the $v_k(T)$ and $v_{k+1}(T)$ curves converge in this zone, wherein $\delta v_k = v_{k+1}(\tilde{T}) - v_k(\tilde{T}) \ll v_k(\tilde{T})$, $|\overline{W}_{kQ}| \approx |\overline{W}_{k+1Q}|$. Then the total effect of this zone in observation on the earth's surface is described by

$$u^{\Sigma}_{k,k+1}(t, r, h) = |\overline{W}_{kQ}(0, h, \tilde{T})| \exp\left[i\tilde{p}\left(t - \frac{r}{v_k}\right)\right] \times$$
$$\times \left[1 \pm \exp\left(-i\tilde{p}\frac{\delta v_k}{v_k^2} r\right)\right],$$

where $\tilde{p} = 2\pi/\tilde{T}$; the positive sign is used if the source is in the surface waveguide and the negative sign is used if the source is in the interior waveguide. For sources outside the waveguides, $\overline{W}_{kQ}(0, h, \tilde{T})$ is close to zero. Hence it is apparent that $r \max\limits_t |u^{\Sigma}_{k,k+1}|$ will vary periodically from double the value of the field of one mode to zero. If δv_k is not too small, such variations will occur at realistic epicentral distances r. For example, for the transition zone of modes k = 2 and k = 3 in the G–B model, with $\tilde{T} = 10$ sec, $v_k = 4.4$ km/sec, and $\delta v_k = 0.025$ km/sec, the half-wavelength of variation of $\max |u^{\Sigma}_{k,k+1}|$ is 4000 km. There are reasons for expecting, therefore, that the interference effects associated with this zone can be observed during fairly strong earthquakes.

The role of the transition zones corresponding to greater k and smaller T is insignificant due to their narrowness and the smallness of δv_k. The penetration of short-period wave fields from one waveguide into another is impossible, and therefore these transition zones are only the formal result of the application of a calculation apparatus that describes two independent fields of a single set of eigenfunctions.

Combined theoretical seismograms of Rayleigh waves in various earth models will be given later for a more complete quantitative description of the interference effect caused by a waveguide in the upper mantle, and the theoretical seismograms will be compared with observations.

LITERATURE CITED

1. Andrianova, Z. S., V. I. Keilis-Borok, A. L. Levshin, and M. G. Neigauz (1965), Surface Love Waves, Moscow, Nauka [English translation: Seismic Love Waves, New York, Consultants Bureau, 1967].
2. Buldyrev, V. S., and L. K. Ozerov (1967), "High-frequency Love waves in an inhomogeneous half-space," in: Computational Seismology, No. 3, Moscow, Nauka, pp. 254-268.
3. Vil'kovich, E. V., A. L. Levshin, and M. G. Neigauz (1966), "Love waves in a vertically

inhomogeneous medium (allowance for sphericity, parameter variations, and absorption)," in: Computational Seismology, No. 2, pp. 130-149, Moscow, Nauka [English translation: Seismic Love Waves, New York, Consultants Bureau, 1967].

4. Levshin, A. L. (1964), "Love waves and the low-velocity layer in the upper mantle," Izv. Akad. Nauk SSSR, Ser. Geofiz., No. 11, pp. 1595-1607.

5. Dorman, J., M. Ewing, and J. Oliver (1960), "Study of shear velocity distribution by mantle Rayleigh waves," in: Bull. Seism. Soc. Amer., 50:87-116.

6. Neigauz, M. G., and G. V. Shkadinskaya, "Method for calculating surface Rayleigh waves in a vertically inhomogeneous half-space," this volume, p. 88.

7. Ivanova, Z. S., V. I. Keilis-Borok, A. L. Levshin, and M. G. Neigauz (1964), "Love waves and the structure of the upper mantle," Geophys. J. Roy. Astr. Soc., 9:1-8.

8. Keilis-Borok, V. I., M. G. Neigauz, and G. V. Shkadinskaya (1965), "Application of the theory of eigenfunctions to the calculations of surface wave velocities," Revs. Geophys., 3:105-110.

9. Kovach, R., and D. L. Anderson (1964), "Higher mode surface waves and their bearing on the structure of the earth's mantle," Bull. Seism. Soc. Amer., 54:161.

10. Oliver, J., and M. Ewing (1957), "Higher modes of continental Rayleigh waves," Bull. Seism. Soc. Amer., 47:187-204.

11. Takeuchi, H., F. Press, and N. Kobayashi (1959), "Rayleigh wave evidence for the low-velocity zone in the mantle," Bull. Seism. Soc. Amer., 49:355-364.

SURFACE-WAVE AMPLITUDE SPECTRA†

V. I. Frantsuzova, A. L. Levshin, and G. V. Shkadinskaya

This paper presents a numerical study of theoretical amplitude spectra of surface Rayleigh and Love waves. These spectra are of obvious interest in any problem connected with surface-wave dynamics: determination of source magnitude or energy, focal depth, and source mechanism and investigation of the resonance properties of layers or of the earth's structure.

The initial theoretical and calculation apparatus is described in [1, 3]. The formulas are given in Section 1, and Section 2 gives graphs with the results of calculations. These graphs give an overall representation of the dynamics of the waves and can be used for additional calculations without a computer. In this it is useful to bear in mind the qualitative description of certain properties of surface waves given in [1].

We shall consider here only the fundamental modes in the Gutenberg–Bullen A model.

1. Working Formulas

Let us consider an elastic half-space $z \geq 0$ in which the density ρ, velocities a and b, and decrements of absorption of longitudinal and transverse waves Δ_p and Δ_s are functions only of depth z. Observations are made on the stress-free surface $z = 0$.

Point Sources. Let a source $F^{\Sigma}(t)$ act in the vicinity of a point with the coordinates $z = h$ and $r = 0$. The source $F^{\Sigma}(t)$ is an arbitrary combination of point sources: simple forces, dipoles (with or without moments), and other multipoles. The following conditions are imposed on the function $F^{\Sigma}(t)$: 1) $F^{\Sigma}(t) = 0$ when $t < 0$; 2) $dF^{\Sigma}(t)/dt = 0$ when $t < 0$; 3) the Fourier transform $\Phi^{\Sigma}(\omega) = \int\limits_{-\infty}^{+\infty} F^{\Sigma}(t)\, e^{-i\omega t}\, dt$ exists.

Let $u^{\Sigma}_{kq}(t, r, \varphi, h)$ be the q-th component of displacement of the free surface when the k-th mode of a Rayleigh or Love wave passes by. The component is vertical when $q = z$, radial when $q = r$, and tangential when $q = \varphi$. For sufficiently large r, u_{kr} and u_{kz} are due only to Rayleigh waves, and $u_{k\varphi}$, only to Love waves.

The corresponding amplitude spectra of the displacements are

$$U^{\Sigma}_{kq}(\omega, r, \varphi, h) = \left| \int\limits_{-\infty}^{+\infty} u^{\Sigma}_{kq}(t, r, \varphi, h)\, e^{-i\omega t}\, dt \right|.$$

The following formulas are valid for U^{Σ}_{kq} for large r:

$$\begin{aligned} U^{\Sigma}_{kz} &= N\,|\overline{W}^{\Sigma}_{k}(\omega, h)\,\Phi^{\Sigma}(\omega)|\,\frac{\exp(-\alpha_{kR}(\omega) r)}{\sqrt{r}}, \\ U^{\Sigma}_{kr} &= N\,|\overline{W}^{\Sigma}_{k}(\omega, h)\,\Phi^{\Sigma}(\omega)\,\nu_k(\omega)|\,\frac{\exp(-\alpha_{kR}(\omega) r)}{\sqrt{r}}, \qquad (1) \\ U^{\Sigma}_{k\varphi} &= N\,|\overline{V}^{\Sigma}_{k}(\omega, h)\,\Phi^{\Sigma}(\omega)|\,\frac{\exp(-\alpha_{kL}(\omega) r)}{\sqrt{r}}. \end{aligned}$$

Here, $N = \sqrt{\pi/2}$; $\alpha_{kR}(\omega)$ and $\alpha_{kL}(\omega)$ are the absorption coefficients for Rayleigh and Love waves; $\nu_k(\omega)$ is the ratio of the radial and vertical components in the Rayleigh wave; and the functions $\overline{W}^{\Sigma}_{k}(\omega, h)$ and $\overline{V}^{\Sigma}_{k}(\omega, h)$ for a given source depth and mechanism are the frequency characteristics of the medium for the k-th modes of Rayleigh and Love waves. We

† Translated from Vychislitel'naya Seismologiya, No. 4, Moscow, Nauka (1968), pp. 197-206.

TABLE 1

Source		A_1	A_2
Point force	↗	$\cos\theta$	$i\cos(\delta-\varphi)\sin\theta$
Dipole without moment	↙↗	$-i\xi_{kR}\frac{\sin 2\theta}{2}\cos(\delta-\varphi)$	$\xi_{kR}\sin^2\theta\cos^2(\delta-\varphi)$
Dipole with moment	↙↗	$-i\xi_{kR}\cos(\alpha-\varphi)\sin\gamma\cos\theta$	$\xi_{kR}\cos(\alpha-\varphi)\cos(\delta-\varphi)\sin\theta\sin\gamma$
Center of expansion	↑ ← · → ↓	0	ξ_{kR}
Center of rotation	↑→ ←↓	0	0

see that these functions determine the dependence of the amplitude spectrum of a surface wave upon the source mechanism and depth of source.

The functions $\overline{W}_k^{\Sigma}(\omega, h)$ and $\overline{V}_k^{\Sigma}(\omega, h)$ have the form

$$\overline{W}_k^{\Sigma} = A_1\overline{W}_{kZ} + A_2\overline{W}_{kT} + A_3\frac{d\overline{W}_{kZ}}{dh} + A_4\frac{d\overline{W}_{kT}}{dh}, \qquad (2)$$

$$\overline{V}_k^{\Sigma} = A_5\overline{V}_k + A_6\frac{d\overline{V}_k}{dh}.$$

Here the coefficients A_1-A_6 are in general complex; they can be functions of the frequency, source mechanism, and geometric parameters of the source. Table 1 gives the values of A_1-A_6 for basic elementary sources: arbitrarily oriented point forces, dipoles with and without moments, and also a volume or expansion source (or an equivalent combination of three mutually perpendicular dipoles without moments) and a torque source (or an equivalent combination of two mutually perpendicular horizontal dipoles with moments).

The angles θ and δ determine the directions of the forces: θ is the angle between the force vector and the vertical, and δ is the angle between the horizontal projection of the force and the polar axis. The angles γ and α determine the parameters of the dipole with a moment: γ is the angle between the dipole axis and the vertical (dip angle of fault plane) and α is the angle between the horizontal projection of the dipole axis and the polar axis (strike of fault plane). Of the four angles θ, δ, α, and γ, only three are independent, since

$$\operatorname{ctg}\theta\operatorname{ctg}\gamma = -\cos(\delta-\alpha). \qquad (3)$$

The wave numbers for Rayleigh and Love waves $\xi_{kM}(\omega) = 2\pi/v_{kM}$ (M = R, L; v_{kM} is the phase velocity) are eigenvalues of the boundary-value problems solved in [1, 4].

The functions $\overline{W}_{kZ}$, $\overline{W}_{kT}$, and $\overline{V}_k$ are normalized eigenfunctions of the same problems. As can be seen from Table 1, they and their derivatives with respect to h have the simple physical meaning of the frequency characteristics of the medium for elementary sources: $\overline{W}_{kZ}$ for a vertical force ($\theta=0$); $\overline{W}_{kT}$ for a horizontal force oriented toward the station ($\theta = \pi/2$; $\delta = \varphi$); and $\overline{V}_k$ for a horizontal force perpendicular to the direction to the station ($\theta = \pi/2$; $\delta = \varphi + \pi/2$). The derivative $d\overline{W}_{kZ}/dh$ describes the reaction of the medium to a vertical dipole without a moment; $d\overline{W}_{kT}/dh$, to a horizontal dipole with a moment acting in the direction toward the station for a horizontal fault plane ($\theta = \pi/2$; $\gamma = 0$; $\delta = \varphi$); and $d\overline{V}_k/dh$, to a horizontal dipole with a moment acting perpendicular to the direction to the station ($\delta = \varphi + \pi/2$).

Volume Sources. If the source is "blurred," i.e., applied not to a small vicinity of the point r = 0, z = h but to some finite volume, then to obtain the displacement spectra we must integrate the corresponding expressions for the spectra U_{kq}^{Σ} over the given volume. In particular, for a field of forces $F^{\dagger} = F(t, r, z)$ that are axisymmetric in absolute value relative to the z axis and act in a fixed direction given by the angles θ and δ, we can derive the expression

$$U_{kq}^{\dagger}(r, \varphi) = \int_0^{\infty} U_{kq}(\omega, z, r)\,|\Psi_M(\omega, z)|\,dz, \qquad (4)$$

where

$$\Psi_M(\omega, z) = \left[\int_{-\infty}^{+\infty} e^{-i\omega t}\int_0^{\infty} F(t, r, z)\,J_0(\xi r)\,r\,dr\,dt\right]_{\xi=\xi_{kM}(\omega)}$$

is the space-time spectrum of the source (M = R when q = r, z; M = L when q = φ); and U_{kq} is the displacement spectrum for a force concentrated at

A_3		A_4	A_5	A_6
0	↗	0	$\sin(\delta-\varphi)\sin\theta$	0
$\cos^2\theta$	↙↗	$i\sin\frac{2\theta}{2}\cos(\delta-\varphi)$	$-i\xi_{\lambda L}\sin^2\theta\frac{\sin 2(\delta-\varphi)}{2}$	$\sin(\delta-\varphi)\frac{\sin 2\theta}{2}$
$\cos\theta\cos\gamma$	↙↗	$i\cos(\delta-\varphi)\sin\theta\cos\gamma$	$-i\xi_{\lambda L}\sin(\delta-\varphi)\cos(\alpha-\varphi)\times$ $\times\sin\theta\sin\gamma$	$\sin(\delta-\varphi)\sin\theta\cos\gamma$
1	↑ ← · → ↓	0	0	0
0	↑→ ←↓	0	$-i\xi_{\lambda L}$	0

the point z = h, r = 0 and acting in the same direction as the force field.

For a field of forces distributed in the plane z = h such that $F^{\dagger} = F(t, r)\,\delta(z - h)$, where $\delta(x)$ is the delta function, we have

$$U^{\uparrow}_{kq} = |U_{kq}\cdot\Psi_M(\omega)|. \qquad (5)$$

2. Results

The principal calculations were made for a Gutenberg–Bullen A model, the parameters of which are given in Table 1 of the preceding paper. The fundamental modes of Love and Rayleigh waves in the period range T = 1-300 sec were calculated. A program that took into account the earth's sphericity was used for Love waves and a program for a flat earth was used for Rayleigh waves.

Figure 1 shows a graph of the wavenumbers $\xi_{1L}(T)$ (solid line) and $\xi_{1R}(T)$ (broken line), and also of the ratio of the components $\nu_1(T)$. It is apparent that the wavenumbers are in the range of 0.004-2.0 (the phase velocities, accordingly, are within 5.5-3 km/sec). It follows from Table 1 that the absolute values of the coefficients A_1-A_6 are within the limits of 0 to 2. The function $\nu_1(T)$ varies within the limits of 0.65-0.85 and is slightly dependent upon frequency. Therefore, the amplitude spectra of the horizontal and vertical components of the fundamental mode of Rayleigh waves must be similar in form.

Figures 2-4 show graphs of $|\bar{V}_1|$ and $|d\bar{V}_1/dh|$,† $|\bar{W}_{1Z}|$ and $|d\bar{W}_{1Z}/dh|$, and $|\bar{W}_{1T}|$ and $|d\bar{W}_{1T}/dh|$ in arbitrary units for various h from 0 to 200 km as functions of T.

† The functions $\bar{V}_1$ for two-dimensional models without the additional factor $\sqrt{\xi_{1L}}$ are given in [1, 3].

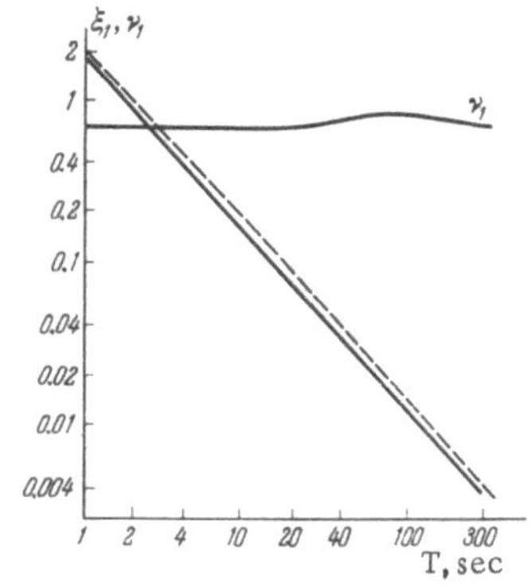

Fig. 1. Graph of ν_1, ξ_{1L} (solid lines), and ξ_{1R} (broken line) as functions of period.

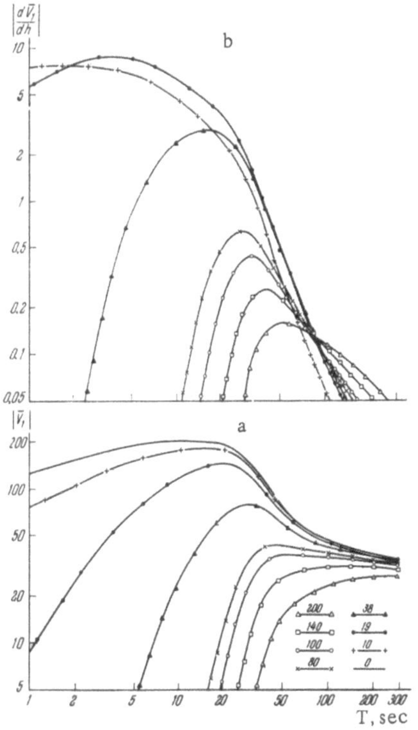

Fig. 2. Frequency characteristics of Love waves for various h: a) $|\bar{V}_1(T)|$: b) $|d\bar{V}_1(T)/dh|$.

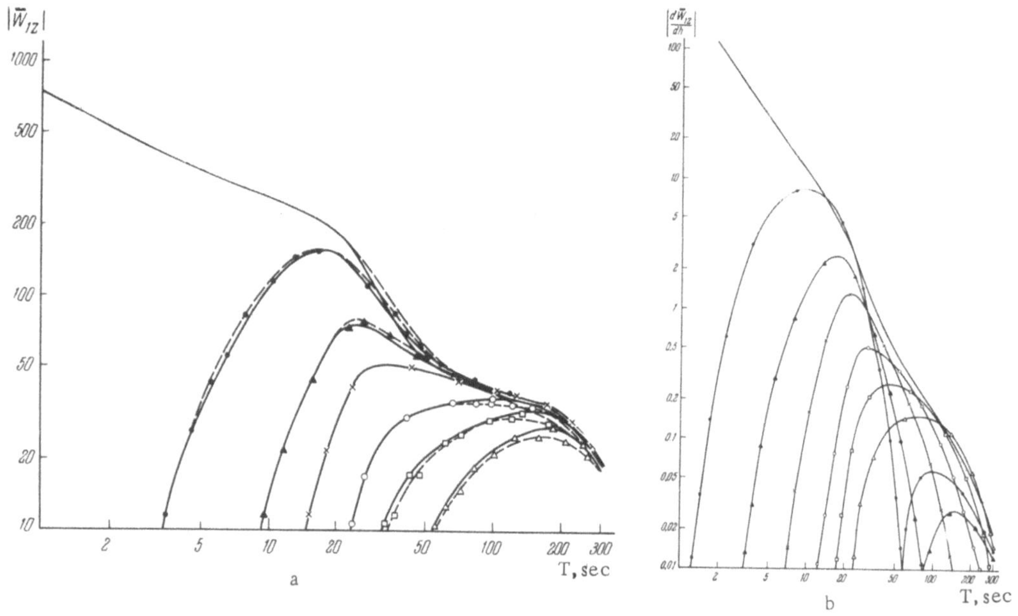

Fig. 3. Frequency characteristics of Rayleigh waves for vertical force for various h: a) $|\bar{W}_{1Z}(T)|$; b) $|d\bar{W}_{1Z}(T)/dh|$. The characteristics $\bar{W}_{1Z}$ for the Jeffreys model are shown by broken lines.

Let us consider the functions $\overline{W}_{1Z}$ and $\overline{V}_1$ for various h. For a surface source (h = 0), $|\overline{W}_{1Z}|$ describes, as it were, a short-period filter with a comparatively flat right-hand cutoff; $\overline{V}_1$ has a left-hand cutoff at small T < 5 sec. This difference is due to the nature of the waves: the Rayleigh surface wave exists even at infinitely small periods; the Love wave is always an interference wave and vanishes as the period approaches zero.

As the source depth increases, both characteristics have left-hand cutoffs. The steepness of the cutoffs and the boundary period increase with h.

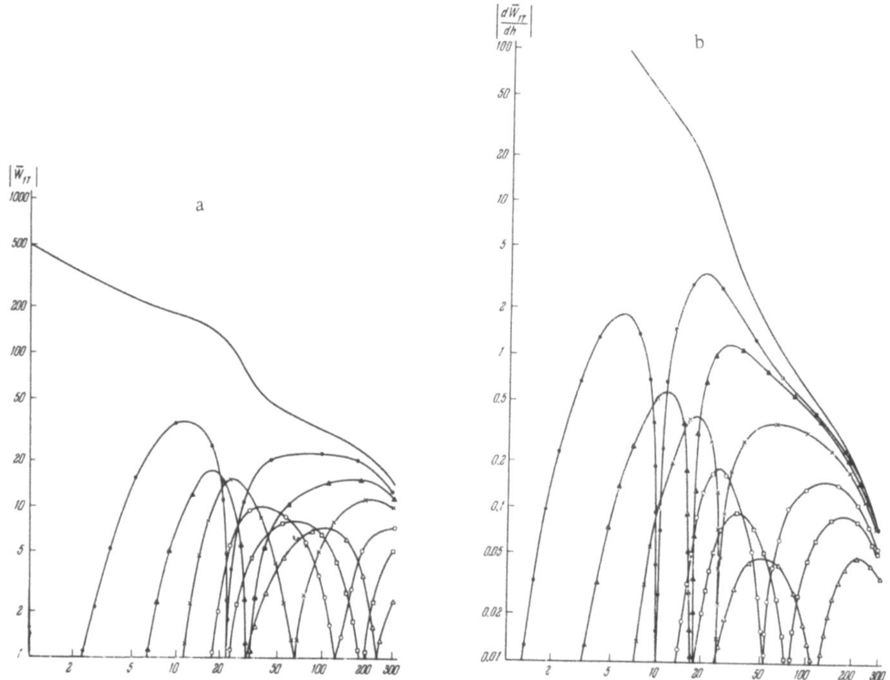

Fig. 4. Frequency characteristics of Rayleigh waves for horizontal force for various h: a) $|\overline{W}_{1T}(T)|$; b) $|d\overline{W}_{1T}(T)/dh|$.

When $h > 0$ the graphs of $\overline{W}_{1T}$ and their derivatives are more complex, since these functions have nodes (zeros) on the z axis at fixed T. For every h there is a narrow zone of intermediate periods where these functions are very small. As h increases, these zones are shifted in the direction of greater T. At small T, however, these functions behave similarly to $\overline{W}_{1Z}$ and $\overline{V}_1$ – the steepness and boundary period of the left-hand cutoff increases rapidly with h.

The functions $\overline{W}_1^{\Sigma}$ and $\overline{V}_1^{\Sigma}$ for any compound sources must behave similarly, since they are only linear combinations of the elementary frequency characteristics. As the focal depth increases, therefore, the short periods must be relatively attenuated in the amplitude spectra of surface waves. This effect has been described in a number of papers [3, 5, 6].

In the range of periods $T < 100$ sec, the frequency characteristics are markedly affected by the properties of the upper part of the profile – in particular, the thickness and the transverse wave velocity in the crust. The parameters of the mantle have a considerably smaller influence. In Fig. 3a, for example, the broken line shows the frequency characteristics $\overline{W}_{1Z}$ for the Jeffreys model (see Table 1 in preceding paper) without a waveguide in the upper mantle. It is apparent that for periods less than 50 sec, there are no substantial differences in the curves for the two models with and without a waveguide. In particular, since these periods are used in practice for determining magnitude, it is clear that the structure of the upper mantle cannot affect the so-called "magnitude defect" – the apparent decrease in earthquake energy as determined from surface waves when the focus is beneath the crust.

The formulas in Section 1 make it easy to calculate from the graphs in Figs. 1-4 a number of more comprehensive properties of the wave in question: the effect of absorption, the attenuation of spectral amplitudes with depth for any specific mechanism and source spectrum, etc. They also make it possible to take into account the filtering properties of the medium in "amplitude equalization" – in determining the amplitude spectrum of a source from observed amplitude spectra [7].

LITERATURE CITED

1. Andrianova, Z. S., V. I. Keilis-Borok, A. L. Levshin, and M. G. Neigauz (1965), Surface Love Waves, Moscow, Nauka [English translation: Seismic Love Waves, New York, Consultants Bureau, 1967].
2. Vil'kovich, E. V., A. L. Levshin, and M. G. Neigauz (1966), "Love waves in a vertically inhomogeneous medium (allowance for sphericity, parameter variations, and absorption)," Computational Seismology, No. 2, Moscow, Nauka, pp. 130-149 [English translation: Seismic Love Waves, New York, Consultants Bureau, 1967].
3. Keilis-Borok, V. I., and T. B. Yanovskaya (1962), "Surface wave spectrum versus source depth in the earth's crust," Izv. Akad. Nauk SSSR, Ser. Geofiz., No. 11, pp. 1532-1539.
4. Neigauz, M. G., and G. V. Shkadinskaya, "Method for calculating surface Rayleigh waves in a vertically inhomogeneous half-space," this volume, p. 88.
5. Sabitova, T. M. (1964), "Estimation of focal depth in the crust from surface wave spectra," Izv. Akad. Nauk Kirg. SSR, Ser. estestv. i tekhn. nauk, 6:89-96.
6. Levshin, A. L., M. G. Neigauz, and T. M. Sabitova (1965), "Spectra of Love waves and the depth of the normal source," Geophys. J. Roy. Astr. Soc., 9:253-260.
7. Toksöz, M. N., A. Ben-Menahem, and D. G. Harkrider (1964), "Determination of source parameters of explosions and earthquakes by amplitude equalization of seismic surface waves; I," J. Geophys. Res., 69:4355-4366.

PART III

INVERSE PROBLEMS OF SEISMOLOGY

INVERSE PROBLEMS OF SEISMOLOGY (STRUCTURAL REVIEW)†

V. I. Keilis-Borok and T. B. Yanovskaya

1. Introduction

The inverse problem of seismology is formulated as follows.

Given: (a) some combination of observations of body and surface seismic waves and of free oscillation of the earth; (b) postulated data on the earth's constitution, i.e., limitations regarding the velocity and density profiles as well as the relationship between various parameters.

To be found: a set of velocity and density profiles that are in agreement with the given data, with allowance for possible inaccuracy of the data.

The above formulation of the problem is evident but not quite usual. In common practice, a single profile is sought; that is why different single profiles from the above-mentioned may be obtained in different investigations by chance, especially if different waves have been used.

There are two approaches to the solution of the inverse problem of seismology:

(a) to invert the records (or some of their parts) as a function of time, without separating the records into single wave types;

(b) the inversion of some functions characterizing the laws of propagation of single waves. The following functions can be considered: travel-time curves $t(\Delta, h)$; apparent velocities $dt(\Delta, h)/d\Delta$; amplitude distance curves $A(\Delta, h, T)$; phase and group velocities $c_K(T)$ and $C_k(T)$; free oscillation periods T_K; and the decrement of attenuation λ_K of the free oscillations of the earth.

Here, t is the travel time, Δ the epicentral distance, h the depth of focus, A the amplitude, T the oscillation period, and k the mode number.

Two methods are possible in each approach:

(a) Trial-and-error method to find the cross sections for which the theoretically calculated properties of the waves are in agreement with observations.

(b) Direct inversion of data, i.e., solution of problem that is inverse in mathematical terms.

The first approach, the inversion of records, has been developed only in a few papers [15-17]. The technique for solving the one-dimensional inverse problem is well developed in seismic prospecting, but the one-dimensional problem is seemingly of little interest to seismology. This technique is generalized to the three-dimensional case when the medium is horizontally homogeneous [5]. But to use this generalization, one needs unreasonably exact and detailed observations. Only the second approach, i.e., inversion of the propagation characteristics of separate waves, is considered in this paper, and in the following papers of this book.

2. Trial-and-Error Method

The general scheme of the method is shown in Fig. 1a. The solid lines indicate the cycles or loops in the process, and the broken lines indicate paths followed only once.

The box marked "summary of observations" needs little comment. The raw observations are to be introduced here, i.e., the raw clouds of measured dots, without any drawing of fitted curves and without identification of waves. The waves need only be divided into body, Rayleigh, and Love waves and into spheroidal and torsional free oscillations. Only the

† Adapted from Geophys. J. R. Astr. Soc., 13:223-234 (1967).

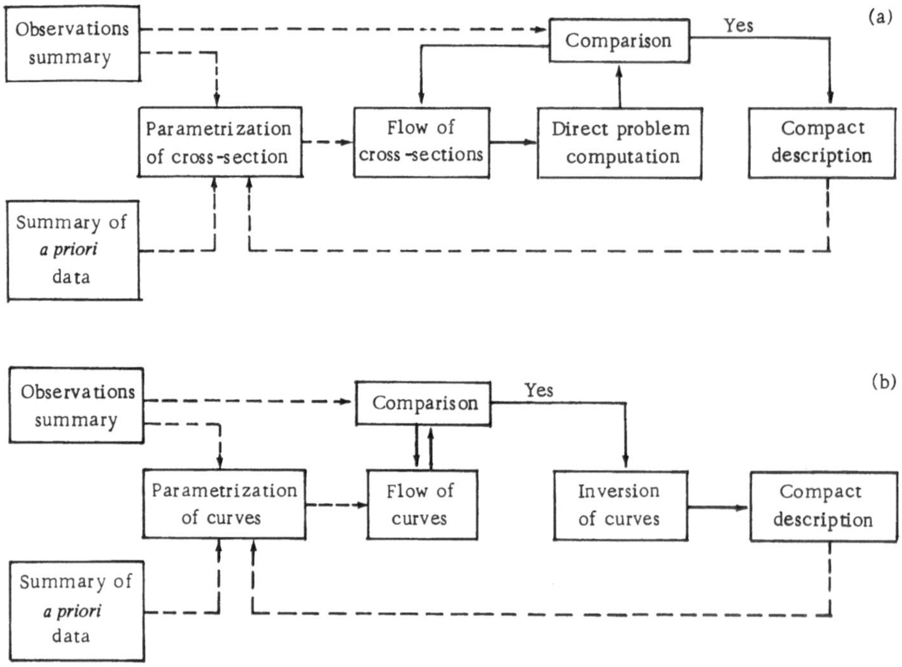

Fig. 1. General schemes of trial-and-error method (a) and direct inversion method (b).

simplest corrections, which make the dots comparable, are allowed here.

For example, the arrival times are given as a cloud of dots in the (t, Δ) plane for each h, without drawing the travel-time curve t(Δ, h), much less loops or triplications. The velocities c_K or C_k are given as a cloud of dots in the (c, T) or (C, T) plane, without drawing the dispersion curve or indicating the mode number k. The error should be estimated for each dot; so perhaps we should speak of "blobs" rather than dots.

Summary of *a priori* data, i.e., of the limitations on possible cross sections which follow from previous investigations of the same cross section, from physical considerations, from mass and moment of inertia of the earth, etc., should also include the possible relationships between various properties – elasticity, density, attenuation, etc. [14].

Parametrization of cross section means representation of the cross section through a finite number of parameters, and the indication of *a priori* limits for them. This operation is based on *a priori* data and on preliminary qualitative analysis of given observations. The pattern of the cloud of observational dots indicates how detailed the representation of the cross section can be.

In particular, the limits inside which the cross-section lies can be determined from (t, Δ, dt/dΔ) in a discrete set of points [14].

Optimal parameters are those which can be determined most precisely from the given observations. The problem of the choice of optimal parameters is not yet solved. Some indications can be found in [24, 25].

In routine practice, the cross section is usually divided into layers with a constant velocity or velocity gradient. Some of the following parameters are used: depth of layer boundaries; velocity, velocity gradient, jump of velocity or jump of velocity gradient at boundaries; and average velocity in some depth interval. To ensure that different combinations of such parameters will correspond to different travel-time curves, one should assume a standard form of $a(z)$ in each waveguide. A set of other possible forms can be calculated later. In the results of this operation, the cross section is represented as a point in the space of its unknown parameters. Their limits determine the region in which this point lies. Our problem is to narrow this region as far as the observations allow. This is done by the following three operations.

Points of the above-mentioned region (i.e., possible cross sections) are tried in succession.

The data known from observations are computed theoretically for each point, and the discrepancy between the theoretical and observed data is estimated.

The solution of our inverse problem is a region in which this discrepancy is sufficiently small. Our problem is reduced, therefore, to the determination of the region of the minimum of some function (the discrepancy mentioned) in a multidimensional space of the unknown parameters of the cross section.

The flow of cross sections is organized in accordance with the accepted method for finding a minimum. Four methods are widely employed: the Monte Carlo method, descent along the gradient, the random search method, and the hedgehog method. The first three are compared in [4].

Descent along the gradient is effective only for a small number of unknown parameters.

The random search method is effective far from the minimum, but the search gives way to casual wandering in the vicinity of the minimum.

Strange as it may seem, the Monte Carlo method is preferable, but it has a great disadvantage in that the results of previous trials are not used in the next trial. On gaining the minimum, we leave it at once. For this reason, the hedgehog method [29] seems to be the most promising.

Direct Problem Computation. The theoretical computation of the data that are known from observations needs no comment. In practice it can be done for any kind of data, but only if the model of the cross section is horizontally homogeneous.

The programs used are described in [1, 14, 21].

Comparison of Computations and Observations. The choice of the measure of discrepancy between computation and observation is a most important but inadequately developed problem. This measure should answer the purpose of the investigation: it must depend heavily on those properties of the cross section that we wish to study and depend little on other properties. Unfortunately, we do not always know in advance what we are investigating the cross section for.

Ideally, the measure should be the probability that the discrepancy between observations and computations is random. This probability, after proper normalization, could be considered the probability of the corresponding cross section. Then the solution of our inverse problem could be represented in a most natural form – as the density distribution of the unknown cross section (more precisely, of the corresponding vector in the space of the unknown parameters). Afterwards a confidence region could be determined. A method of determining this distribution has not yet been developed, but a pioneering paper has been published [10]. This paper gives a method of drawing a variant of a travel-time curve with the maximum statistical likelihood ratio. The determination of a distribution function and confidence region for a curve through clouds of dots remains the key problem, along with the optimal parametrization. In routine practice, we have followed the usual method: comparison of theoretical curves with corresponding curves obtained by smoothing of observations. The curves were represented by a set of ordinates. The measure of discrepancy was chosen by numerical experiments as some combination of the usual measures, standard and maximal deviations, correlation coefficient, etc. One such experiment is described in [2].

This practice is far from optimal, but it gives comparatively good results.

Compact description of the found cross sections means determination of their common geometrical properties. A trivial example is the average velocity; it is sometimes approximately the same for all cross sections [1, 20].

If the parametrization were optimal in the sense indicated above, we would automatically obtain the compact description. It would be given by the intervals of values of some parameters for all cross sections found. And conversely, the common properties of the cross sections are the optimal parameters which should be introduced initially. Thus cross-section parametrization can be improved after this operation.

3. Inversion Method

The general scheme for this method is shown in Fig. 1b.

Four of the operations are the same as in the trial-and-error method described above and only the three others need be described.

Parametrization of curves means the representation of each curve by a finite number of parameters and the indication of *a priori* limits for them. This operation is based on the general theory of these curves, on *a priori* data, and on the pattern of observational dots.

The general theoretical properties of travel-time curves are given in [8, 9], but such properties have not yet been formulated for dispersion curves.

The problem of optimal parametrization of curves is equivalent to that of parametrization of cross sections and is not yet solved. In routine practice we represent each curve by a set of ordinates or by the parameters of the straight lines and parabolas that approximate individual parts of the curve.

The flow of curves means here the successive selection of various combinations of corresponding parameters. Of course, all possible variants of wave identification should be considered for each variant of the curve.

Comparison of a given variant of the curve with observations is the same operation as in the trial-and-error method. The only difference is that we now compare with the observations not the curve computed for a given cross section but the curve drawn directly in the cloud of observed dots.

Inversion of the curve into a cross section (or of the parameters of the curve into the parameters of the cross section) is the next operation. It is applied to all variants of the curve that fit the observations.

The inversion method has been developed only for body wave travel-time curves [7]. Only an approximate "partial derivative" method is known for the dispersion of surface waves and free oscillations. Examples of application of the inversion method can be found in [10, 22].

4. Comparison of Methods

The advantage of the trial-and-error method is the possibility of joint analysis of the largest amount of different data – any data for which the direct problem can be calculated theoretically. To add any data – gravitational, thermal, etc. – one merely includes in the "direct problem computation" operation a subroutine for computing the corresponding data theoretically for a given cross section, and then one formulates an algorithm for comparison. Examples of application of the trial-and-error method can be found in [1, 2, 12, 19, 20, 23, 27]. Other examples are the joint interpretation of the first-arrival delay and of the Bouguer anomaly [21] and the joint analysis of seismic and temperature data [13]. Temperature distributions in the crust were calculated in [13] for various crustal cross sections obtained in [12].

Temperature differences amount to 100° at a depth of 20 km and to 500° at the bottom of the crust. Some cross sections give a strong negative temperature gradient and may possibly be rejected.

The main advantage of the inversion method is the smaller amount of computation time required.

At present, since the inversion method is developed only for body wave travel times, the combined method may be of use. A set of cross sections is obtained by inversion of travel times and then cross sections that fit the other observations are selected by trial and error.

LITERATURE CITED

1. Adams, R. D., and M. J. Randall (1964), "The fine structure of the earth's core," Bull. Seism. Soc. Amer., 54:1299-1314.
2. Andrianova, Z. S., V. I. Keilis-Borok, A. L. Levshin, and M. G. Neigauz (1965), Surface Love Waves, Moscow, Nauka [English translation: Seismic Love Waves, New York, Consultants Bureau, 1967].
3. Anderson, D. L., and R. Kovach (1964), "Attenuation in the mantle and rigidity of the core from multiply reflected core phases," Proc. Nat. Acad. Sci. Wash., 51:168-172.
4. Azbel', I. Ya., V. I. Keilis-Borok, and T. B. Yanovskaya (1966), "A technique of joint interpretation of travel time and amplitude–distance curves in upper mantle studies," Geophys. J. Roy. Astr. Soc., 11:25-55.
5. Blagoveshchenskii, A. S. (1967), "The inverse problem in the theory of wave propagation," in: (M. Sh. Birman, ed.), Topics in Mathematical Physics; Vol. 1: Spectral Theory and Wave Processes, New York, Consultants Bureau (1967).
6. Fedotov, S. A., N. N. Matveeva, R. Z. Tarakanov, and T. B. Yanovskaya (1964), "On longitudinal wave velocities in the upper mantle in the area of the Japanese and Kurile Islands," Izv. Akad. Nauk SSSR, Ser. Geofiz., No. 8, p. 1185.
7. Grudeva, N. P. (1966), "Dependence of earthquake magnitude on focal depth," in: Computational Seismology, No. 2, Moscow, Nauka, pp. 104-120.
8. Gerver, M. L., and V. M. Markushevich, "Determining seismic-wave velocity from travel-time curves," this volume, p. 123.
9. Gerver, M. L., and V. M. Markushevich, "Properties of surface-source travel-time curves," this volume, p. 148.
10. Gutenberg, B. (1953), "Travel times of longitudinal waves from surface foci," Proc. Nat. Acad. Sci. Wash., 39:849-853.

11. Hannon, W. J., and R. L. Kovach (1966), "Velocity filtering of seismic core phases," Bull. Seism. Soc. Amer., 56:441-454.
12. Jeffreys, H., and K. E. Bullen (1940), Seismological Tables, London, British Association for the Advancement of Science, Gray-Milne Trust.
13. Knopoff, L., and T. L. Teng (1965), "Analytical calculation of the seismic travel time problem," Revs. Geophys., 3:11-24.
14. Knopoff, L. (1967), "Density–velocity relations for rocks," Geophys. J. Roy. Astr. Soc., 13:1-6.
15. Levshin, A. L., T. M. Sabitova, and V. P. Valyus (1966), "Joint interpretation of body and surface waves for a district in Central Asia," Geophys. J. Roy. Astr. Soc., 11:57-66.
16. Lubimova, E. A. (1966), "Temperature profiles under oceanic and continental crust," in: The Crust and Upper Mantle of the Pacific Area, Washington, D.C., American Geophysical Union (1968), pp. 17-21.
17. Neigauz, M. G., and G. V. Shkadinskaya, "Method for calculating surface Rayleigh waves in a vertically inhomogeneous half-space," this volume, p. 88.
18. Phinney, R. A. (1966), "Theoretical calculation of the spectrum, phase velocity, and decay of first arrivals in layered elastic systems," Geophys. J. Roy. Astr. Soc., 11:260.
19. Alexander, S. S., and R. A. Phinney (1966), "A study of the core–mantle boundary using P waves diffracted by the earth's core," J. Geophys. Res., 71:5943-5958.
20. Phinney, R. A., and S. S. Alexander (1966), "P wave diffraction theory and the structure of the core–mantle boundary," J. Geophys. Res., 71:5959-5976.
21. Press, F., and S. Biehler (1964), "Inference on crustal velocities and densities from P wave delays and gravity anomalies," J. Geophys. Res., 69:2979-2996.
22. Yanovskaya, T. B. (1963), "Determination of velocity distribution in the upper mantle as an inverse mathematical problem," Izv. Akad. Nauk SSSR, Ser. Geofiz., No. 8, pp. 1171-1178.
23. Yanovskaya, T. B., and I. Ya. Azbel' (1964), "The determination of velocities in the upper mantle from observations of P waves," Geophys. J. Roy. Astr. Soc., 8:313.
24. Yanovskaya, T. B. (1966), "Program for calculation of travel-time curves and amplitudes of body waves in layered media," in: Problems of Quantitative Study of Seismic-Wave Propagation, No. 8, pp. 87-92.
25. Yanovskaya, T. B. (1968), "Inversion of travel times related to the inner core boundary," paper presented at the Fifth International Symposium on Geophysical Theory and Computers, Tokyo.
26. Press, F. (1967), "Earth models obtained by Monte Carlo inversion," paper presented at the Fourth International Symposium on Geophysical Theory and Computers, Trieste.
27. Backus, G. E., and F. Gilbert (1968), "The resolving power of gross earth data," Geophys. J. Roy. Astr. Soc., 16:169-205.
28. Markushevich, V. M., "Characteristic properties of deep-focus travel-time curves," this volume, p. 172.
29. Valyus, V. P., "Determining seismic profiles from a set of observations," this volume, p. 114.
30. Tuve, M. A., H. E. Tatel, and P. J. Hart (1954), "Crustal structure from seismic exploration," J. Geophys. Res., 59:415-422.

DETERMINING SEISMIC PROFILES FROM A SET OF OBSERVATIONS†

V. P. Valyus

Introduction

This paper considers the inverse problem of seismology: to find a set of profiles for which the calculated characteristics of seismic wave propagation (travel-time curves, amplitude–distance curves, dispersion curves) agree with observations.

This problem was examined in [1-4] as applied to travel-time and amplitude–distance curves of body waves. This paper describes a method of interpreting any set of data. A flow diagram for our solution of the problem is shown in Fig. 1. A similar scheme is described in [1, 3]. Let us discuss the elements that differ substantially from those described in [1, 3].

The experimental curves are any of the following functions as determined from observations:

1. The travel-time curve $t(\Delta)$ and/or the amplitude–distance curve $A(\Delta)$ of any noninterference body wave for any focal depth (t is travel-time and Δ is epicentral distance. The focal depth can be unknown, to be determined along with the parameters of the profile. The wave type (P, SP, etc.) is assumed to be known. In the presence of multiple-valued travel-time curves, however, it is not necessary to know beforehand which branch corresponds to a given segment of the travel-time curve.

2. The dispersion curves of various modes of the phase and/or group velocities of surface waves. The wave type (Rayleigh or Love) and the mode number are assumed known. Any other experimental data can also be included in this scheme, if there are programs for solving the corresponding direct problems. The total number of experimental curves is n. They can be given in any combination.

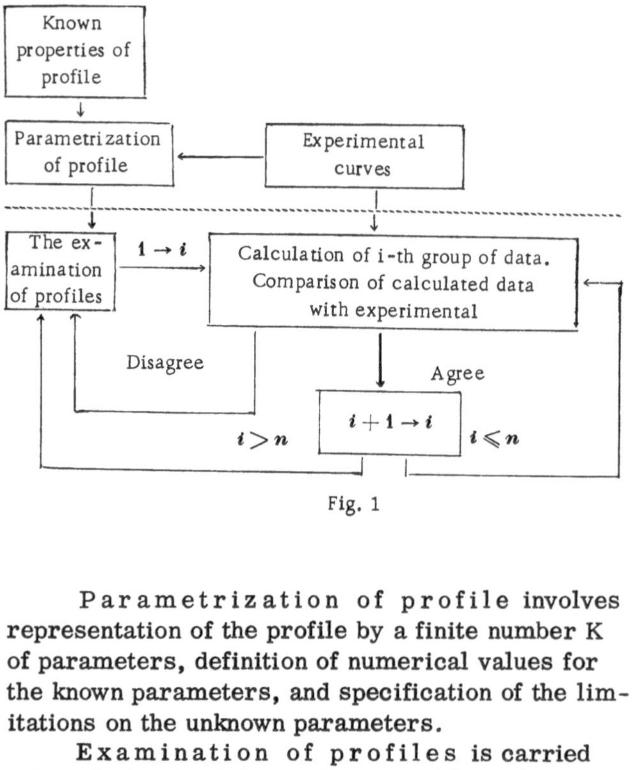

Fig. 1

Parametrization of profile involves representation of the profile by a finite number K of parameters, definition of numerical values for the known parameters, and specification of the limitations on the unknown parameters.

Examination of profiles is carried out on the basis of the mathematical formulation of the problem: finding in the space of unknown parameters of the profile the region of the minimum – the region where some function f (the measure of disagreement between the calculated and experimental curves) is below the allowable level. In such a problem, the conventional methods for finding the minimum of a multivariate

† Translated from Vychislitel'naya Seismologiya, No. 4, Moscow, Nauka (1968), pp. 3-14.

function (Monte Carlo, steepest descent, etc. [1,5]) have a considerable drawback: they are suited for finding the minimum of a function but are poorly equipped for describing the entire region of the minimum as we have just defined it.

For this reason, a new method was suggested for studying the minimum region of a multivariate function – the hedgehog method, which consists of the following procedure: the nodes of the coordinate grid in the space of the unknown parameters are examined by one of the conventional methods (in practice, the Monte Carlo method is used). At each node the function to be minimized f is calculated, and it is checked whether the node belongs to the desired minimum region.

After a node that belongs to this region is found, the function f is calculated at the nodes adjacent to that node. It is determined which of these nodes also belongs to the minimum region. The adjacent nodes are, in turn, considered for each of these nodes, etc. This procedure is continued until all nodes adjacent to each node in the minimum region have been studied. In other words, the nodes of the coordinate grid within the minimum region are examined successively as long as it is possible to move continuously from one node to another.

Then, a new part of the minimum region of f is sought by one of the conventional methods. Of course, each node is examined only once. It is best to select as adjacent nodes those that differ from the initial by not more than one step along each coordinate. But this is too tedious: each node has $3^K - 1$ such neighbors, where K is the number of unknown parameters (usually not less than 6-8). Therefore the program provides for consideration of only adjacent points of the "lower ranks," for which there are not more than m coordinates ($m \leq K$) that differ by one step from the coordinates of the main point.

This is illustrated in Fig. 2, in which the straight lines show the coordinate grid (two-dimensional case), the intersections of the straight lines are the nodes, the heavy lines show the coarse grid, and the fine lines (along with the heavy) show the finer grid in which one step along a coordinate is twice as fine as a step in the coarse network. The minimum region of f consists of two isolated parts A and B. Point 1 is found by the Monte Carlo method, and point 2 is the nearest node; it belongs to A, and after it is found, f is calculated at the nodes adjacent to it. In the case of the coarse grid and m = 1, these are nodes 3-6. Node 3 is in A. Therefore, f is also calculated in the nodes adjacent to it (7, 8, and 9).

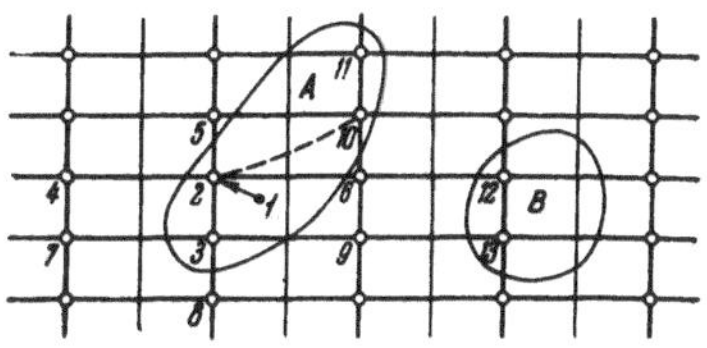

Fig. 2

Examination of profile consists of calculating theoretically for the next profile a group of functions that have been determined from observations. The calculations are then compared with the observations. If the comparison is sufficiently good, the next group of functions is calculated. If not, the next profile is taken. Generally speaking, all functions could be calculated immediately, but this would increase computer time.

1. Parametrization of Profile

Specification Profile Parameters. A limitation on the profile is the possibility of calculating theoretical curves for it. There are programs (see below, in Section 3) for calculating $t(\Delta)$, $dt(\Delta)/d\Delta$, $A(\Delta)$, and the dispersion curves for a layered spherically symmetric earth model. Within each layer the parameters of the medium are monotonic, continuous, and can be interpolated from the values at the layer boundaries. Parameter discontinuities at the boundaries are permitted. By increasing the number of layers, we can approximate various piecewise continuous profiles.

Let v_{i0} ($i = 0, 1, \ldots$) be the distances of the layer boundaries from the center of the earth, and let v_{00} be the earth's radius. We have the following parameters at boundary v_{i0}: the longitudinal wave velocity v_{i1}, the transverse wave velocity v_{i2}, the density v_{i3}, the longitudinal wave damping factor v_{i4}, and the transverse wave damping factor v_{i5}.

To introduce a discontinuity of a parameter v_{ij} at any boundary, we should let $v_{i0} = v_{i+1,0}$, $v_{ij} \neq v_{i+1,j}$.

Thus the profile is represented by a set of parameters v_{ij}. Of course, v_{ij} that do not affect the data to be interpreted (for example, v_{i1} in Love wave interpretation) are ignored.

Setting Limitations on Profile Parameters. In solving the inverse problem by the method described here, the variation limits are indicated for each v_{ij}. They are the input data for the profile-sorting units and they limit the region in which we seek the solution. We let V_{ij}^1 and V_{ij}^0 be the upper and lower limits of v_{ij}, so that $V_{ij}^0 \leq v_{ij} \leq V_{ij}^1$.

The following methods of setting the limits are programmed (below, C_1 and C_2 are given numbers, and k = 0 or 1):

1. $V_{ij}^k = C_1$ is a constant limit.
2. $V_{ij}^k = v_{i-1,j} + C_1$ is the limitation on the increment of v_{ij}.
3. $V_{ij}^k = C_1 + C_2 v_{ij'}$ is the linear dependence of the limit upon one of the parameters $v_{ij'}$ (for example, depth).
4. $V_{ij}^k = C_1(v_{i-1,0} - v_{i0}) + v_{i-1,j}$ is the limitation on the mean value of dv_{ij}/dr, where r is the distance from the earth's center (note that v_{i0} are the r values at the layer boundaries).
5. $V_{ij}^k = \left[C_1 + \frac{v_{i-1,j} - v_{i-2,j}}{v_{i-2,0} - v_{i-1,0}}\right](v_{i-1,0} - v_{i0}) + v_{i-1,j}$ is the limitation on the increment of dv_{ij}/dr in the i-th layer compared to the preceding layer.

Several limitations can be set for one v_{ij}. For example, we can require that V_{i2}^0 be in a definite relationship to v_{i1} (limitation of type 3) and that in this, $V_{i2}^0 \geq v_{i-1,2}$ (limitation of type 4).

Some parameters may be known beforehand or may be expressed practically uniquely in terms of other unknown parameters. In these cases, $V_{ij}^0 = V_{ij}^1$.

The requirements on the setting of V_{ij}^k are as follows: V_{ij}^k must not be inconsistent; a sequence of v_{ij} such that the value of the preceding must be known to calculate the limitation on one of these parameters is called a chain. The limitations must not contain closed chains.

When profiles are examined by the hedgehog method (see Section 2), another requirement is imposed on V_{ij}^k: for each v_{ij} the upper limits V_{ij}^0 and lower V_{ij}^1 must be attainable. Let, for example, the limitations on v_{ij} and $v_{i'j'}$ be

$$a < v_{ij} < b, \quad (1)$$

$$c < v_{i'j'} < v_{ij}. \quad (2)$$

If $a < c$, then $v_{ij} = a$ is unattainable, since a $v_{i'j'}$ that satisfies (2) does not exist when $v_{ij} = a$.

If the latter requirement is not satisfied for the selected V_{ij}^k, it is simple to rearrange V_{ij}^k so that they bound the same region and satisfy this requirement.

2. Examination of Profiles

As was stated in the introduction, we are considering profiles that correspond to the nodes of a coordinate network in the space of the unknown parameters. These nodes are first examined by the Monte Carlo method, and then by moving successively from one node to a number of adjacent nodes. The set of nodes that belong to the region of the minimum is the solution of the problem in question. Let us consider the steps in examining profiles by the proposed method.

Grid of Parameters. The grid is set up in the space of parameters for which $V_{ij}^0 \neq V_{ij}^1$. Their number K is equal to the dimensionality of the space in which the profiles are sorted. Let V_H be the maximum of the lower limits on the l-th variable parameter (l = 1, 2, ..., k) and let V_B be the minimum of the upper limits; l can be considered the coordinate number. The nodes are selected, starting from V_H with a step Δ_l. Even if it does not come into the node, V_B is always considered. The node that immediately precedes V_B (let us call it V) can be very close to V_B: we check the condition

$$V_B - V < \xi_l \cdot \Delta_l;\ \xi_l < 1$$

and if the condition is satisfied, then V is excluded from consideration. Figure 3 shows how the grid is set up. The triangle is the region of allowable values of the variable parameters v_1 and v_2. The points are the nodes of the grid. A node for which the condition $V_B - V < \xi_2\Delta_2$ is satisfied is denoted by a small circle. The function is not computed at this node.

We shall count the node number n_l from V_H, beginning with zero. If $V_B - V_H < \xi_l\Delta_l$, there is only one node V_B with the number 0. Thus each node corresponds to a vector $n = \{n_l\}$ of dimensionality k, where n_l are integers. The profile can be set up from n.

We shall call the two profiles v_1 and v_2 "adjacent with rank m" under the following three conditions: they satisfy the limits V_{ij}^k; they are nodes of the grid; and at the corresponding n_1 and n_2, m of the coordinates differ by 1 while the others are equal. In other words, the coordinates of the vector $n_3 = n_2 - n_1$ can be 0, −1, or 1, wherein the number of coordinates that differ from zero is m. The number of adjacent nodes increases rapidly as K and m increase. Let us give the corresponding formulas.

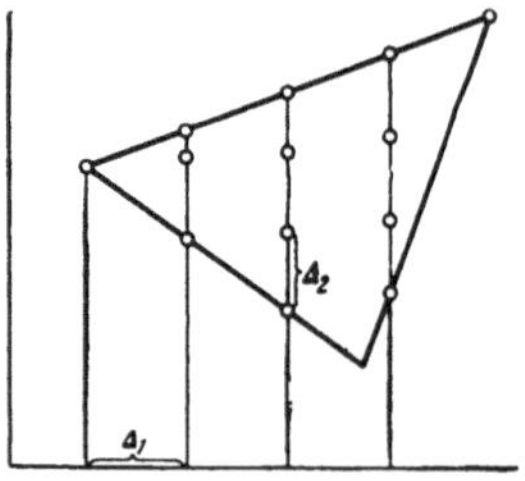

Fig. 3

Let $\varphi(K, m)$ be the number of adjacent nodes of a rank not greater than m and let $\varphi^*(K, m)$ be the number of adjacent nodes of rank m. If none of the adjacent nodes is cut off by the limits V_{ij}^k, then

$$\varphi(K, 1) = \varphi^*(K, 1) = 2K,$$
$$\varphi(K, m) = \varphi(K, m-1) + \varphi^*(K, m),$$
$$\varphi^*(K, m) = 2[\varphi^*(K-1, m-1) + \varphi^*(K-2, m-1) + \ldots$$
$$\ldots + \varphi^*(m-1, m-1)],$$
$$\varphi(K, 2) = 2K^2,$$
$$\varphi(K, K) = 3^k - 1.$$

Examination of Profiles Over a Grid. For calculations by the hedgehog method, we must assign the limits V_{ij}^k and set up a subroutine F that examines the profile: calculates for it the theoretical curves, compares them with the corresponding experimental curves, and makes a decision. If the calculated curves agree with the experimental, the profile is "good," i.e., it is in the minimum region; otherwise the profile is "poor."

The hedgehog routine transfers control to subroutine F, with indication of the profile, i.e., of the coordinates of the node in question. Subroutine F examines this profile to determine whether or not it belongs to the minimum region. Then the hedgehog program selects the next profile for examination by subroutine F. At the same time, a list of profiles (nodes) for which calculations by program F have been made is compiled. This list includes the parameters that determine the profile (all parameters are listed in one computer word), and the "poor" profiles are indicated.† In addition, the program remembers two current addresses: the address ω of the end of the list and the address α of the last node for which all adjacent nodes have been considered. At the beginning of operation, $\alpha = \omega = \alpha_0 - 1$ (α_0 is the address of the beginning of the list). The algorithm of the hedgehog program is basically as follows:

1) The profile $\{v_{ij}^k\}$ is selected by the Monte Carlo method, taking into account the limits $\{V_{ij}^k\}$; it is rounded off to the nearest node.

2) It is determined whether or not this profile is in the list of profiles examined (cells from α_0 to ω). If so, control is transferred to 1; if not, it proceeds.

3) A calculation by subroutine F is made for the selected node and the node is entered in the list; $\omega + 1 \rightarrow \omega$.

4) If $\alpha = \omega$, nodes can no longer be found in the given part of the minimum region, and control is transferred to 1. We seek another isolated part of the minimum region. If $\alpha < \omega$, we go on.

5) We consider a node from cell α and increase the current address $\alpha + 1 \rightarrow \alpha$. If the node is indicated as a "poor" profile, control is transferred to 4. Otherwise we construct all adjacent nodes of rank $\leq$ m (m is preassigned). The nodes for which a calculation by program F has not yet been made are selected, i.e., those not in the list, in cells from α_0 to ω. For these profiles, we calculate by F, enter them in the list, increase ω in doing so, and transfer control to 4.

An essential aspect of the hedgehog method is the fact that adjacent nodes are constructed only for "good" profiles, and it is determined whether or not each newly constructed profile has been studied before.

Comments on Choice of m and Δ_l. It is obvious that the volume of calculations increases greatly with an increase in m and a decrease in Δ_l.

But Δ_l should not be too large or m too small, since this would make it possible to miss the minimum region or to mistakenly divide it into isolated parts. In Fig. 2, for example, points 2, 3 and 10, 11 belong to one region A, but with a coarser grid and m = 1, A breaks down into two regions,

Cases have been encountered in which several large parts (containing many nodes) of the minimum region were found, but it was difficult to decide whether or not these parts were in fact isolated.

It is inadvisable to reduce Δ_l in such a case: the count must be started all over again, and many internal nodes that are not of particular interest will be counted for each part. It is better to attempt to increase m. It can also be useful to rotate the coordinate axes: we take in each of the two parts of the minimum region we have found the profile with the minimum value of the function f. Let these be the profiles $\{v_{ij}\}$ and $\{v'_{ij}\}$. We calculate f by program F for the profiles $\{v_{ij} + \alpha(v'_{ij} - v_{ij})\}$, where α varies from 0 to 1 with a small step. If not one of these profiles is "poor," the parts of the minimum region were joined. In Fig. 2, this corresponds to detailed checking of f along the dotted line.

3. Estimation of Profile by Subroutine F

Subroutine F must be set up individually for each problem, depending upon the available observations, the criterion for comparing calculations

† E. Nyland and J. Gardner (Institute of Geophysics, UCLA) have indicated that it is unnecessary to include "poor" structures produced by the hedgehog program (but not by the Monte Carlo method) in the list: they can be recognized as neighbors of "good" structures already estimated. This provides considerable memory economy.

with observations, the order in which the comparison is made, etc. This program consists essentially of a number of subroutines, which are as follows:

1. Subroutines for Solving Direct Problems – Calculation of Theoretical Curves for Given Profile. Subroutine L calculates the phase and group velocities of Love waves of any mode [6, 7].

Subroutine R calculates the phase and group velocities of Rayleigh waves for any mode [8, 9].

Subroutine AG calculates the travel times and amplitudes of the first five arrivals of any body wave (using a ray approximation) [10].

Subroutine G calculates the times of the first arrivals of P or S for a surface source [11].

Subroutine GB calculates the times and amplitudes of the first five arrivals of refracted P or S, and also waves reflected from internal interfaces at an angle greater than the critical angle; only waves that have emerged downward from the focus are considered.

Programs G and GB are not as universal as AG, but they operate much more rapidly.

2. Subroutines for Calculating a Measure of Divergence between Theoretical and Experimental Curves. These subroutines provide for the case when it is unknown to which branch of the travel-time curve the experimental curves correspond. However, the type of body wave (P, PP, etc.) and the mode number of the surface wave are assumed to be known.

3. Subroutine for Comparison with Given Thresholds. This subroutine separates the profiles that agree and disagree with the observations according to the criteria adopted.

LITERATURE CITED

1. Azbel', I. Ya., V. I. Keilis-Borok, and T. B. Yanovskaya (1966), "Method for joint interpretation of travel-time and amplitude curves in upper mantle studies," Computational Seismology, No. 2, Moscow, Nauka, pp. 3-45.
2. Yanovskaya, T. B. (1963), "Calculation of velocity profiles of the upper mantle as an inverse mathematical problem," Izv. Akad. Nauk SSSR, Ser. Geofiz., No. 8, pp. 1171-1178.
3. Keilis-Borok, V. I., and T. B. Yanovskaya (1967), "The inverse problem of seismology (structural review)," Geophys. J. Roy. Astr. Soc., 13:223-234 (see also this volume, p. 109).
4. Yanovskaya, T. B., and I. Ya. Azbel' (1964), "The determination of velocities in the upper mantle from observations of P waves," Geophys. J. Roy. Astr. Soc., 8:313.
5. Rasstrigin (1965), Random Search Methods, Izd. Zinatne, Riga.
6. Andrianova, Z. S., V. I. Keilis-Borok, A. L. Levshin, and M. G. Neigauz (1965), Surface Love Waves, Moscow, Nauka [English translation: Seismic Love Waves, New York, Consultants Bureau, 1967].
7. Vil'kovich, E. V., A. L. Levshin, and M. G. Neigauz (1966), "Love waves in a vertically inhomogeneous medium (allowance for sphericity, parameter variations, and absorption)," Computational Seismology, No. 2, Moscow, Nauka, pp. 130-149 [English translation in: Seismic Love Waves, New York, Consultants Bureau (1967)].
8. Neigauz, M. G., and G. V. Shkadinskaya, "Method for calculating surface Rayleigh waves in a vertically inhomogeneous half-space," this volume, p. 88.
9. Keilis-Borok, V. I., M. G. Neigauz, and G. V. Shkadinskaya (1965), "Application of the theory of eigenfunctions to the calculation of surface wave velocities," Revs. Geophys., 3:105-110.
10. Yanovskaya, T. B., and G. V. Golikova (1964), "Method and program for calculating P wave dynamics in the earth's mantle," Problems of the Dynamic Theory of Seismic Wave Propagation, No. 7, pp. 123-129.
11. Azbel', I. Ya., and T. B. Yanovskaya, "Approximation of velocity distributions for calculation of P-wave times and amplitudes," this volume p. 62.

EXPEDITING BODY WAVE CALCULATIONS†

V. P. Valyus

This paper describes several methods that considerably speed up calculations of the travel times and amplitude of body waves by the ray method.

1. Interpolation of Velocity in Layers. Formulas for x(u) and t(u).

Here and below, the apparent velocity at the surface u will be taken as the ray parameter.

We introduce the notation:

$$v = \begin{cases} a, & \text{the compressional wave velocity, if we compute } t(x) \text{ and } A(x) \text{ for P waves} \\ b, & \text{the shear wave velocity, if we compute } t(x) \text{ and } A(x) \text{ for S waves.} \end{cases}$$

Let us reduce the problem for a sphere to the problem for half-space symmetry [1]:

$$v^* = \frac{vR}{r}, \qquad y = R \ln \frac{R}{r}, \tag{1}$$

where R is the radius of the earth.

For each spherical velocity model v(r), there exists a corresponding plane-layered model $v^*(y)$ which gives the same travel-time curve. The boundaries r_i and velocities v_i in the sphere correspond to the depths y_i and velocities v_i^* in the flat model. For the flat model, x(u) and t(u) are given by:

$$x(u) = \int_0^{y_{fd}} \frac{v^*(y)}{\sqrt{u^2 - v^*(y)^2}} dy + 2 \int_{y_{fd}}^{Y(u)} \frac{v^*(y)}{\sqrt{u^2 - v^*(y)^2}} dy, \tag{2}$$

$$t(u) = \int_0^{y_{fd}} \frac{u}{v^*(y)\sqrt{u^2 - v^*(y)^2}} dy + 2 \int_{y_{fd}}^{Y(u)} \frac{u}{v^*(y)\sqrt{u^2 - v^*(y)^2}} dy. \tag{3}$$

The first terms in these formulas give x and t in the layers above the source and the second terms give x and t in the layers below the source. The factor 2 means that the ray intersects the lower layers twice. Y(u) is the maximum depth of penetration of the ray with ray parameter u, and y_{fd} is the focal depth.

Each layer makes the contribution

$$\int_{y_{i-1}}^{y_i} \frac{v^*(y)}{\sqrt{u^2 - v^*(y)^2}} dy, \tag{4}$$

and

$$\int_{y_{i-1}}^{y_i} \frac{u}{v^*(y)\sqrt{u^2 - v^*(y)^2}} dy, \tag{5}$$

respectively, to the right sides of equations (2) and (3).

First of all we shall find the roots u(x) of Eq. (2), and then from the obtained u calculate t from formula (3). To find the roots u(x), we must compute x(u) many times. Therefore, our problem is to find an interpolation of v^* such that Eq. (4) can be computed quickly.

For linear interpolation of $v^*(y)$, integration of (4) results in $\alpha_i\sqrt{u^2 - v_i^{*2}}$, where α_i is the layer parameter.

If v(r) is interpolated by Bullen's law, v^* is an exponential function of y. Then, integration of (4) leads to an expression such as $C_i \arcsin v_i^*/u$, where C_i is the layer parameter. We shall use a linear interpolation of $v^*(y)$, since the square root can be computed more quickly than arcsin, so that linear interpolation saves machine time.

†Adapted from Vychislitel'naya Seismologiya, Vol. 5, Moscow, Nauka (1971), pp. 206-213.

Another way to speed up the calculations is the possibility of using a limited number of layers to approximate v(r). We shall not go into the question of what is meant by "a good approximation of the structure v(r)," but we shall compare linear interpolation in the flat model $v^*(y)$ with linear interpolation in the sphere v(r). For small depths, these interpolations are practically the same. For large depths, there is a significant difference. We shall linearly interpolate $v^*(y)$, i.e., the structure in the flat model. This gives for each layer

$$x(u) = \alpha_K \sqrt{u^2 - v_K^{*2}} + \sum_{i=0}^{K-1} \alpha_i \left(\sqrt{u^2 - v_i^{*2}} - \sqrt{u^2 - v_{i+1}^{*2}}\right), \tag{6}$$

$$t(u) = \alpha_K \ln \frac{u + \sqrt{u^2 - v_K^{*2}}}{v_K^*} + \sum_{i=0}^{K-1} \alpha_i \ln \left(\frac{v_{i+1}^*}{v_i^*} \frac{u + \sqrt{u^2 - v_i^{*2}}}{u + \sqrt{u^2 - v_{i+1}^{*2}}} \right). \tag{7}$$

Here, K is the number of the layer; it coincides with the number of the upper boundary of the layer. We consider a boundary as a place where one of the parameters has a discontinuity, or a layer of zero thickness. At the layer boundaries one of the parameters has a discontinuity. For such a layer, instead of a diffracted wave, we have a wave that is reflected at an angle greater than the critical. The sum is taken over all layers that intersect the ray. The layer parameter

$$\alpha_i = \begin{cases} (y_{i+1} - y_i)/(v_{i+1}^* - v_i^*) & \text{for layers above the source;} \\ 2\,(y_{i+1} - y_i)/(v_{i+1}^* - v_i^*) & \text{for layers below the source.} \end{cases}$$

The layer in which the source is located is divided into two parts: below and above the source.

2. Finding the Roots u(x)

Let us consider the behavior of the function x(u) for a ray that passes through the K-th layer. The ray parameter u (the apparent velocity on the surface) is equal to the velocity v^* at the deepest point of the ray. For a ray that has its deepest point in the K-th layer, therefore, $u \le v_{K+1}^*$, since otherwise the ray would go deeper. On the other hand, $u > \max\{v_0^*, \ldots, v_K^*\}$, because otherwise the ray would not reach that deep. If $v_{K+1}^* \le \max\{v_0^*, \ldots, v_K^*\}$ (the condition for the existence of a waveguide), there are no rays that have their deepest point in the K-th layer.

Example. Assume a structure as indicated in Fig. 1, with the source at depth y_1. Then for a

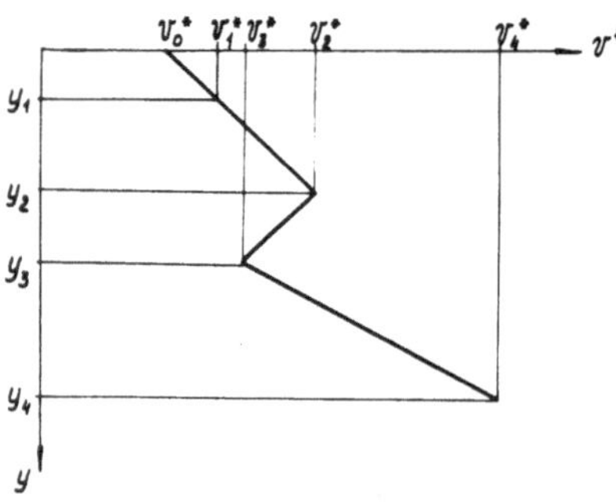

Fig. 1

ray refracted in the first layer, u changes from v_1^* to v_2^*. There are no rays with deepest points in the second layer, because $\max\{v_0^*, v_1^*, v_2^*\} = v_2^* < v_3^*$. For a ray whose deepest point is in the third layer, u changes from $\max\{v_0^*, v_1^*, v_2^*, v_3^*\} = v_2^*$ to v_4^*.

Thus in formula (6), u varies from $\max\{v_0^*, \ldots, v_K^*\}$ to v_{K+1}. We shall denote $\max\{v_0^*, \ldots, v_K^*\}$ as v_{HK}^*.

Near the point v_{HK}^*, x(u) has large first and second derivatives, because of $\alpha_i\sqrt{u^2 - v_{HK}^{*2}}$, whose derivatives are infinite when $u \to v_{HK}^*$. Hence, it is not convenient to seek roots u(x) in steps along the u axis (even if we use linear interpolation), since in that case very small steps would have to be used.

Let us represent x(u) as

$$x(u) = x^*(u) + \alpha \sqrt{u^2 - v_{HK}^{*2}}. \tag{8}$$

Here all terms such as $\alpha_i\sqrt{u^2 - v_{HK}^{*2}}$ are lumped together; α is the sum of the factors before them; $x^*(u)$ are the remaining terms; and $x^*(u)$ has bounded derivatives everywhere on $[v_{HK}^*, v_{K+1}^*]$. Thus, in our example (Fig. 1), for rays refracted in the first layer,

$$x(u) = \alpha_1 \sqrt{u^2 - v_1^{*2}} + \alpha_0 \left(\sqrt{u^2 - u_0^{*2}} - \sqrt{u^2 - v_1^{*2}}\right);$$

$$v_{HK}^* = v_1^*$$

can be reduced to (8) if we let

$$\alpha = \alpha_1 - \alpha_0, \quad x^*(u) = \alpha_0 \sqrt{u^2 - v_0^{*2}}.$$

For rays whose deepest point is in the third layer,

$$x(u) = \alpha_3 \sqrt{u^2 - v_3^{*2}} + \alpha_2\left(\sqrt{u^2 - v_2^{*2}} - \sqrt{u^2 - v_3^{*2}}\right) + \alpha_1\left(\sqrt{u^2 - v_1^{*2}} - \sqrt{u^2 - v_2^{*2}}\right) + \alpha_0\left(\sqrt{u^2 - v_0^{*2}} - \sqrt{u^2 - v_1^{*2}}\right),$$

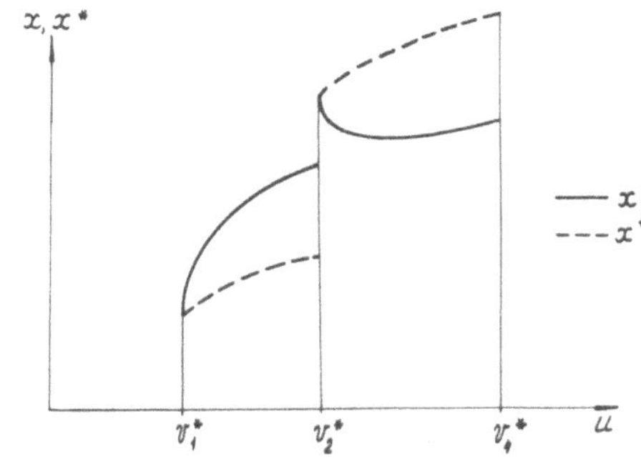

Fig. 2

$$\overset{*}{v}_{HK} = \overset{*}{v}_2,$$

$$\alpha = \alpha_2 - \alpha_1,$$

$$x^*(u) = \alpha_3 \sqrt{u^2 - \overset{*}{v}_3{}^2} - \alpha_2 \sqrt{u^2 - \overset{*}{v}_3{}^2} + \alpha_1 \sqrt{u^2 - \overset{*}{v}_1{}^2} +$$

$$+ \alpha_0 \left(\sqrt{u^2 - \overset{*}{v}_0{}^2} - \sqrt{u^2 - \overset{*}{v}_1{}^2} \right).$$

Graphs of x(u) and $x^*(u)$ for our example are shown in Fig. 2. We sought roots u(x) in the following way. We represent $x^*(u)$ as a broken line. This is done by bisection, as follows (Fig. 3): we compute x^* at the extreme points $\overset{*}{v}_{HK}$ and $\overset{*}{v}_{K+1}$, and then divide the interval $[\overset{*}{v}_{HK}, \overset{*}{v}_{K+1}]$ by two and compute x^* in the middle. We compare the difference

$$x^*\left(\frac{\overset{*}{v}_{HK} + \overset{*}{v}_{K+1}}{2}\right) - \frac{x^*(\overset{*}{v}_{HK}) + x^*(\overset{*}{v}_{K+1})}{2}$$

with some σ error of hitting a given point on the travel-time curve. If the difference is less than σ, we assume that x^* on the interval $[\overset{*}{v}_{HK}, \overset{*}{v}_{K+1}]$ is, with sufficient accuracy, replaced by two straight-line segments. If it is greater than σ, we divide both halves of our interval by two, calculate x^* at their middles, and so on. The case when the difference is smaller than σ but x^* is not accurately approximated by a broken line is theoretically possible but has a low probability (Fig. 4). As a result of the procedure described above, on each part of the interval $\overset{*}{v}_{HK}{}^2$, $\overset{*}{v}_{K+1}$, where x^* is approximated by a straight line, x(u) has the form

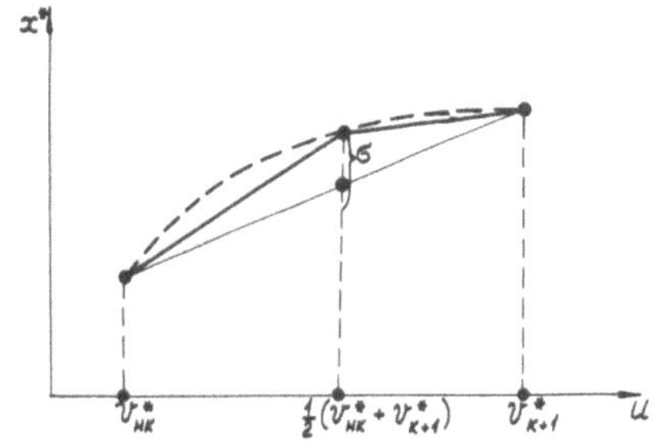

Fig. 3

$$x(u) = au + b + \alpha \sqrt{u^2 - \overset{*}{v}_{HK}{}^2}, \tag{9}$$

where a and b are the parameters of the straight line, which can be determined from the x^* values at the ends of the interval.

Then for each interval we determine whether x(u), as defined by formula (9), has an extremum. It is easy to see that there can be only one extremum. The conditions under which an extremum exists are that a and α have different signs and $|a| > |\alpha|$. If an extremum exists and the extreme point u is inside our interval, we divide the interval into two parts, from the beginning to u_{extr} and from u_{extr} to the end. We can calculate u_{extr} as

$$u_{extr} = \left| \frac{a\overset{*}{v}_{HK}}{\sqrt{a^2 - \alpha^2}} \right|. \tag{10}$$

As a result of this procedure, the entire interval $[\overset{*}{v}_{HK}, \overset{*}{v}_{K+1}]$ is divided into segments $[u_j, u_{j+1}]$, (j = 0, 1, . . .), within which x(u) has the form (9), is unique, and has maximum and minimum values at the ends of the segment.

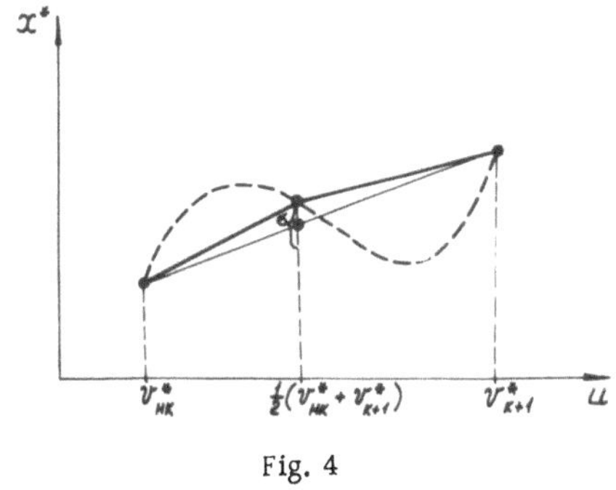

Fig. 4

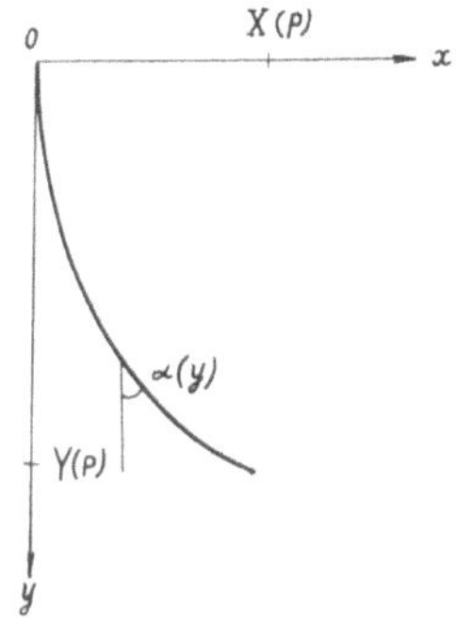

Fig. 5

For each segment $[u_j, u_{j+1}]$, we successively try all points x that are between $x(u_j)$ and $x(u_{j+1})$ and at which travel times and amplitudes must be calculated. If we substitute these x values into (9) and solve for u, we obtain for u(x)

$$u = \frac{a(x-b) - \alpha^* \sqrt{(x-b)^2 - (a^2-\alpha^2)\, v_{HK}^{*2}}}{a^2 - \alpha^2}, \tag{11}$$

where

$$\alpha^* = \begin{cases} -|\alpha|, & \text{if } a \text{ and } \alpha \text{ have different signs and } |a| > |\alpha| \text{ and } u_{extr} \geq u_j, \\ |\alpha| & \text{in the other cases.} \end{cases}$$

3. Calculation of A(u)

We can calculate A(u) by the following equation [5]:

$$A(u) = \sqrt{\frac{\rho_{fd}\, v_{fd}\, v_{fd}^{*2} \left|\frac{du}{dx}\right|}{\rho_{surf} v_{surf} R \sin\frac{x}{R}\, u \sqrt{(u^2 - v_{fd}^{*2})(u^2 - v_{surf}^2)}}} \times$$

$$\times \prod_m |\varkappa_m| \sqrt{\frac{v_{m2}\sqrt{u^2 - v_{m_2}^2}}{v_{m1}\sqrt{u^2 - v_{m1}^2}}}\, |w|. \tag{12}$$

Here, the first square root is the ray factor, which allows for geometrical spreading. The subscripts "surf" and "fd" indicate the earth's surface and the focal depth of the source, respectively. The derivative du/dx is obtained by differentiation of (9):

$$\frac{du}{dx} = \frac{\sqrt{u^2 - v_{HK}^{*2}}}{a\sqrt{u^2 - v_{HK}^{*2}} - \alpha u}, \tag{13}$$

and

$$\prod_m |\varkappa_m| \sqrt{\frac{v_{m2}\sqrt{u^2 - v_{m2}^2}}{v_{m1}\sqrt{u^2 - v_{m1}^2}}}$$

is the product of the refractive indices at all discontinuities intersected by the ray. In the case of reflection at more than the critical angle, one of the factors in ρ is the absolute value of the reflection coefficient, m is the index of discontinuity, V_{m_1} is the value of v^* at the m-th boundary, V_{m_2} is the value of v^* at the next boundary, and $|W|$ is the absolute value of a coefficient that allows for the free boundary.

LITERATURE CITED

1. Gerver, M. L., and V. M. Markushevich, "Determining seismic-wave velocity from travel-time curves," this volume, p. 123.
2. Azbel', I. Ya., and T. B. Yanovskaya, "Approximation of velocity distributions for calculation of P-wave times and amplitudes," this volume, p. 62.
3. Yanovskaya, T. B. (1966), "Algorithm for calculating reflection and refraction coefficients of head waves," Computational Seismology, No. 1, Moscow, Nauka, pp. 107-111.
4. Yanovskaya, T. B., "Program for calculating travel-time and amplitude curves of body waves in a layered medium," Problems of the Dynamic Theory of Seismic Wave Propagation, No. 6, pp. 87-93.
5. Alekseev, A. S., and B. Ya. Gel'chinskii (1959), "Ray method for calculating wave fields for inhomogeneous media with curvilinear interfaces," Problems of the Dynamic Theory of Seismic Wave Propagation, No. 3, pp. 107-160.

DETERMINING SEISMIC-WAVE VELOCITY FROM TRAVEL-TIME CURVES†

M. L. Gerver and V. M. Markushevich

INTRODUCTION

1. Statement of Problem

1. One of the main problems of geophysics is the determination of seismic-wave propagation velocity from travel-time curves. The problem is as follows: let a transient deformation occur on the earth's surface at some point A at time t = 0. This produces an elastic wave, and its arrival time is observed at each point B on the surface of the earth. It is required to determine the propagation velocity of the wave from the known dependence of time upon the coordinates of the points B.

2. As is usual in seismology we assume that: 1) the earth is spherically symmetric; 2) the propagation velocity of seismic waves varies only with depth; and 3) the propagation of the wave obeys the laws of geometric optics.

The latter assumption allows us to study, instead of a seismic wave, the propagation of a transient pulse along rays emerging from the source. Since a ray in a spherically symmetric medium is a plane curve, instead of the sphere we can consider only its central section.

3. The problem is now as follows: let K be a circle of radius R with center C and let A be a point on its circumference (Fig. 1). We introduce the polar coordinates r, θ, taking point C as the pole and line CA as the polar axis.

Seismic rays emerge from the point A. The one that emerges at the angle α to the radius CA ($\alpha \in [0, \pi/2]$) returns to the circumference at point B.

The pulse is propagated along ray AB with the velocity v(r), $r \in [0, R]$. This can be imagined as follows: a point M slides along ray AB with ve-

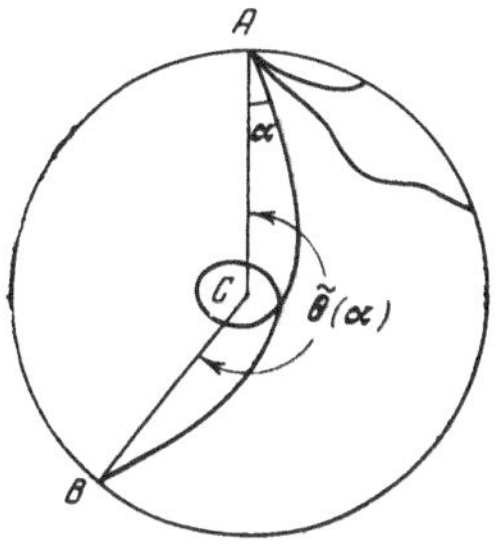

Fig. 1

locity v(r); it emerges from A at time t = 0 and arrives at B at time $t = t(\alpha)$. The arrival time $t(\alpha)$ is recorded at point B.

Let $\theta(\alpha)$ be the total rotation angle of the radius vector CM as M moves from A to B, and let $\tilde{\theta}(\alpha)$ be the angle between the radii CA and CB. It is clear that $\theta(\alpha)$ and $\tilde{\theta}(\alpha)$ are congruent modulo 2π: the difference $\theta(\alpha) - \tilde{\theta}(\alpha) = 2\pi k$, where k is the number of revolutions made by ray AB about the center.

Let us consider two curves in the plane θ, t:

$$\Gamma_0 \{\theta = \theta(\alpha),\ t = t(\alpha)\}$$

and

$$\tilde{\Gamma}_0 \{\tilde{\theta} = \tilde{\theta}(\alpha),\ t = t(\alpha)\}, \alpha \in [0, \tfrac{\pi}{2}].$$

The assumption that time $t(\alpha)$ is recorded at point B means that we know the curve $\tilde{\Gamma}_0$, the true travel-time curve proper. The problem consists of finding the velocity v(r) from the curve $\tilde{\Gamma}_0$.

Let us divide this problem into two parts: determining Γ_0 from $\tilde{\Gamma}_0$, and finding v(r) from Γ_0. We

† Translated from Vychislitel'naya Seismologiya, No. 3, Moscow, Nauka (1967), pp. 3-51.

shall devote a special paper to determining Γ_0 from $\tilde{\Gamma}_0$. Here we consider only the problem of finding v(r) from Γ_0.

4. This problem (determination of v(r) from Γ_0) was solved by Herglotz and Wiechert in 1907 for an elementary particular case under the condition that $dv/dr \leq v(r)/r$. Physically, this condition means the absence of waveguides; in the real earth there is evidence that this condition is not satisfied.

In this paper we assume the possible existence of waveguides. In this case, the solution is not unique. We shall study fully this ambiguity, and in particular we shall show the extent to which this indefiniteness is reduced if, along with the travel-time curve from a surface source, we consider the travel-time curves for sources at depth.

2. Summary of Work

In Chapter 1, to simplify all the formulas we reduce the problem of ray propagation in a circle to the problem of rays in a half-plane. In this transformation, the velocity v(r) becomes velocity u(y) and curves Γ_0 and $\tilde{\Gamma}_0$ become Γ and $\tilde{\Gamma}$.

Rays emerge into the lower half-space from a point O of plane x, y. Let L be one of the rays, which forms an angle α with the y axis, $0 < \alpha < \pi/2$ (Fig. 2). We denote sin α by p, which we shall call the ray parameter.

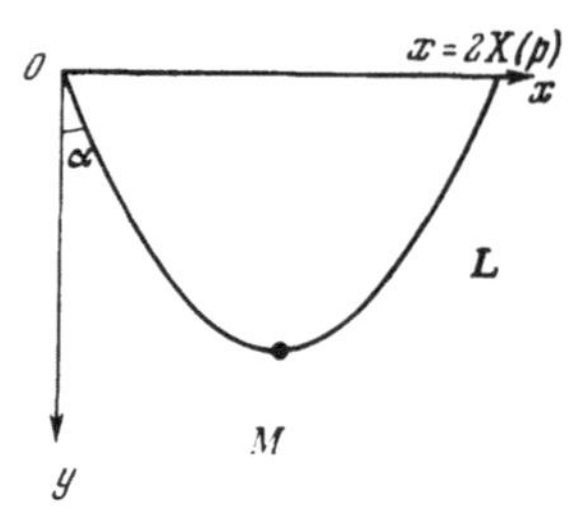

Fig. 2

Let [X(p), Y(p)] be the coordinates of the deepest point of the ray L. At the point $x = 2X(p)$, $y = 0$ the ray returns to the line $y = 0$. The segments of the ray between points 0, [X(p), Y(p)] and between points [2X(p), O], [X(p), Y(p)] (descending and ascending branches of ray) are symmetric with respect to the line $x = X(p)$.

Point M moves along ray L. At the time origin it emerges from point O, and at the point $(x, y) \in L$ it has velocity u(y); the velocities are normalized so that $u(0) = 1$.

Certain limitations that are natural from a physical point of view (piecewise second-order differentiability, only a finite number of waveguides, etc.) are imposed on velocity u(y).

Let T(p) be the travel time of M along the descending branch of L. Then the arrival time of M at point [2X(p), O] is 2T(p).

Ray L takes the form such that time T(p) is a minimum. As is known from geometric optics, this is equivalent to the following condition: the descending branch of L forms at the point (x, y) an angle $\alpha(y)$ with the y axis such that $\sin \alpha(y) = pu(y)$.

Let Γ be a curve in the plane x, t$\{x = 2X(p), t = 2T(p)\}$, $p \in (0, 1)$. Henceforth we shall call it the travel-time curve.

Let $\tilde{X}(p)$ be a function in the interval $0 < p < 1$ such that $0 \leq \tilde{X}(p) < \pi R/v(R)$, $\tilde{X}(p) \equiv X(p) \pmod{\pi R/v(R)}$. Let $\tilde{\Gamma}$ be the curve $\{x = 2\tilde{X}(p), t = 2T(p)\}$, $p \in (0, 1)$.

The problem of determining v(r) from $\tilde{\Gamma}_0$ or Γ_0 are reduced to the problems of determining u(y) from $\tilde{\Gamma}$ or from Γ, respectively.

In Chapter II, the functions X(p) and T(p) are studied for a given velocity distribution u(y).

In Chapter III we solve the problem of determining the velocity distribution from given functions X(p) and T(p). It turns out that this problem has no unique solution. There exists a set of functions $\{u^\nu(y)\}$ that have the following property: let L be a ray with parameter p that is propagated with velocity $u^\nu(y)$. Then the abscissa of its deepest point is X(p) and the motion time along the descending branch is T(p).

We shall completely describe the set of solutions $\{u^\nu(y)\}$. Analytic conditions are formulated that are necessary and sufficient for a function $u^\nu(y)$ to be a solution of this problem. Then we construct a set G in the plane y, u such that: 1) the graph of any solution $u^\nu(y)$ is located in $\bar{G}$, and 2) the graph of some solution passes through any point of G.

In Chapter IV, we study the geometric properties of the travel-time curve Γ and explain how to find the functions X(p) and T(p) from curve Γ, i.e., how to find among all possible parametrizations of the travel-time curve that in which the ray parameter p is taken as the parameter. As a rule, the latter problem has a unique solution. In this case, comparison of the results of Chapters III and IV leads to solution of the problem of finding the velocity from the travel-time curve Γ. For some Γ, however, it is impossible to find X(p) and T(p) without some additional information. As additional information we shall use travel-time curves from sources below the free surface.

In Chapter V, by joint investigation of the travel-time curves for surface and deep sources, two problems are solved:

a) Determination of the functions X(p) and T(p) when only from curve Γ are they determined nonuniquely.

b) Reduction of the ambiguity in determining velocity from a travel-time curve.

In particular, the following theorem is proved.

Theorem. If the travel-time curves are known for several deep sources that are situated such that at least one source is between two adjacent waveguides, then the velocity outside of the waveguides is uniquely determined.

Chapter I

PROBLEM OF RAYS IN A HALF-PLANE

We reduce the problem of ray propagation in a circle K, that was formulated in Section 1, to the problem of rays in a half-plane.

3. Transformation of Circle to Half-Plane

1. We assume that the curve Γ_0 (not just the curve $\tilde{\Gamma}_0$) is known. It is therefore more convenient to consider seismic rays not in one circle K (see Section 1) but on a Riemann surface Z consisting of a denumerable set of such circles (cut along the radius CA): having made a revolution about the center C, ray AB moves to the next sheet of the Riemann surface.

Let us consider the conformal mapping of the surface Z (omitting the point C) onto a half-plane: $w = \ln R/z$. The modulus of expansion of this transformation is $|w'| = 1/|z| = 1/r$. Therefore, if we assume that the velocity along the image of the ray AB is $v(r)/r$, then the travel time along the ray AB and along its image of the plane w will be the same.

Let us make another transformation – one that normalizes velocity: we reduce the velocity by a factor of $v(R)/R$ and compress the plane w by the same factor. This does not change the travel time along the ray, but the velocity along the real axis becomes equal to 1.

The total transformation is written as:

$$x = \frac{R}{v(R)}\,\theta,\quad y = \frac{R}{v(R)}\ln\frac{R}{r},\quad u(y) = \frac{v\left(Re^{-\frac{v(R)}{R}y}\right)}{v(R)e^{-\frac{v(R)}{R}y}}.$$

2. In this transformation, each seismic ray AB that does not pass through the center C is mapped into a ray L on the plane x, y.

Let $M = (r, \theta)$ be an arbitrary point on the descending branch of AB, and let (x, y) be its image on ray L. Let $\alpha(x, y)$ be the angle between ray L and the y axis at the point (x, y). Since the transformation under consideration is conformal, $\alpha(x, y)$ is equal to the angle between ray AB and the radius CM at the point (r, θ). It is known from geometric optics that the ratio $\sin\alpha(x, y)/u(y)$ is constant along L. That is, along the descending branch of L the angle $\alpha(x, y)$ is a function only of y. Therefore, we shall henceforth write $\alpha(y)$ instead of $\alpha(x, y)$. Let $\sin\alpha(0) = p$ for the ray in question. Since by definition $u(0) = 1$, then

$$\sin\alpha(y) = pu(y).$$

The number p we shall call the ray parameter.

3. We now prove the following lemma.

Lemma. Among the seismic rays that pass through the center C, there is at most one that returns to the circumference of circle K in a finite time.

Proof. According to its physical meaning, the function v(r) is positive and bounded in the segment [O, R]. Let us consider two cases:

a) The ratio $v(r)/r$ is not bounded when $r \in (0, R]$. Thus the function u(y) is not bounded on the semiaxis $y \geq 0$. On the strength of the condition

$$\sin\alpha(y) = pu(y),$$

only ray AB, which moves along the diameter of circle K, passes through the center C.

b) The ratio $v(r)/r$ is bounded when $r \in (0, R]$. Thus, $u(y) < u_0 = \text{const}$ when $y \geq 0$. The motion time along any ray that passes through the center C is infinite.

According to this lemma, the shape of curves Γ_0 and $\tilde{\Gamma}_0$ does not depend upon whether or not the rays that pass through the center are considered.

4. Limitations Imposed on Velocity u(y)

We shall assume that the seismic-wave propagation velocity v(r) satisfies certain physically reasonable requirements. These requirements will be formulated directly for the transformed function u(y).

1. The function u(y) is positive and scaled so that $u(0) = 1$.

2. The function u(y) is everywhere twice continuously differentiable, with the exception of a finite number of points at which it or its derivatives either do not exist or are discontinuous.

3. In every finite segment, u(y) is bounded, but on the entire semiaxis $y \geq 0$ it is not bounded

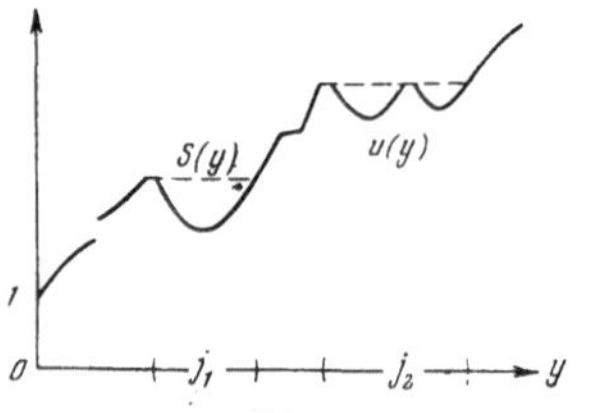

Fig. 3

(note that consideration of a function u(y) that is bounded on the entire semiaxis does not present any additional difficulties).

4. There exists a finite number of waveguides. To formulate this requirement exactly, we introduce the function $s(y) = \sup\{u(y^0), y^0 \in [0, y]\}$. In Fig. 3, the graph of s(y) is given by a broken line where it does not coincide with the graph of u(y).

The waveguides $j_1, \ldots, j_n$ are intervals of the y axis that have the following properties (Fig. 3):

1) s(y) is constant in each of them;

2) Each of them contains points where $u(y) < s(y)$;

3) Outside of these intervals, $u(y) = s(y)$.

We shall denote the ends of waveguide j_k by y_k and $\bar{y}_k$, $1 \leq k \leq n$:

$$y_1 < \bar{y}_1 < y_2 < \bar{y}_2 < \ldots < y_n < \bar{y}_n.$$

For definiteness, we shall assume that y_1 is strictly greater than zero. So that some formulas can be written more compactly, we shall let

$$y_0 = \bar{y}_0 = 0.$$

5. Refinement of Assumptions about Seismic Rays

In defining curves Γ_0 and $\tilde{\Gamma}_0$ in Section 1, we stipulated that the travel time $t(\alpha)$ was recorded for each seismic ray AB that emerged at the surface. Let us refine this condition.

Some rays can be divided into two or more rays. For example, where the velocity undergoes a discontinuity, the ray can be divided into reflected and refracted continuation. We shall agree not to consider such divided rays.

Let us explain in more detail which rays are thus eliminated from consideration.

1) We shall not consider rays whose images on the plane x, y contain a segment parallel to the x axis.

2) We shall agree that the deepest point of a ray is determined by the following condition: the ordinate of the deepest point of a ray L with parameter p on the plane x, y is

$$Y(p) = \inf\{y,\ pu(y) \geqslant 1\}, \quad p \in (0, 1).$$

According to this condition, when a ray with a nonzero angle of incidence is divided into reflected and refracted continuations, only the refracted ray is considered. When a ray with a zero angle of incidence is divided into several rays with different deepest points (see Fig. 4), the ray whose deepest point is closest to the surface is considered.

Points $\theta(\alpha)$ and $t(\alpha)$ corresponding to rays that we agreed not to consider are eliminated from curve Γ_0. We treat curve $\tilde{\Gamma}_0$ similarly.

By virtue of these agreements, each ray L with parameter $p \in (0, 1)$ on the plane x, y is situated in a band $0 \leq y \leq Y(p)$. The descending branch of L is uniquely projected on the y axis. If the abscissa of the deepest point $X(p) < \infty$, then L has an ascending branch that is symmetric with the descending branch with respect to the line $x = X(p)$.

6. Curves Γ and $\tilde{\Gamma}_0$

A ray with parameter p that has an ascending branch terminates at the point $x = 2X(p)$, $y = 0$; the travel time along it (with velocity u(y)) is $2T(p)$.

With mapping onto the plane x, y, curve Γ_0 becomes the curve $\Gamma\ \{x = 2X(p),\ t = 2T(p)\}$, $p \in (0, 1)$. The problem of finding v(r) from Γ_0 reduces to the problem of determining velocity u(y) from travel-time curve Γ.

If curve Γ_0 is not known and only $\tilde{\Gamma}_0$ is given, then travel-time curve Γ cannot be considered given, but only the curve

$$\tilde{\Gamma}\ \{x = 2\tilde{X}(p), \quad t = 2T(p)\}, \quad p \in (0, 1),$$

where

$$0 \leqslant \tilde{X}(p) < \pi R/v(R), \quad \tilde{X}(p) \equiv X(p) \pmod{\pi R/v(R)}.$$

In this paper, we shall be concerned with determining u(y) from Γ. As was already stated, a separate paper will be devoted to determining u(y) from $\tilde{\Gamma}$.

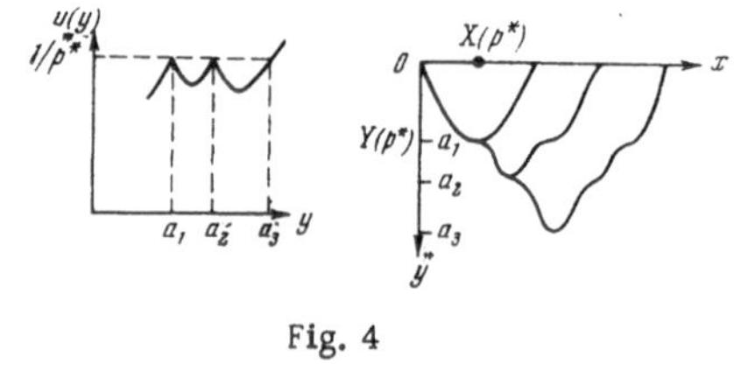

Fig. 4

Chapter II

THE FUNCTIONS X(p) AND T(p)

In this chapter, we study the properties of the functions X(p) and T(p) for a given velocity u(y).

7. Formulas Expressing X(p) and T(p) in Terms of u(y)

It is clear (Fig. 5) that the equation of the descending branch of ray L with parameter p has the form: $x=\int_0^y \operatorname{tg} \alpha(y)\, dy$ or (on the strength of the relation $\sin \alpha(y) = pu(y)$)

$$x=\int_0^y \frac{pu(y)\,dy}{\sqrt{1-p^2u(y)}}.$$

The travel time along L from 0 to (x, y) is obviously

$$\int_0^y \frac{dy}{u(y)\cos\alpha(y)} = \int_0^y \frac{dy}{u(y)\sqrt{1-p^2u^2(y)}}.$$

Therefore,

$$X(p)=\int_0^{Y(p)} \frac{pu(y)\,dy}{\sqrt{1-p^2u^2(y)}},\quad T(p)=\int_0^{Y(p)} \frac{dy}{u(y)\sqrt{1-p^2u^2(y)}},$$

$$p\in(0,1).$$

We shall begin the study of these formulas with a discussion of the properties of the function Y(p).

8. The Function Y(p)

1. By definition

$$Y(p)=\inf\{y,\ pu(y)\geqslant 1\},\quad p\in(0,\ 1).$$

2. Closely related to Y(p) is the function

$$f(y)=s^{-1}(y)=[\sup\{u(y^0),\ \ y^0\in[0,\ y]\}]^{-1},\quad y\geqslant 0.$$

It is obvious that the function $f(y)$ is nonincreasing, and $0 < f(y) \leq 1$.

Let us examine the graph of $p = f(y)$ on the plane y, p (Fig. 6). It is easily obtained from the graph of u(y); outside of the waveguides $j_1, \ldots, j_n$, $f(y)$ coincides with $u^{-1}(y)$, and in each of the intervals of j_k it is constant ($f(y) = p_k$ when $y \in j_k$, $1 \leq k \leq n$).

Let y^0 be a discontinuity point of $f(y)$; we connect the points $\{y^0, f(y^0 - 0)\}$ and $\{y^0, f(y^0 + 0)\}$,

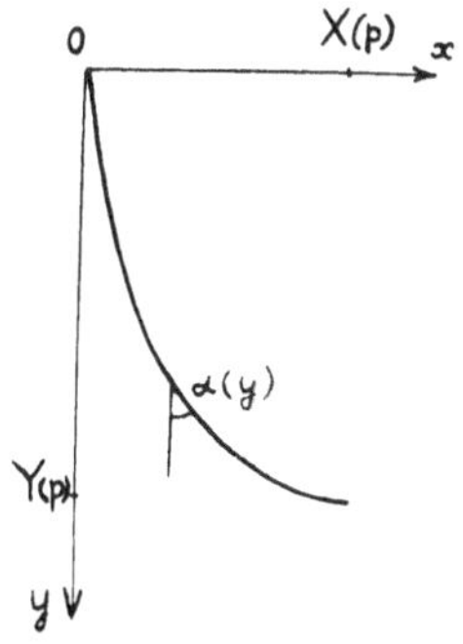

Fig. 5

by a straight-line segment and include it in the graph of $f(y)$. The graph of Y(p) is treated similarly. Then, by virtue of the relation

$$\inf\{y,\ pu(y)\geqslant 1\}=\inf\{y,\ p\geqslant f(y)\}$$

the graphs of $f(y)$ and Y(p) (as curves on the plane y, p) will coincide (in regions of monotonicity, these are inverse functions).

Below in determining velocity u(y), we shall repeatedly use the property that from the known function Y(p) we can determine the function u(y) outside of the waveguides.

3. The function Y(p) obviously does not increase. The physical interpretation of this property is that the more steeply the ray emerges from the source O, the lower its deepest point.

4. If for some p, $Y(p - 0) > Y(p + 0)$, then in the interval $(Y(p + 0), Y(p - 0))$, the function $f(y)$ is constant and equal to p. This is a necessary and sufficient condition for a point $p \in (0, 1)$ to be a discontinuity point of Y(p).

In particular, the values p_k of the function $f(y)$ when $y \in j_k = (y_k, \bar{y}_k)$, $1 \leq k \leq n$, are discontinuity points of Y(p), wherein $y_k = Y(p_k + 0)$, $\bar{y}_k = Y(p_k - 0)$. Thus, from the known function Y(p) and the known numbers $p_1, \ldots, p_n$ we can determine the location of the waveguides $j_1, \ldots, j_n$. Because of this property, the numbers $p_1, \ldots, p_n$ play an important role.

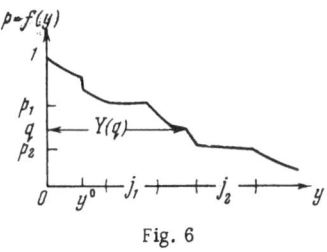

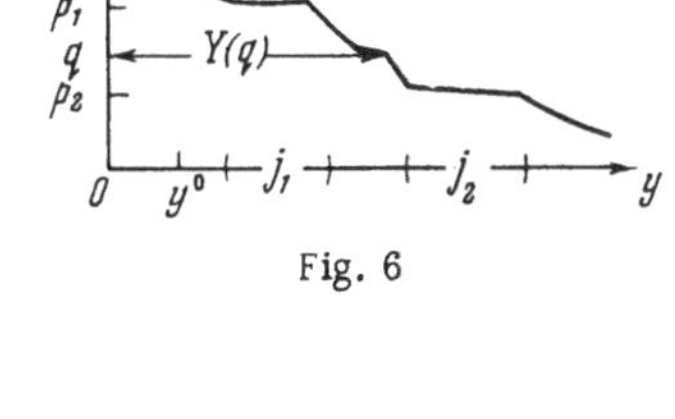

Fig. 6

Let $p_0 = 1$, $p_{n+1} = 0$. Since

$$0 = y_0 = \bar{y}_0 < y_1 < \bar{y}_1 < \ldots < y_n < \bar{y}_n < \infty,$$

then

$$1 = p_0 > p_1 > \ldots > p_n > p_{n+1} = 0.$$

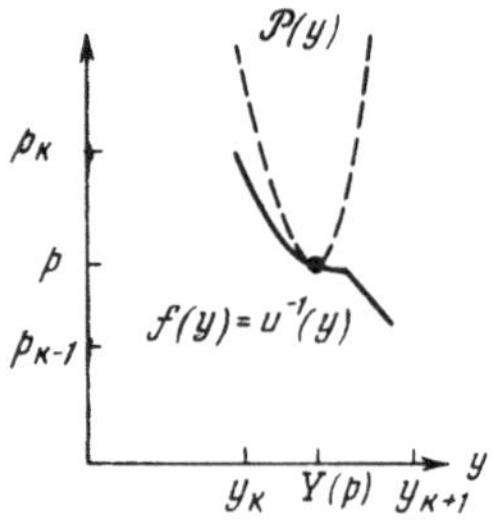

Fig. 7

9. The Function X(p)

The behavior of the function X(p) is determined by the behavior of Y(p) and by the differential properties of u(y).

1. Note that the interval (0, 1) of the p axis has a finite number of points: points $f(y - 0)$ and $f(y + 0)$ if $f(y - 0) > f(y + 0)$ and points $f(y^0)$ if $f(y)$ is continuous, and $f'(y)$ or $f''(y)$ either does not exist or is discontinuous when $y = y^0$.

The remaining points $p \in (0, 1)$ are divided into two nonoverlapping sets: Π and Π', and we assign to Π' those where $f'(Y(p)) = 0$.

We shall show that at points $p \in \Pi'$

$$X(p) = X(p \pm 0) = +\infty,$$

while at points $p^0 \in \Pi$, the function X(p) is continuously differentiable.

2. Let $p \in \Pi'$. We write X(p) as

$$X(p) = p \int_0^{Y(p)} \frac{dy}{\sqrt{u^{-2}(y) - p^2}}.$$

Among the numbers $k = 0, 1, \ldots, n$ we select one such that $p \in [p_{k+1}, p_k)$. The equation $u^{-1}(y) = f(y)$ is satisfied when $y \in (\bar{y}_k, y_{k+1}]$ (see Fig. 7).

To verify the relation $X(p) = X(p \pm 0) = +\infty$, it is sufficient, therefore, to establish that

a) the integral $\int_{\bar{y}_k}^{Y(p)} \frac{dy}{\sqrt{f^2(y) - p^2}}$ becomes $+\infty$,

b) for any sequence q_n that converges to $p + 0$

$$\int_{\bar{y}_k}^{Y(q_n)} \frac{dy}{\sqrt{f^2(y) - q_n^2}} \to +\infty,$$

and c) for any sequence q_n that converges on $p - 0$

$$\int_{\bar{y}_k}^{Y(p)} \frac{dy}{\sqrt{f^2(y) - q_n^2}} \to +\infty$$

(we maintain a constant upper limit, using the fact that the function Y(q) does not increase).

Since when $y \in (\bar{y}_k, y_{k+1}]$ and when $p \in [p_{k+1}, p_k)$,

$$\frac{1}{\sqrt{f(y) + p}} > \frac{1}{\sqrt{2p_k}},$$

then assertions a), b), and c) can be replaced by simpler ones:

a') the integral $\int_{\bar{y}_k}^{Y(p)} \frac{dy}{\sqrt{f(y) - p}}$ becomes $+\infty$,

b') for any sequence q_n that converges on $p + 0$

$$\int_{\bar{y}_k}^{Y(q_n)} \frac{dy}{\sqrt{f(y) - q_n}} \to +\infty,$$

c') for any sequence q_n that converges on $p - 0$

$$\int_{\bar{y}_k}^{Y(p)} \frac{dy}{\sqrt{f(y) - q_n}} \to +\infty.$$

The function $f(y)$ is twice continuously differentiable at the point Y(p). Therefore, there exists a parabola P(y) such (see Fig. 7) that $P(Y(p)) = f(Y(p)) = p$ and $P(y) > f(y)$ for $y \in (\bar{y}_k, Y(p))$. Assertions a'), b'), and c') obviously follow from:

a'') $$\int_{\bar{y}_k}^{Y(p)} \frac{dy}{\sqrt{P(y) - p}} = +\infty,$$

b'') $$\int_{\bar{y}_k}^{Y(q_n)} \frac{dy}{\sqrt{P(y) - p}} \to +\infty \quad \text{for } q_n \to p + 0,$$

c'') $$\int_{\bar{y}_k}^{Y(p)} \frac{dy}{\sqrt{P(y) - q_n}} \to +\infty \quad \text{for } q_n \to p - 0.$$

Since a''), b''), and c'') are obviously satisfied, the relation $X(p) = X(p \pm 0) = +\infty$ when $p \in \Pi'$ is proved.

3. Now let $p^0 \in \Pi$. It must be proved that when $p = p^0$ the derivative $X'(p)$ exists and is continuous.

To prove the continuous differentiability of the function $X(p)$ at the point $p = p^0$, it is, of course, sufficient to verify that the function

$$\frac{X(p)}{p} = \int_0^{Y(p)} \frac{dy}{\sqrt{u^{-2}(y) - p^2}}$$

when $p = p^0$ has a continuous derivative with respect to p.

Let us simplify the symbols. Let $p^2 = k$, $u^{-2}(y) = h(y)$, and $g(k) = \inf\{y, h(y) \le k\}$. Then $\int_0^{Y(p)} \frac{dy}{\sqrt{u^{-2}(y) - p^2}}$ can obviously be rewritten as

$$\int_0^{g(k)} \frac{dy}{\sqrt{h(y) - k}}.$$

The problem is solved if it is shown that the function $\int_0^{g(k)} \frac{dy}{\sqrt{h(y)-k}}$ has a continuous derivative with respect to k when $k = k_0 = (p^0)^2$.

Let us consider two cases.

Case I. The function h(y) is discontinuous at the point $g_0 = g(k_0)$ (Fig. 8) and the function g(k) is constant in some vicinity of the point k_0: $g(k) = g_0$ (it would be so if $Y(p) = Y(p_0)$ for p close to p^0).

The derivative with respect to k of the function $\int_0^{g_0} \frac{dy}{\sqrt{h(y)-k}}$ when $k = k_0$ obviously exists and is continuous.

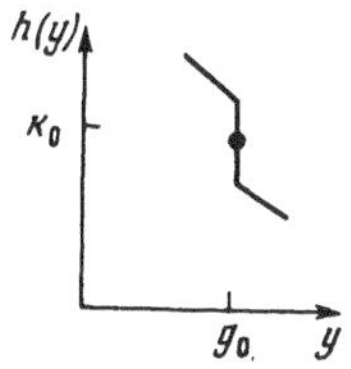

Fig. 8

Case II. The function h(y) is continuous at the point $g_0 = g(k_0)$ (Fig. 9).

In some neighborhood of the point g_0 the function h(y) is monotone decreasing and is the inverse of g(k).

Within this neighborhood we select an interval (g_1, g_2) containing the point g_0 and let $h(g_1) = k_1$ and $h(g_2) = k_2$. The function $\int_0^{g_1} \frac{dy}{\sqrt{h(y)-k}}$ is obviously continuously differentiable when $k = k_0$, and

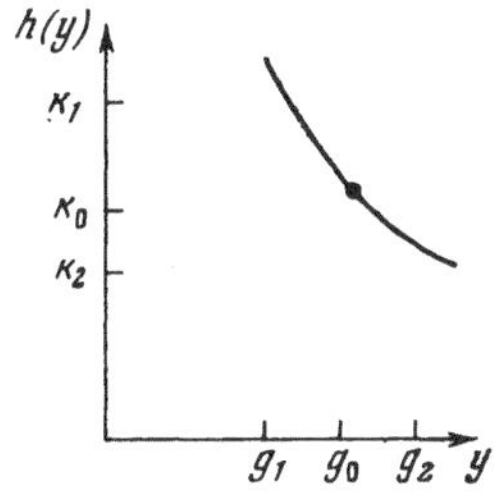

Fig. 9

it remains to check whether

$$\int_{g_1}^{g(k)} \frac{dy}{\sqrt{h(y) - k}} = I(k)$$

has the same property.

This cannot be checked by direct differentiation, since the integrand has a singularity when $k = k_0$. Let $h(y) = h$. Then $y = g(h)$, $dy = g'(h)\,dh$, $h(g_1) = k_1$, and $h(g(k)) = k$, so that

$$I(k) = \int_{k_1}^{k} \frac{g'(h)\,dh}{\sqrt{h-k}}.$$

Integrating by parts, we obtain

$$I(k) = -2g'(k_1)\sqrt{k_1 - k} - 2\int_{k_1}^{k} g''(h)\sqrt{h-k}\,dh.$$

Now we can find dI/dk:

$$I'(k) = \frac{g'(k_1)}{\sqrt{k_1 - k}} + \int_{k_1}^{k} \frac{g''(h)\,dh}{\sqrt{h-k}}.$$

We recall that the function u(y) is twice continuously differentiable in the vicinity of the point $y = Y(p^0)$ and that $u'(Y(p^0)) \ne 0$.

Therefore, the function $g''(h)$ is continuous in some interval containing (k_2, k_1), if the interval $g_1 < y < g_2$ was sufficiently small.

It remains to verify that the function $\int_{k_1}^{k} \frac{g''(h)\,dh}{\sqrt{h-k}}$ is continuous when $k = k_0$. Since

$$\int_{k}^{k_1} \frac{g''(h)\,dh}{\sqrt{h-k}} = \int_{k_0}^{k_1+k_0-k} \frac{g''(h+k-k_0)\,dh}{\sqrt{h-k_0}},$$

then

$$\int_{k_1}^{k} \frac{g''(h)\,dh}{\sqrt{h-k}} - \int_{k_1}^{k_0} \frac{g''(h)\,dh}{\sqrt{h-k_0}} =$$

$$= \int_{k_0}^{k_1} \frac{g''(h) - g''(h+k-k_0)}{\sqrt{h-k_0}}\,dh + \int_{k_1+k_0-k}^{k_1} \frac{g''(h+k-k_0)}{\sqrt{h-k_0}}\,dh.$$

When k is sufficiently close to k_0, the integrals on the right side of the latter equation are arbitrarily small:

$$\left|\int_{k_0}^{k_1} \frac{g''(h) - g''(h+k-k_0)}{\sqrt{h-k_0}}\, dh\right| <$$

$$< \max_{h\in(k_0,\, k_1)} |g''(h) - g''(h+k-k_0)| \int_{k_0}^{k_1} \frac{dh}{\sqrt{h-k_0}}$$

$$\left|\int_{k_1+k_0-k}^{k_1} \frac{g''(h+k-k_0)}{\sqrt{h-k_0}}\, dh\right| <$$

$$< |k-k_0| \max_{h\in(k_1+k_0-k,\, k_1)} \left|\frac{g''(h+k-k_0)}{\sqrt{h-k_0}}\right|.$$

Thus, $I'(k)$ is continuous when $k = k_0$: Q.E.D.

Note. Set Π' is a set of measure zero, and set Π, therefore, is a set of full measure. This follows from the definition of Π' and from the following theorem.

Let a function f be smooth in an interval, and its derivative f' be zero on some set of points. Then the measure of the set of values taken by f in these points is equal to zero.

4. At points $p \in \Pi \cup \Pi'$,

$$X(p) = X(p \pm 0).$$

At a finite number of points that do not belong to $\Pi \cup \Pi'$, it may not be satisfied, as the following shows.

For example, let (see Fig. 10)

$$u(y) = \begin{cases} 1+y, & 0 \leqslant y \leqslant 1 \\ \dfrac{1}{\sqrt{\frac{1}{4}-(y-1)^2}}, & 1 \leqslant y \leqslant \frac{5}{4}. \end{cases}$$

Thereby,

$$Y(p) = \begin{cases} p^{-1}-1, & {}^1/_2 \leqslant p \leqslant 1 \\ 1+\sqrt{{}^1/_4 - p^2}, & \frac{\sqrt{3}}{4} \leqslant p \leqslant \frac{1}{2}. \end{cases}$$

We shall show that

$$X({}^1/_2) = X({}^1/_2 + 0) \neq X({}^1/_2 - 0).$$

When $\frac{1}{2} \leq p \leq 1$

$$X(p) = \int_0^{p^{-1}-1} \frac{(1+y)\,dy}{\sqrt{p^{-2}-(1+y)^2}} =$$

$$= \sqrt{p^{-2}-(1+y)^2}\,\Big|_{p^{-1}-1}^{0} = \sqrt{p^{-2}-1}.$$

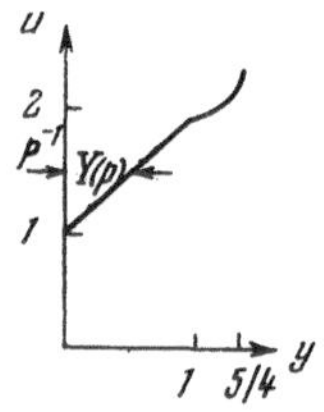

Fig. 10

When $\sqrt{\frac{3}{4}} < p < \frac{1}{2}$

$$X(p) = \int_0^1 \frac{(1+y)\,dy}{\sqrt{p^{-2}-(1+y)^2}} +$$

$$+ p \int_1^{1+\sqrt{{}^1/_4 - p^2}} \frac{dy}{\sqrt{({}^1/_4 - p^2)-(y-1)^2}} = \sqrt{3} + p\,\frac{\pi}{2}.$$

Therefore, $X(\frac{1}{2}) = X(\frac{1}{2}+0) = \sqrt{3} \neq \sqrt{3} + \pi/4 = X(\frac{1}{2}-0)$.

It can be shown that at each point $p \in (0, 1)$

$$\varliminf_{q\to p-0} X(q) \geqslant X(p), \qquad \varlimsup_{q\to p-0} X(q) \geqslant \varliminf_{q\to p+0} X(q).$$

In the present paper, however, these properties of X(p) are not needed. We shall therefore omit their proofs.

10. The Auxiliary Function $\tau(p)$. The numbers $\sigma_1, \ldots, \sigma_n$. The Relationship between X(p), T(p), and $\tau(p)$. The Properties of T(p)

We shall introduce the auxiliary function

$$\tau(p) = \int_0^{Y(p)} \sqrt{u^{-2}(y) - p^2}\,dy, \qquad p \in (0, 1).$$

1. It is clear that $\tau(p)$ is monotone decreasing: as p increases, the integration interval does not increase, and the integrand decreases. At the points of continuity of Y(p), the function $\tau(p)$ is continuous. At the points of discontinuity of Y(p), it can be either discontinuous or continuous, depending upon the behavior of the integrand. If $Y(p^0 - 0) > Y(p^0 + 0)$ at some point p^0 but $u^{-1}(y) = f(y) = p^0$ when $y \in (Y(p^0 + 0), Y(p^0 - 0))$, then $\tau(p)$ is continuous when $p = p^0$. And only at points $p_1, \ldots, p_n$ (corresponding to waveguides $j_1, \ldots, j_n$ does the function $\tau(p)$ have negative discontinuities:

$$\tau(p_k+0) - \tau(p_k-0) = \int_{Y(p_k-0)}^{Y(p_k+0)} \sqrt{u^{-2}(y) - p_k^2}\,dy =$$

$$= -\int_{v_k}^{\bar{v}_k} \sqrt{u^{-2}(y) - p_k^2}\,dy, \quad k = 1, 2, \ldots, n.$$

We introduce

$$\sigma_k = \int_{y_k}^{\bar{y}_k} \sqrt{u^{-2}(y) - p_k^2}\, dy, \quad 1 \leqslant k \leqslant n.$$

The numbers $\sigma_1, \ldots, \sigma_n$ are an important characteristic of waveguides, just as are the numbers $p_1, \ldots, p_n$.

It is convenient not to determine $\tau(p)$ at all at points $p_1, \ldots, p_n$; all the assertions we make about $\tau(p)$ are valid only when $p \neq p_i$. We shall not expressly stipulate this below.

2. It is easy to see that $\tau(p) = T(p) - pX(p)$. In fact,

$$T(p) - pX(p) = \int_0^{Y(p)} \frac{dy}{u(y)\sqrt{1 - p^2u^2(y)}} - $$
$$- p \int_0^{Y(p)} \frac{pu(y)\, dy}{\sqrt{1 - p^2u^2(y)}} = \int_0^{Y(p)} \frac{1 - p^2u^2(y)}{u(y)\sqrt{1 - p^2u^2(p)}} = \tau(p).$$

Thus, the integrals expressing X(p) and T(p) converge and diverge simultaneously. In particular, $T(p) < \infty$ when $p \overline{\in} \Pi$. That is, the function $\tau(p)$ can be considered known if the functions X(p) and T(p) are given: in the set of full measure Π the function $\tau(p)$ is equal to the difference $T(p) - pX(p)$, and when $p \in \Pi$, $p \neq p_k$, and $1 \leq k \leq n$, its definition can be extended by continuity. In particular, from the known functions X(p) and T(p) we can determine the numbers p_k and σ_k, $k = 1, 2, \ldots, n$. Further, at the continuity points of $\tau(p)$,

$$T(p \pm 0) - pX(p \pm 0) = T(p) - pX(p).$$

That is, $T(p) = T(p \pm 0)$ everywhere, except for a finite number of points, and $T(p) = T(p \pm 0) = +\infty$ for $p \in \Pi'$.

3. Being monotonic, the function $\tau(p)$ has a finite derivative almost everywhere in (0, 1). It can be verified by direct calculation that when $p \in \Pi$ the derivative $\tau'(p)$ exists and is equal to $-X(p)$.

In fact,

$$\tau'(p) = -\int_0^{Y(p)} \frac{p\,dy}{\sqrt{u^{-2}(y) - p^2}} + Y'(p)\sqrt{u^{-2}(y) - p^2}\,\Big|_{y=Y(p)}.$$

If u(y) is continuous when $y = Y(p)$, then $u^{-2}(Y(p)) = p^2$, but if $u^{-1}(Y(p) - 0) > p > u^{-1}(Y(p) + 0)$, then $Y'(p) = 0$. Therefore, the second term vanishes when $p \in \Pi$, i.e., $\tau'(p) = -X(p)$.

When $p \in \Pi$, therefore, the function $T(p) = \tau(p) + pX(p)$ is continuously differentiable and $T'(p) = pX'(p)$. This property will be used in Chapter IV in determining X(p) and T(p) from Γ.

4. Note another important consequence that follows from the equation

$$\tau'(p) = -X(p), \quad p \in \Pi.$$

From the known function $\tau(p)$ we can determine X(p) and T(p) everywhere, except, perhaps, for a finite number of points. As a proof, let us consider the set $\hat{\Pi}$, in which the derivative $\tau'(p)$ exists and is finite and continuous. Since $\tau'(p) = -X(p)$ when $p \in \Pi$, then $\hat{\Pi} \supseteq \Pi$, and $\hat{\Pi} \cap \Pi' = 0$. Thus if $\tau(p)$ is known, we can let

$$X(p) = \begin{cases} -\tau'(p), & p \in \hat{\Pi} \\ +\infty, & p \overline{\in} \hat{\Pi}\,, \end{cases} \qquad T(p) = \tau(p) + pX(p).$$

In this case, incorrect values can arise only at a finite number of points.

Thus, specifying the pair of functions X(p) and T(p) is equivalent to specifying the function $\tau(p)$.

Chapter III

DETERMINING THE VELOCITY u(y) FROM GIVEN FUNCTIONS X(p) AND T(p)

The problem of determining the velocity u(y) from the given functions X(p) and T(p) will be solved in this chapter.

11. Exact Statement of Problem

We must find a function u(y) that has the following property: Let L be a ray with parameter p that is propagated in a medium with velocity u(y). Then, the abscissa of its deepest point will be X(p), and the travel time along the descending branch will be T(p).

We shall assume that the given functions X(p) and T(p) are not arbitrary but are such that there exists at least one solution u(y) of the problem:

$$X(p) = \int_0^{Y(p)} \frac{pu(y)\,dy}{\sqrt{1 - p^2u^2(y)}}, \quad T(p) = \int_0^{Y(p)} \frac{dy}{u(y)\sqrt{1 - p^2u^2(y)}},$$
$$p \in (0, 1),$$

where $Y(p) = \inf\{y, pu(y) \geq 1\}$, and u(y) is a function that satisfies all requirements of Section 4, Chapter I (here, of course, the solution u(y) is unknown – only its existence is assumed).

In accordance with Section 10, Chapter II, this problem is equivalent to the following. Given the function

$$\tau(p)=\int_0^{Y(p)}\sqrt{u^{-2}(y)-p^2}\,dy,\quad p\in(0,\,1),$$

find u(y).

We shall seek the velocity u(y) in the class of piecewise twice continuously differentiable functions that are positive, equal to unity when y = 0, bounded in any finite segment, and unbounded over the entire semiaxis $y \geq 0$.

Beforehand, it was unclear whether or not the solution of our problem was unique: perhaps there exists a function $u^*(y)$ that does not equal u(y) but is such that when $Y^*(p) = \inf\{y,\ pu^*(y) \geq 1\}$

$$\int_0^{Y^*(p)}\sqrt{[u^*(y)]^{-2}-p^2}\,dy\equiv\tau(p),\quad p\in(0,\,1).$$

We shall retain the designation u(y) for the solution whose existence was assumed above. The other solutions (it turns out that they exist) will be denoted by various superscripts: $u^0(y)$, $u^1(y)$, $u^*(y)$, etc.

12. Derivation of Main Formula Linking X(p), Y(p), and u(y)

In this section, we shall derive the following basic formula relating the functions X(p), Y(p), and u(y).

When $q \in (p_{k+1}, p_k)$, $k = 0, 1, \ldots, n$

$$Y(q)=\frac{2}{\pi}\int_q^1\frac{X(p)\,dp}{\sqrt{p^2-q^2}}+\sum_{i=0}^{k}\frac{2}{\pi}\int_{y_i}^{\bar y_i}\operatorname{arctg}\sqrt{\frac{u^{-2}(y)-p_i^2}{p_i^2-q^2}}\,dy.$$

1. Consider the following sets on the plane y, p (Fig. 11). Some of them are functions of the parameter q, which takes values in the interval (0, 1).

Domain D^q consists of those points (y, p) that satisfy the inequalities

$$0<y<Y(q),\quad q<p<u^{-1}(y).$$

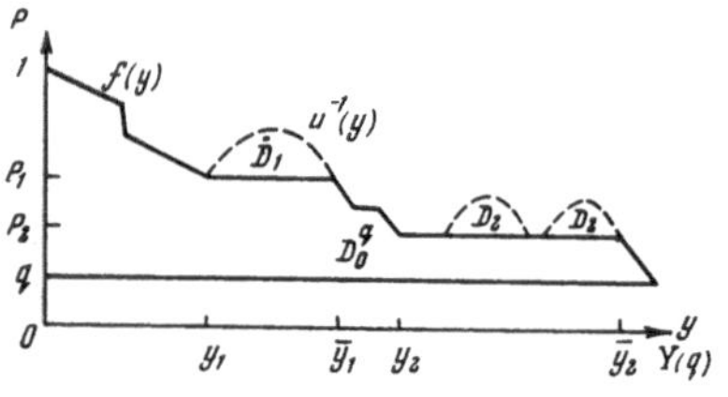

Fig. 11

Domain D_0^q is a subset of D^q that is defined by the inequalities

$$0<y<Y(q),\quad q<p<f(y).$$

Finally, D_k is a set of points (y, p) such that

$$y_k<y<\bar y_k,\quad p_k\leqslant p<u^{-1}(y),\quad k=0,\,1,\,2,\ldots,\,n.$$

For any fixed q = (0, 1), sets $D_0, D_1, \ldots, D_n$ are disjoint.

2. Let us use the relation

$$\int_a^b\frac{p\,dp}{\sqrt{(b^2-p^2)(p^2-a^2)}}=\frac{\pi}{2}.$$

This relation is valid for any a and b for which $0 < a < b$. In particular, for any $q \in (0, 1)$ and for any $y \in (0, Y(q))$

$$\int_q^{u^{-1}(y)}\frac{p\,dp}{\sqrt{(u^{-2}(y)-p^2)(p^2-q^2)}}=\frac{\pi}{2}.$$

Therefore,

$$\frac{\pi}{2}Y(q)=\int_0^{Y(q)}\frac{\pi}{2}\,dy=\int_0^{Y(q)}\left[\int_q^{u^{-1}(y)}\frac{p\,dp}{\sqrt{(u^{-2}(y)-p^2)(p^2-q^2)}}\right]dy.$$

We shall denote the integrand $\frac{2}{\pi}\frac{p}{\sqrt{(u^{-2}(y)-p^2)(p^2-q^2)}}$ by $F_q(p, y)$.

3.

a) The function $F_q(p, y)$ is a positive and measurable function of the variables p and y in domain D^q, and one of the iterated integrals exists. Therefore, $F_q(p, y)$ is integrable in D^q, and

$$Y(q)=\iint_{D^q}F_q(p,\,y)\,dp\,dy.$$

b) According to the Fubini theorem

$$\iint_{D_0^q}F_q(p,\,y)\,dp\,dy=\int_q^1\left[\int_0^{Y(p)}F_q(p,\,y)\,dy\right]dp=$$

$$=\frac{2}{\pi}\int_q^1\left[\int_0^{Y(p)}\frac{p\,dy}{\sqrt{u^{-2}(y)-p^2}}\right]\frac{dp}{\sqrt{p^2-q^2}}.$$

Recalling that

$$\int_0^{Y(p)}\frac{p\,dy}{\sqrt{u^{-2}(y)-p^2}}=X(p),$$

we obtain

$$\iint\limits_{D_0^q} F_q(p, y)\,dp\,dy = \frac{2}{\pi}\int\limits_q^1 \frac{X(p)\,dp}{\sqrt{p^2 - q^2}}.$$

The function on the right side of the latter equation will be denoted by $\Phi(q)$, $q \in (0, 1)$.

c) It is not difficult to verify that when $0 < q < a < b$

$$\int\limits_a^b \frac{p\,dp}{\sqrt{(b^2 - p^2)(p^2 - q^2)}} = -\operatorname{arctg}\sqrt{\frac{b^2 - p^2}{p^2 - q^2}}\Bigg|_a^b =$$

$$= \operatorname{arctg}\sqrt{\frac{b^2 - a^2}{a^2 - q^2}},$$

so that when $q < p_k$

$$\int\limits_{p_k}^{u^{-1}(y)} F_q(p, y)\,dp = \frac{2}{\pi}\operatorname{arctg}\sqrt{\frac{u^{-2}(y) - p_k^2}{p_k^2 - q^2}}.$$

Therefore, for any $k = 0, 1, 2, \ldots, n$ and for any $q < p_k$

$$\iint\limits_{D_k} F_q(p, y)\,dp\,dy = \frac{2}{\pi}\int\limits_{y_k}^{\bar{y}_k} \operatorname{arctg}\sqrt{\frac{u^{-2}(y) - p_k^2}{p_k^2 - q^2}}\,dy.$$

The function on the right side of this equation will be denoted by $A_k(q)$. For $q \in (p_{k+1}, p_k)$, $k = 0, 1, 2, \ldots, n$, let

$$\Psi(q) = \sum_{i=0}^{k} A_i(q).$$

d) Since when $q \in (p_{k+1}, p_k)$

$$D^q = D_0^q \cup \bigcup_{i=1}^{k} D_i,$$

then we finally have

$$Y(q) = \frac{2}{\pi}\int\limits_q^1 \frac{X(p)\,dp}{\sqrt{p^2 - q^2}} + \sum_{i=0}^{k} \frac{2}{\pi}\int\limits_{y_i}^{\bar{y}_i} \operatorname{arctg}\sqrt{\frac{u^{-2}(y) - p_i^2}{p_i^2 - q^2}}\,dy,$$

$$q \in (p_{k+1}, p_k), \quad 0 \leqslant k \leqslant n$$

or (using the symbols introduced above)

$$Y(q) = \Phi(q) + \Psi(q), \quad q \in (0, 1).$$

13. Properties of the Functions $\Phi(q)$ and $\Psi(q)$

We shall point out some properties of the functions $\Phi(q)$ and $\Psi(q)$ introduced above.

1. Each of the functions

$$A_k(q) = \frac{2}{\pi}\int\limits_{y_k}^{\bar{y}_k} \operatorname{arctg}\sqrt{\frac{u^{-2}(y) - p_k^2}{p_k^2 - q^2}}\,dy, \quad 1 \leqslant k \leqslant n$$

is obviously monotone increasing in the interval $0 < q < p_k$. Thus, the function $\Psi(q)$ is monotone increasing in the interval (p_{k+1}, p_k) for any $k = 1, \ldots, n$.

In the interval $(p_1, 1)$, the function $\Psi(q)$ is identically equal to zero.

2. The function $Y(q)$ is nonincreasing in $(0, 1)$. This means that $\Phi(q) = Y(q) - \Psi(q)$ does not increase when $q \in (p_1, 1)$ and is strictly decreasing when $q \in (p_{k+1}, p_k)$, $1 \le k \le n$.

3. In the interval $0 < q < p_k$, the function $A_k(q)$ is analytic. That is, $\Psi(q)$ is analytic when $q \in (p_{k+1}, p_k)$, $0 \le k \le n$.

4. Let us consider $\lim\limits_{q \to p_k - 0} A_k(q)$. It is clear that it equals the measure of the set of points in the interval $j_k = (y_k, \bar{y}_k)$, where $u^{-1}(y) > p_k$

$$A_k(p_k - 0) = \operatorname{mes}\{y \in j_k, \quad u^{-1}(y) > p_k\}.$$

Generally speaking, the interval j_k contains points where $u^{-1}(y) = p_k$. Thus, $A_k(p_k - 0)$ can be strictly less than the length of interval j_k. In any case, $0 < A_k(p_k - 0) \le \bar{y}_k - y_k$.

At point p_k, the function $\Psi(q)$ has a negative discontinuity:

$$\Psi(p_k + 0) - \Psi(p_k - 0) = -A_k(p_k - 0) < 0.$$

Therefore, the graph of $\Psi(q)$ has approximately the same shape as that in Fig. 12.

5. If $A_k(p_k - 0) = \bar{y}_k - y_k$, then the function $\Phi(q)$ is continuous at point p_k. In fact,

$$\Phi(p_k - 0) - \Phi(p_k + 0) = [Y(p_k - 0) -$$

$$- Y(p_k + 0)] - [\Psi(p_k - 0) - \Psi \times$$

$$\times (p_k + 0)] = (\bar{y}_k - y_k) - A_k(p_k - 0) = 0.$$

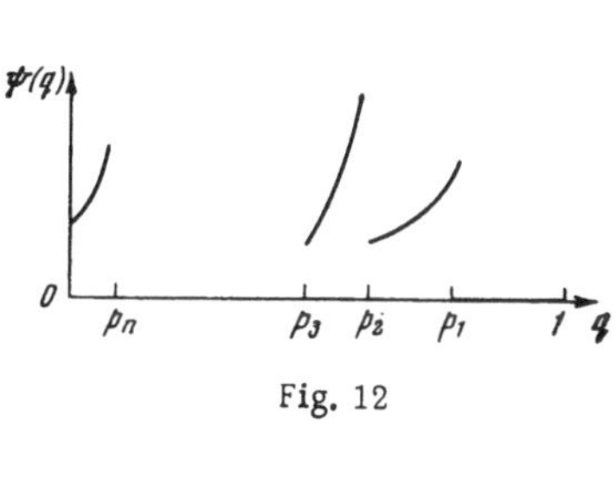

Fig. 12

If $A_k(p_k - 0) < \bar{y}_k - y_k$, then $\Phi(q)$ has at point p_k a negative discontinuity:

$$\Phi(p_k + 0) - \Phi(p_k - 0) = A_k(p_k - 0) - (\bar{y}_k - y_k) < 0.$$

Thus, $\Phi(q)$ does not increase over the entire interval $(0, 1)$ and strictly decreases when $q \in (0, p_1]$. The discontinuities of $\Phi(q)$ and $Y(q)$ coincide at the points of discontinuity of $Y(q)$ that differ from $p_1, \ldots, p_n$.

14. Study of Main Formula

Analysis of the formula $Y(p) = \Phi(p) + \Psi(p)$, $p \in (0, 1)$ leads to the following conclusions.

1. When $p \in (p_1, 1)$, $\Psi(p) \equiv 0$, so that $Y(p) = \Phi(p)$ in the interval $(p_1, 1)$. Thus, if we know the function $X(p)$ when $p \in (p_1, 1)$, we can also determine the function $Y(p)$:

$$Y(q) = \frac{2}{\pi} \int_q^1 \frac{X(p)\,dp}{\sqrt{p^2 - q^2}}, \quad q \in (0, 1).$$

The functions $Y(q)$, $q \in (p_1, 1)$ and $u^{-1}(y)$, $y \in (0, y_1)$ are inverses. Thus, from the function $X(p)$, $p \in (p_1, 1)$, we can also find the function $u(y)$ in the interval $(0, y_1)$. This assertion is approximately equivalent to the above-mentioned result of Herglotz and Wiechert.

2. Let the function $X(p)$ when $p \in (p_{k+1}, 1)$, the intervals $j_1 = (y_1, \bar{y}_1), \ldots, j_k = (y_k, \bar{y}_k)$, and the $u(y)$ values in these intervals be known. Then we can find the function $Y(p)$ when $p \in (p_{k+1}, 1)$ and thereby the function $u(y)$ when $y \in (0, y_{k+1})$.

In particular, if we know $X(p)$ for all $p \in (0, 1)$ and all intervals $j_1, \ldots, j_n$, we can find the function $u(y)$ over the entire semiaxis $y > 0$ if its values in the intervals $j_1, \ldots, j_n$ are known.

The physical meaning of this assertion is as follows. If all waveguides and the propagation velocities in them are known, we can from $X(p)$ find the propagation velocity of a wave at any depth.

Questions arise about how to find the waveguides and how to determine the velocities in them. Before considering this problem, we recall the properties of waveguides that are already known.

15. Waveguide Properties (Résumé of Results Obtained Above)

The number of waveguides n is uniquely determined by the function $\tau(p)$: it is equal to the number of discontinuities of $\tau(p)$.

It is not evident whether or not the locations of the waveguides and the propagation velocities in them could be uniquely determined. Perhaps there exists a solution $u^*(y)$ that does not equal $u(y)$. Let $j_1^*, \ldots, j_n^*$ be waveguides corresponding to $u^*(y)$. It may happen than an interval j_k^* coincides with interval j_k but $u^*(y) \neq u(y)$ when $y \in j_k$. And it may happen that intervals j_k^* and intervals j_k do not coincide.

The following conditions are necessarily satisfied, however. Let $j_k^* = (y_k^*, \bar{y}_k^*)$, $1 \leq k \leq n$, $y_0^* = \bar{y}_0^* = 0$, and let $Y^*(p) = \inf\{y, pu^*(y) \geq 1\}$. Let

$$\sigma_k^* = \int_{y_k^*}^{\bar{y}_k^*} \sqrt{[u^*(y)]^{-2} - p_k^2}\,dy,$$

$$A_k^*(q) = \frac{2}{\pi} \int_{y_k^*}^{\bar{y}_k^*} \operatorname{arctg} \sqrt{\frac{[u^*(y)]^2 - p_k^2}{p_k^2 - q^2}}\,dy, \quad q < p_k,$$

$\Psi^*(q) = \sum_{i=0}^{k} A_i^*(q)$ when $q \in (p_{k+1}, p_k)$, $0 \leq k \leq n$. Then 1) $y_k^* = Y^*(p_k + 0)$, $\bar{y}_k^* = Y^*(p_k - 0)$; 2) $\sigma_k^* = \sigma_k$, $k = 1, \ldots, n$; and 3) $Y^*(p) = \Phi(p) + \Psi^*(p)$, $p \in (0, 1)$. Here, the numbers $p_1, \ldots, p_n$ and $\sigma_1, \ldots, \sigma_n$ are uniquely determined from $\tau(p)$: when $p \neq p_k$, $1 \leq k \leq n$, $\tau(p)$ is continuous, $\tau(p_k - 0) - \tau(p_k + 0) = \sigma_k$.

Note 1. Conditions 1), 2), and 3) were proved above for $u(y)$, i.e., it was shown that

$$y_k = Y(p_k + 0), \quad \bar{y}_k = Y(p_k - 0),$$

$$\int_{y_k}^{\bar{y}_k} \sqrt{u^{-2}(y) - p_k^2}\,dy = \sigma_k, \quad k = 1, \ldots, n,$$

and $Y(p) = \Phi(p) + \Psi(p)$, $p \in (0, 1)$. Only the fact that $u(y)$ is a solution, however, was used in the proof. Thus these conditions are valid for any other solution $u^*(y)$.

Note 2. We emphasize that condition 3) contains, among other things, the following assertion: if $u^*(y)$ is a solution, then $\Phi(p) + \Psi^*(p)$ does not increase in the interval $(0, 1)$.

16. Necessary and Sufficient Conditions for $u^*(y)$ to be a Solution

It turns out that the conditions formulated in the preceding section that are necessary for the function to be a solution are also sufficient. More precisely, we have the following theorem.

Theorem. Let it be possible to select intervals $j_k^* = (y_k^*, \bar{y}_k^*)$, $1 \leq k \leq n$ and assign $u^*(y)$ in them such that a) the function $Y^*(q)$, which, when

$q \in (p_{k+1}, p_k)$, is equal to the sum

$$\Phi(q) + \sum_{i=0}^{k} \frac{2}{\pi} \int_{y_i^*}^{\bar{y}_i^*} \operatorname{arctg} \sqrt{\frac{[u^*(y)]^{-2} - p_i^2}{p_i^2 - q^2}}\, dy,$$

$$y_0^* = \bar{y}_0^* = 0, \quad 0 \leqslant k \leqslant n,$$

does not increase in the interval $0 < q < 1$;

b) $Y^*(p_k + 0) = y_k^*, \quad Y^*(p_k - 0) = \bar{y}_k^*$

c) $\int_{y_k^*}^{\bar{y}_k^*} \sqrt{[u^*(y)]^{-2} - p_k^2}\, dy = \sigma_k$

for any $k = 1, \ldots, n$.

From $Y^*(q)$ we define the function $u^*(y)$, nondecreasing outside of $\bigcup_{k=1}^{n} j_k^*$ such that

$$Y^*(p) = \inf\{y,\ pu^*(y) \geqslant 1\}.$$

Thus, $u^*(y)$ is defined over the entire semiaxis $y \geq 0$.

Then $u^*(y)$ is a solution.

Note. In the formulation of this theorem, it is understood that all the integrals involved exist. In order that this be clearly satisfied, we shall assume that $0 < u^*(y) \leq p_k^{-1}$ when $y \in j_k^*$ and that $u^*(y)$ is twice continuously differentiable in intervals j_k^* everywhere, with the exception of a finite number of points. This excessively strong limitation ensures the piecewise double differentiability of $u^*(y)$ over the entire semiaxis. Moreover, since by definition $u^*(y)$ does not decrease outside of the intervals j_k^*, then all the results obtained above for $u(y)$ can be applied to $u^*(y)$.

Proof

1. By virtue of the relation $Y^*(p) = \inf\{y, pu^*(y) \geq 1\}$, the function $Y^*(p)$ is equal to the ordinate of the deepest point of a ray with parameter p that is propagated with velocity $u^*(y)$. The abscissa of this point will be denoted by $X^*(p)$. The integral $\int_0^{Y^*(p)} \sqrt{[u^*(y)]^{-2} - p^2}\, dy$ will be denoted by $\tau^*(p)$. We must establish that $\tau^*(p) \equiv \tau(p)$, $p \in (0, 1)$.

2. Let

$$\Phi^*(q) = \frac{2}{\pi} \int_q^1 \frac{X^*(p)\, dp}{\sqrt{p^2 - q^2}}, \quad q \in (0, 1).$$

According to the main formula of Section 12 (see note on $u^*(y)$), when $q \in (p_{k+1}, p_k)$

$$Y^*(q) = \Phi^*(q) + \sum_{i=1}^{k} \int_{y_i^*}^{\bar{y}_i^*} \operatorname{arctg} \sqrt{\frac{[u^*(y)]^{-2} - p_i^2}{p_i^2 - q^2}}\, dy.$$

Therefore, $\Phi^*(q) \equiv \Phi(q)$ when $q \in (0, 1)$.

3. We recall that $\tau'(p) = -X(p)$. If r is an arbitrary point in $r \in (0, 1)$, then since $\tau(p)$ is a monotonic and bounded function in the interval $(r, 1)$, the integral $\int_r^1 X(p)\, dp$ exists.

We shall use the equation

$$\frac{\pi}{2} \int_r^1 X(p)\, dp = \int_r^1 X(p) \left[\int_r^p \frac{q\, dp}{\sqrt{(p^2 - q^2)(q^2 - r^2)}} \right] dp.$$

In the triangle $\Delta_r \{r < p < q < 1\}$ (see Fig. 13) in the plane p, q, the function $qX(p)/\sqrt{(p^2 - q^2)(q^2 - r^2)}$ is measurable and positive, and one of the iterated integrals exists. Therefore, this function is integrable in Δ_r. By changing the order of integration we obtain

$$\frac{\pi}{2} \int_r^1 X(p)\, dp = \int_r^1 \frac{q \int_q^1 \frac{X(p)\, dp}{\sqrt{p^2 - q^2}}}{\sqrt{q^2 - r^2}}\, dq = \frac{\pi}{2} \int_r^1 \frac{q\Phi(q)\, dq}{\sqrt{q^2 - r^2}}.$$

4. In quite the same way, it is proved that

$$\int_r^1 X^*(p)\, dp = \int_r^1 \frac{q\Phi^*(q)\, dq}{\sqrt{q^2 - r^2}}.$$

Thus, for any $r \in (0, 1)$,

$$\int_r^1 X^*(p)\, dp = \int_r^1 X(p)\, dp.$$

5. According to the theorem, $\tau^*(p)$ is continuous when $p \neq p_k$, $1 \leq k \leq n$; and at points $p_1, \ldots, p_n$ it has discontinuities $-\sigma_1, \ldots, -\sigma_n$.

Thus both the continuous components and the discontinuity functions of the monotonic functions $\tau^*(p)$ and $\tau(p)$ coincide. Therefore, to prove that $\tau^*(p) \equiv \tau(p)$ it is sufficient to be sure of the absolute continuity of the difference $\tau^*(p) - \tau(p)$.

6. Let us transform the expression $\tau(p) = \int_0^{Y(p)} \sqrt{u^{-2}(y) - p^2}\, dy$. We recall that $u^{-1}(y) = f(y)$ if $y \overline{\in} \bigcup_{k=1}^{n} j_k$ and that $f(y) = p_k$ when $y \in j_k$. When $p \in$

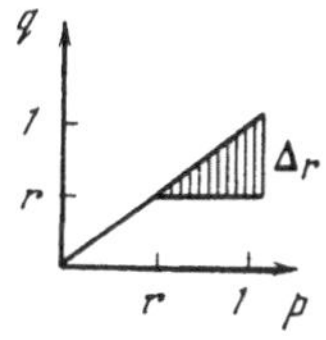

Fig. 13

(p_{k+1}, p_k), $0 \le k \le n$

$$\tau(p) = \sum_{i=1}^{k}\left[\int_{y_i}^{\bar{y}_i} \sqrt{u^{-2}(y) - p^2}\,dy + \int_{\bar{y}_{i-1}}^{y_i} \sqrt{f^2(y) - p^2}\,dy\right] +$$

$$+ \int_{y_k}^{Y(p)} \sqrt{f^2(y) - p^2}\,dy = \int_0^{Y(p)} \sqrt{f^2(y) - p^2}\,dy +$$

$$+ \sum_{i=1}^{k}\left[\int_{y_i}^{\bar{y}_i} \sqrt{u^{-2}(y) - p^2}\,dy - \int_{y_i}^{\bar{y}_i} \sqrt{p_i^2 - p^2}\,dy\right].$$

We denote by $\tilde{\tau}(p)$ the integral $\int_0^{Y(p)} \sqrt{f^2(y) - p^2}\,dy$. It is clear that the function $\tau(p) - \tilde{\tau}(p)$ is continuously differentiable for any $p \in (p_{k+1}, p_k)$.

Let

$$f^*(y) = [\sup_{y^0 \le y} u^*(y^0)]^{-1}, \quad \tilde{\tau}^*(p) = \int_0^{Y^*(p)} \sqrt{[f^*(y)] - p^2}\,dy.$$

In order to verify that the difference $\tau^*(p) - \tau(p)$ is not a singular function, it is sufficient to see that $\tilde{\tau}^*(p) - \tau(p)$ is absolutely continuous.

7. Let us write $\tilde{\tau}(p)$ as a Stieltjes integral:

$$\tilde{\tau}(p) = \int_0^{Y(p)} \sqrt{f^2(y) - p^2}\,dy = \int_1^p \sqrt{z^2 - p^2}\,dY(z).$$

In exactly the same manner, $\tilde{\tau}^*(p) = \int_1^p \sqrt{z^2 - p^2}\,dY^*(z)$.

Thus,

$$\tilde{\tau}^*(p) - \tilde{\tau}(p) = \int_1^p \sqrt{z^2 - p^2}\,d\,[Y^*(z) - Y(z)].$$

Since the difference $Y^*(z) - Y(z)$ is obviously continuously differentiable for any $z \ne p_k$, $k = 1, 2, \ldots, n$, the function $\tilde{\tau}^*(p) - \tilde{\tau}(p)$ is absolutely continuous.

Thus, $\tau^*(p) \equiv \tau(p)$: Q.E.D.

17. Construction of a Solution $u^*(y)$ not Equal to $u(y)$

Using the theorem in the preceding section, we can show that the velocity is determined nonuniquely from the given functions $X(p)$ and $T(p)$ or, what is the same, from the given function $\tau(p)$.

Now we construct a velocity function $u^*(y)$ that does not equal $u(y)$ and yet is a solution.

1. The solution will have the following form (see Fig. 14).

Any interval j_k^* will be divided into two parts:

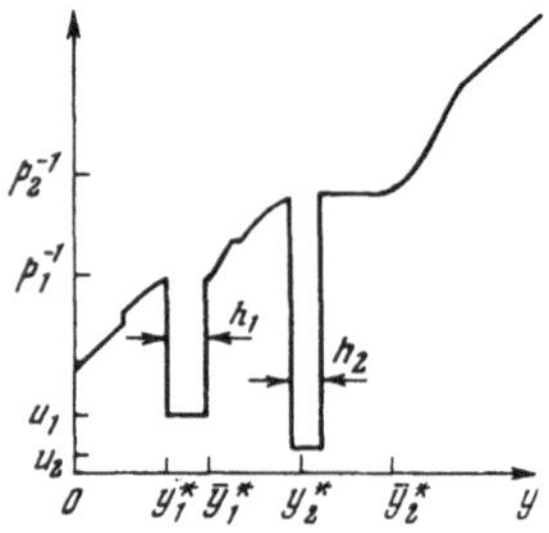

Fig. 14

numbers $h_1, \ldots, h_n$ are constructed such that

$$y_k^* < y_k^* + h_k \leqslant \bar{y}_k^*, \quad k = 1, 2, \ldots, n.$$

The function $u^*(y)$ will be a step function in each interval j_k^*:

$$u^*(y) = \begin{cases} u_k = \text{const} > 0, & y_k^* < y < y_k^* + h_k \\ p_k^{-1}, & y_k^* + h_k < y < \bar{y}_k^*. \end{cases}$$

2. The function $A_k^*(q)$, which in general is defined by the formula

$$A_k^*(q) = \frac{2}{\pi}\int_{y_k}^{\bar{y}_k^*} \operatorname{arctg} \sqrt{\frac{[u^*(y)]^{-2} - p_k^2}{p_k^2 - q^2}}\,dy, \quad q < p_k,$$

for such a special solution takes the form

$$A_k^*(q) = \frac{2}{\pi} h_k \operatorname{arctg} \sqrt{\frac{u_k^{-2} - p_k^2}{p_k^2 - q^2}}.$$

It will be a function of h_k and not of the location of the interval j_k^* (of y_k^* and $\bar{y}_k^*$). The following function will have a similar property:

$$\Psi^*(q) = \sum_{i=1}^{k} A_i^*(q), \quad q \in (p_{k+1}, p_k).$$

3. The numbers h_k and constants u_k are related as

$$h_k \sqrt{[u_k^{-2} - p_k^2} = \sigma_k, \quad 1 \leqslant k \leqslant h.$$

Thus, condition c) of the theorem of Section 16

$$\int_{y_k^*}^{\bar{y}_k^*} \sqrt{u^*(y)]^{-2} - p_k^2}\,dy = \sigma_k, \quad 1 \leqslant k \leqslant n,$$

will be satisfied for any choice of the numbers y_k^*, $\bar{y}_k^*$.

4. Let us try to satisfy condition a) of the theorem of Section 16: that the function $Y^*(q) = \Phi(q) + \Psi^*(q)$ does not increase in the interval (0, 1). We shall use the monotonicity of $Y(q)$ and the fact that $Y^*(q) = Y(q) + [\Psi^*(q) - \Psi(q)]$.

Let us differentiate $A_k(q)$ and $A_k^*(q)$:

$$A_k'(q) = \frac{2}{\pi}\int_{y_k}^{\bar{y}_k} \frac{\sqrt{u^{-2}(y) - p_k^2}}{u^{-2}(y) - q^2}\frac{q\,dy}{\sqrt{p_k^2 - q^2}},$$

$$[A_k^*(q)]' = \frac{2}{\pi}h_k\frac{\sqrt{u_k^{-2} - p_k^2}}{u_k^{-2} - q^2}\cdot\frac{q}{\sqrt{p_k^2 - q^2}} =$$

$$= \frac{2}{\pi}\frac{\sigma_k}{u_k^{-2} - q^2}\frac{q}{\sqrt{p_k^2 - q^2}},$$

$$0 < q < p_k, \quad k = 1, 2, \ldots, n.$$

We make the constant $u_k > 0$ so small that

$$\frac{\sigma_k}{u_k^{-2} - p_k^2} < \int_{y_k}^{\bar{y}_k} \frac{\sqrt{u^{-2}(y) - p_k^2}}{u^{-2}(y)}dy.$$

Then, for any $q \in (0, p_k)$

$$\frac{\sigma_k}{u_k^{-2} - q^2} < \frac{\sigma_k}{u_k^{-2} - p_k^2} <$$

$$< \int_{y_k}^{\bar{y}_k} \frac{\sqrt{u^{-2}(y) - p_k^2}}{u^{-2}(y)}dy < \int_{y_k}^{\bar{y}_k} \frac{\sqrt{u^{-2}(y) - p_k^2}}{u^{-2}(y) - q^2}dy.$$

Let the constants u_k be chosen as indicated for any $k = 1, 2, \ldots, n$ and let $h_k = \sigma_k/\sqrt{u_k^{-2} - p_k^2}$. Then $[A_k^*(q) - A_k(q)]' < 0$ for any $q \in (0, p_k)$, $1 \le k \le n$, so that $\Psi^*(q) - \Psi(q)$ does not increase in any of the intervals (p_{k+1}, p_k), $0 \le k \le n$.

To verify that the function $Y^*(q) = Y(q) + [\Psi^*(q) - \Psi(q)]$ does not increase over the entire interval (0, 1), we must show the validity of the inequalities $Y^*(p_k - 0) - Y^*(p_k + 0) \ge 0$, $1 \le k \le n$. They are obviously satisfied. In fact, on the strength of Section 13, for any $k = 1, \ldots, n$

$$\Phi(p_k - 0) - \Phi(p_k + 0) \geqslant 0,$$

$$\Psi^*(p_k - 0) - \Psi^*(p_k + 0) = A^*(p_k - 0) =$$

$$= \lim_{q \to p_k - 0}\frac{2}{\pi}h_k \operatorname{arctg}\sqrt{\frac{u_k^{-2} - p_k^2}{p_k^2 - q^2}} = h_k.$$

Thus, the even stronger inequalities

$$Y^*(p_k - 0) - Y^*(p_k + 0) \geqslant h_k, \quad 1 \leqslant k \leqslant n$$

are satisfied.

That is, condition a) of the theorem of Section 16 is satisfied.

5. It remains to select the intervals j_k^* such that condition b) of the theorem of Section 16 is satisfied. We shall use this very condition as a definition of the numbers y_k^* and $\bar{y}_k^*$: the function $Y^*(q)$ is already constructed. Let

$$y_k^* = Y(p_k + 0), \quad \bar{y}_k^* = Y(p_k - 0).$$

6. As we planned, let

$$u^*(y) = \begin{cases} u_k, & y_k^* < y < y_k^* + h_k, \\ p_k^{-1}, & y_k^* + h_k < y < \bar{y}_k^*. \end{cases}$$

Outside the intervals $j_1^*, \ldots, j_n^*$ we define $u^*(y)$ such that it does not decrease outside $\bigcup_{k=1}^{n} j_k^*$ and satisfies the relation

$$Y^*(p) = \inf\{y, \; pu^*(y) \geqslant 1\}.$$

All three conditions a), b), and c) of the theorem of Section 16 are satisfied. That is, $u^*(y)$ is a solution. Actually, not one solution but a whole set was constructed here: if all constants $(u_1, \ldots, u_n)$ are reduced, the function $u^*(y)$ remains a solution, and the different solutions correspond to different sets $(u_1, \ldots, u_n)$.

Therefore, the solution $u(y)$ is not unique.

In the following sections, we shall continue our study of the set of solutions. But now we shall make one remark (which will soon be necessary).

Remark. For any $\varepsilon > 0$ when $u_1, \ldots, u_n$ are sufficiently small

$$0 < Y^*(q) - \Phi(q) < \varepsilon, \quad q \in (0, 1).$$

To verify the latter assertion, it is sufficient to note that

$$|A_k^*(q)| \leqslant h_k = \frac{\sigma_k}{\sqrt{u_k^{-2} - p_k^2}}.$$

18. Continuous Transformation of One Solution to Another

We have shown that velocity is not uniquely determined from the given functions $X(p)$ and $T(p)$. The question of whether or not a specific function is a solution can be answered with the aid of the theorem of Section 16.

Now we have the problem of characterizing the entire set of solutions. We will be able to describe compactly the set G in the plane y, u, which contains the graphs of all solutions.

The following fact will be used in studying the set G.

Let $u^0(y)$ and $u^1(y)$ be any two solutions. Then one of them can be continuously transformed to the other.

More accurately: we construct a cone-parameter set of solutions $\{u^\omega(y)\}$ that is a continuous function of the parameter $\omega \in [0, 1]$ and such that when $\omega = 0$ and when $\omega = 1$ the initial functions $u^0(y)$ and $u^1(y)$, respectively, are obtained.

1. Let $Y^i(p) = \inf\{y, pu^i(y) \geq 1\}$, i = 0, 1. We arbitrarily fix $\omega \in [0, 1]$. Let

$$Y^\omega(p) = (1-\omega) Y^0(\omega) + \omega Y^1(p).$$

The function $Y^\omega(p)$ is obviously a monotonic function of p in the interval $0 < p < 1$.

Let $y_k^\omega = Y^\omega(p_k - 0)$, $\bar{y}_k^\omega = Y^\omega(p_k - 0)$.

We shall construct a function $u^\omega(y)$, nondecreasing outside the intervals $j_k^\omega = (y_k^\omega, \bar{y}_k^\omega)$, such that $Y(p) = \inf\{y, pu^\omega(y) \geq 1\}$.

2. We shall show that in the intervals j_k^ω the definition of the function $u^\omega(y)$ can be extended so that it is a solution.

We shall fix any one of the intervals j_k^ω, k = 1, 2, ..., n; the construction is similar in the remaining intervals.

We shall also consider the intervals $j_k^i = (y_k^i, \bar{y}_k^i) = (Y^i(p_k + 0), Y(p_k - 0)$, i = 0, 1 (see Fig. 15a and 15b). We divide j_k^ω into two parts by a point s_k^ω such that the length of the interval $j^0 = (y_k^\omega, s_k^\omega)$ is $(1-\omega)(\bar{y}_k^0 - y_k^0)$ and the length of the interval $j^1 = (s_k^\omega, \bar{y}_k^\omega)$ is $\omega(\bar{y}_k^1 - y_k^1)$.

Let (see Fig. 15c)

$$u^\omega(y) = \begin{cases} u^0\left[\dfrac{y - y_k^\omega}{1-\omega} + y_k^0\right], & y \in j^0 = (y_k^\omega, s_k^\omega) \\ u^1\left[\dfrac{y - s_k^\omega}{\omega} + y_k^1\right], & y \in j^1 = (s_k^\omega, \bar{y}_k^\omega). \end{cases}$$

3. According to Section 16, the function $u^\omega(y)$ is a solution. In fact, let

$$A_k^i(q) = \frac{2}{\pi}\int_{y_k^i}^{\bar{y}_k^i} \operatorname{arctg}\sqrt{\frac{[u^i(y)]^{-2} - p_k^2}{p_k^2 - q^2}}\,dy,$$

$$\sigma_k^i = \int_{y_k^i}^{\bar{y}_k^i} \sqrt{[u^i(y)]^{-2} - p_k^2}\,dy,$$

$$i = 0, 1, \omega.$$

Because of the linearity of the transformations used in defining $u^\omega(y)$ in the interval j_k^ω, we have

$$\sigma_k^\omega = (1-\omega)\sigma_k^0 + \omega\sigma_k^1, \quad A_k^\omega(q) = (1-\omega) A_k^0(q) + \omega A_k^1(q).$$

Since $u^0(y)$ and $u^1(y)$ are solutions, then $\sigma_k^0 = \sigma_k^1 = \sigma_k$ and

$$Y^i(q) = \Phi(q) + \sum_{k=1}^{m} A_k^i(q), \quad i = 0, 1,$$

$$q \in (p_{m+1}, p_m), \quad m = 1, 2, \ldots, n.$$

That is, $\sigma_k^\omega = \sigma_k$, and

$$Y^\omega(q) = \Phi(q) + \sum_{k=1}^{m} A_k^\omega(q), \quad q \in (p_{m+1}, p_m), \ 1 \leqslant m \leqslant n.$$

Thus, all conditions of the theorem of Section 16 are satisfied, and $u^\omega(y)$ is, in fact, a solution.

4. When the parameter ω passes through the interval [0, 1], the graphs of the solutions $u^\omega(y)$ move continuously from position $u = u^0(y)$ to position $u = u^1(y)$.

19. The Functions V(p) and N(p)

Let $\{u_\nu(y)\}$ be the set of all solutions; $Y_\nu(p) = \inf\{y, pu_\nu(y) \geq 1\}$. We arbitrarily fix $p \in (0, 1)$. We shall let $V(p) = \sup Y_\nu(p)$, $N(p) = \inf Y_\nu(p)$.

According to the remark at the end of Section 17, $N(p) \equiv \Phi(p)$, $p \in (0, 1)$. Thus the properties of N(p) are already known (see Section 13). Let us investigate the properties of V(p).

1. In the interval $(p_1, 1)$, all functions $Y_\nu(p)$ coincide with one another (see Section 14), so that $V(p) \equiv N(p)$ when $p \in (p_1, 1)$.

2. The inequality $Y_\nu(p) \leq T(p)/p$ obviously follows from the physical meaning of the function $Y_\nu(p)$: $Y_\nu(p)$ is the ordinate of the deepest point of a ray with parameter p that is propagated with ve-

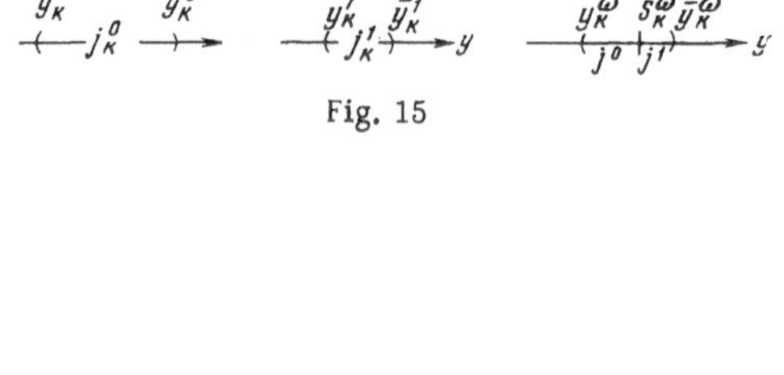

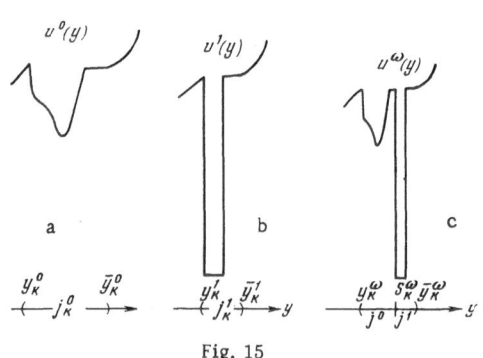

Fig. 15

locity $u_\nu(y)$. The travel time along the descending branch of this ray is T(p) (since $u_\nu(y)$ is a solution). Velocity $u_\nu(y)$ does not exceed 1/p for $y \in (0, y_\nu, (p))$. That is, $Y_\nu(p) \le T(p)/p$ (the path length does not exceed the product of the upper bound of velocity and travel time).

Since $Y_\nu(p_k - 0) \le Y_\nu(p)$ for any $p \in (0, p_k)$, then $V(p_k - 0) \le \inf\{T(p)/p, p < p_k\}$. Thus, $V(p_k - 0)$ is finite for any $k = 1, \ldots, n$.

3. Let $H_k = V(p_k - 0) - N(p_k - 0)$, $1 \le k \le n$. We shall verify that the inequality

$$V(p) \leqslant N(p) + H_k.$$

is satisfied when $p \in (p_{k+1}, p_k)$, $1 \le k \le n$. It is a simple consequence of Section 13. In fact, let

$$\Psi_\nu(p) = Y_\nu(p) - \Phi(p) = Y_\nu(p) - N(p).$$

It is clear that

$$\Psi_\nu(p_k - 0) = Y_\nu(p_k - 0) - N(p_k - 0) \leqslant H_k.$$

Moreover, in the interval (p_{k+1}, p_k) the function $\Psi_\nu(p)$ is less than $\Psi_\nu(p_k - 0)$, i.e., when $p \in (p_{k+1}, p_k)$, $1 \le k \le n$,

$$Y_\nu(p) = N(p) + \Psi_\nu(p) < N(p) + \\ + \Psi_\nu(p_k - 0) \leqslant N(p) + H_k,$$

from which the required inequality follows.

4. For any solution $u_\nu(y)$, the function $Y_\nu(p)$ is included between two nondecreasing functions with a bounded difference:

$$N(p) \leqslant Y_\nu(p) \leqslant V(p), \quad V(p) - N(p) \leqslant \text{const}, \\ p \in (0, 1)$$

According to the construction in the preceding paragraph, for any $p \in (0, 1)$ and any $y \in (N(p), V(p))$, there is a solution $u_\nu(y)$ such that $Y_\nu(p) = y$.

20. The Set G Populated by the Graphs of the Solutions

We relate the point (y, u) to the set G (see Fig. 16) if $1 \le u \le p_1^{-1}$, $Y(u^{-1} + 0) < y < Y(u^{-1} - 0)$, or if $u > p_1^{-1}$, $N(u^{-1}) < y < V(u^{-1})$, or if $0 < u \le p_k^{-1}$, $N(p_k + 0) < y < V(p_k - 0)$.

It is clear that the graph of any solution $u_\nu(y)$ is located in the closure of set G: for any $u_\nu(y)$ and for any $y \ge 0$, the point $(y, u_\nu(y)) \in \overline{G}$.

We shall show that the graph of some solution passes through each point $(y, u) \in G$. This property

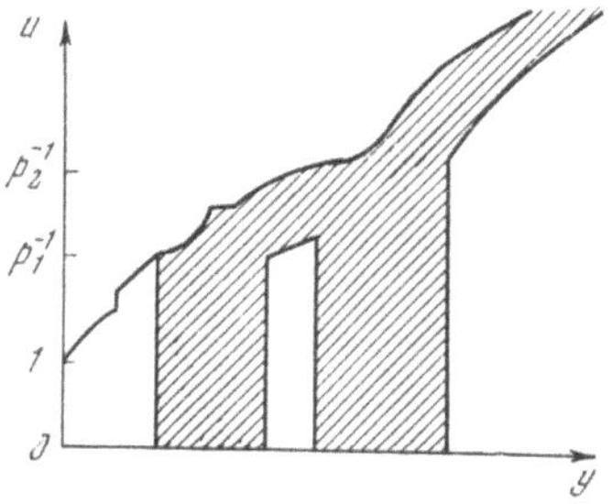

Fig. 16

remains unverified only for points of G that are defined by the inequalities $0 < u \le p_k^{-1}$, $N(p_k + 0) < y < V(p_k - 0)$.

Let (y*, u*) be one such point (see Fig. 17).

By determination of the lower bound a solution $u_1(y)$ is found such that $N(p_k + 0) \le Y_1(p_k + 0) < y^*$ (the function equal to $\inf\{y, pu_1(y) \ge 1\}$ is denoted by $Y_1(p)$).

We shall construct the solution $u^1(y)$ as follows. When $y \in (0, Y_1(p_k + 0))$, the function $u^1(y) = u_1(y)$. The k-th waveguide of velocity $u^1(y)$ will be denoted by j_k^1. In the interval $j_k^1 = (y_k^1, \bar{y}_k^1)$, $u^1(y)$ will be a step function

$$u^1(y) = \begin{cases} u_k, & y_k^1 < y < y_k^1 + h_k \\ p_k^{-1}, & y_k^1 + h_k < y < \bar{y}_k^1. \end{cases}$$

The positive constants u_k and h_k are made small enough to satisfy the inequalities $u_k < u^*$, $y_k^1 + h_k < y^*$ (see Fig. 17). The numbers u_k and h_k must be interrelated: $h_k \sqrt{u_k^{-2} - p_k^2} = \sigma_k$

According to the results of Section 17, we can select the number $\bar{y}_k^1$ and extend the definition of $u^1(y)$ when $y > \bar{y}_k^1$ such that it is a solution (for this, it is necessary only that the constant u_k be sufficiently small).

By determination of the lower bound a solution $u^0(y)$ is found such that $V(p_k - 0) \ge Y^0(p_k - 0) > y^*$ (the function equal to $\inf\{y, pu^0(y) \ge 1\}$ is denoted by $Y^0(p)$).

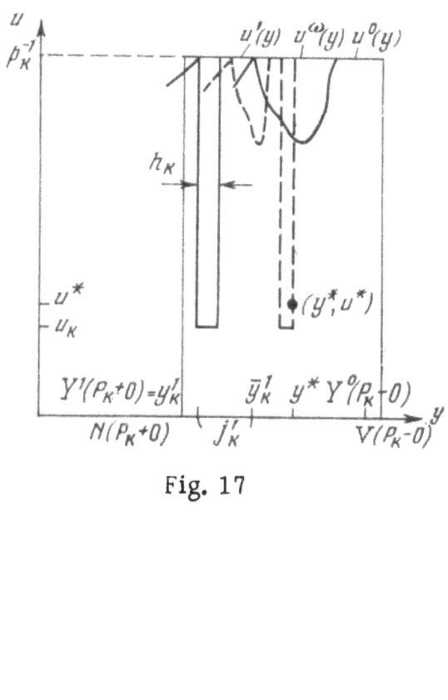

Fig. 17

Let us apply the construction considered in Section 18 to solutions $u^0(y)$ and $u^1(y)$. With continuous movement from $u^0(y)$ to $u^1(y)$, the graph of some solution $u^\omega(y)$, $\omega \in (0, 1)$, will obviously pass through the point (y^*, u^*) (Fig. 17).

Let us investigate the physical meaning of the set G.

If the solution u(y) were uniquely defined, we would, from the given functions X(p) and T(p), find exactly for any $y \geq 0$ the value $u = u(y)$ that the velocity takes at the depth y.

In the presence of waveguides, the velocity is not uniquely determined. We can find exactly the depth y_1 at which the first waveguide begins, and we can determine exactly the velocity u(y) for $y \in (0, y_1)$. When $y > y_1$, it is impossible to determine the velocity at depth y from the functions X(p) and T(p).

The following is valid, however:

Let u be any number greater than 1, and let $p = u^{-1}$. Then: 1) the velocity must take the value u in the interval $N(p) \leq y \leq V(p)$ and 2) when $y < N(p)$, the velocity is strictly less than u. In other words, we can indicate approximately the depth at which a given velocity is reached for the first time. The precise meaning of the words "velocity u(y) takes the value u at the point y^0" is

$$\lim_{y \to y^0 - 0} u(y) \leqslant u \leqslant \lim_{y \to y^0 - 0} u(y)$$

and it would be formally more correct to say that the velocity must take a value greater than or equal to u in the interval [N(p), V(p)].† The number $N(p) = \Phi(p)$ is exactly defined by X(p), and the upper bound was indicated for V(p).

So far we have been talking about the regions between waveguides. The locations of the waveguides j_k (the numbers $y_2, \ldots, y_n, \bar{y}_1, \ldots, \bar{y}_n$) are determined from the function $\tau(p)$ also approximately. But the integral velocity characteristic σ_k, the critical value p_k^{-1} at which the velocity ceases to increase monotonically (for waveguide j_k, $1 \leq k \leq n$), and the number of waveguides n are known. It is impossible to determine the minimum velocity in a waveguide from $\tau(p)$.

21. Relative Positions of Graphs of Two Solutions

The graphs of the solutions populate the set G, and the graphs of many solutions pass through each point. Therefore, indication of the exact velocity at some point y does not allow us to distinguish a unique solution among the set $\{u_\nu(y)\}$.

In this section we shall study the question: What is the set of points for which we must know the values of velocity in order to be able to find a unique solution? The problems of partial coincidence and, in general, of the relative positions of the graphs of two solutions are closely connected with this question.

Let us consider any two solutions: $u^1(y)$ and $u^2(y)$. Let $j_1^i = (y_1^i, \bar{y}_1^i), \ldots, j_n^i = (y_n^i, \bar{y}_n^i)$ be waveguides that correspond to velocity $u^i(y)$; $Y^i(p) = \inf\{y, pu^i(y) \geq 1\}$, $i = 1, 2$.

1. It was noted in the discussion of the main formula that coincidence of waveguides ($j_k^1 = j_k^2$ when $1 \leq k \leq n$) and equality of the solutions in waveguides ($u^1(y) = u^2(y)$ when $y \in j_k^i$, $1 \leq k \leq n$) involves coincidence of the solutions $u^1(y)$ and $u^2(y)$ in the areas between waveguides: $u^1(y) \equiv u^2(y)$.

It is clear that coincidence of the waveguides and equimeasurability of the functions $u^1(y)$ and $u^2(y)$ when $y \in j_k^i$, $1 \leq k \leq n$, are sufficient for coincidence of the solutions between waveguides (the functions $u^1(y)$ and $u^2(y)$ are called equimeasurable in the interval j_k^i if for any r the measure of subset j_k^i, where $u^1(y) \leq r$, is equal to the measure of subset j_k^i, where $u^2(y) \leq r$). To verify the latter property, it is sufficient to note that when

$$A_k^i(q) = \int_{y_k^i}^{\bar{y}_k^i} \operatorname{arctg} \sqrt{\frac{[u^i(y)]^2 - p_k^2}{p_k^2 - q^2}}\, dy, \quad 1 \leqslant k \leqslant n, \quad i = 1, 2$$

the functions $A_k^1(q)$ and $A_k^2(q)$ coincide if $u^1(y)$ and $u^2(y)$ are equimeasurable in the interval $j_k^1 = j_k^2$. Next, we must use the main formula.

2. A more general assertion is also valid. If $\bar{y}_k^1 = \bar{y}_k^2$, and the solutions $u^1(y)$ and $u^2(y)$ are equimeasurable in the interval $(0, \bar{y}_k^i)$, then they coincide between the k-th and (k+1)-th waveguides:

$$y_{k+1}^1 = y_{k+1}^2, \quad u^1(y) = u^2(y) \text{ when } y \in (\bar{y}_k^i, y_{k+1}^i).$$

For a proof, it is sufficient to ensure that $Y^1(q) = Y^2(q)$ when $q \in (p_{k+1}, p_k)$; this follows from the formula

$$Y^i(q) = \frac{2}{\pi} \int_q^{p_k} \frac{X(p)\,dp}{\sqrt{p^2 - q^2}} + \frac{2}{\pi} \int_0^{\bar{y}_k^i} \operatorname{arctg} \sqrt{\frac{[u^i(y)]^{-2} - p_k^2}{p_k^2 - q^2}}\, dy,$$

$$q \in (p_{k+1}, p_k), \quad 1 \leqslant k \leqslant n, \quad i = 1, 2,$$

the proof of which follows almost verbatim the derivation of the main formula of Section 12.

† This also applies to the assertion: "the graph of some solution passes through each point $(y, u) \in G$" (the graphs include intervals that join the limiting velocity values at discontinuity points).

The demonstrated assertion can be formulated as follows: Let

$$F_k^i(r) = \text{mes}\{y,\ y \leqslant y_k^i,\ u^i(y) \leqslant r\},\ i = 1, 2,\ 1 \leqslant k \leqslant n+1$$

($F_k^i(r)$ is the measure of a set situated on the y axis from 0 to the k-th waveguide of velocity $u^i(y)$ and consisting of three points, where $u^i(y)$ is not greater than r). Then, if $F_{k+1}^1(r) = F_{k+1}^2(r)$ when $r \leq p_k^{-1}$, then $Y^1(p) = Y^2(p)$ when $p \in (p_{k+1}, p_k)$.

3. The converse is also true: if $Y^1(p) = Y^2(p)$ when $p \in (p_{k+1}, p_k)$, then $F_{k+1}^1(r) = F_{k+1}^2(r)$ when $r \leq p_k^{-1}$. Let us prove this.

We denote $\bar{y}_k^1 = \bar{y}_k^2$ by $\bar{Y}$ and $y_{k+1}^1 = y_{k+1}^2$ by Y. When $y \in (\bar{Y}, Y)$, the functions $u^1(y)$ and $u^2(y)$ are obviously equal. When $p \in (p_{k+1}, p_k)$,

$$X(p) = \int_0^{\bar{Y}} \frac{pu^i(y)\,dy}{\sqrt{1-[pu^i(y)]^2}} + \int_{\bar{Y}}^{Y^i(p)} \frac{pu^i(y)\,dy}{\sqrt{1-[pu^i(y)]^2}},\quad i = 1,\ 2.$$

That is,

$$\int_0^{\bar{Y}} \frac{u^1(y)\,dy}{\sqrt{1-[pu^1(y)]^2}} = \int_0^{\bar{Y}} \frac{u^2(y)\,dy}{\sqrt{1-[pu^2(y)]^2}},\quad p \in (p_{k+1},\ p_k).$$

If we expand both sides of the latter equation into a series in $\bar{p}$, we obtain

$$\int_0^{\bar{Y}} [u^1(y)]^{2m+1}\,dy = \int_0^{\bar{Y}} [u^2(y)]^{2m+1}\,dy,\quad m = 0, 1, 2, \ldots$$

Any function h(r) that is continuous in the interval $0 \leq r \leq p_k^{-1}$ can be approximated by a polynomial that contains r only in odd powers. Thus,

$$\int_0^{p_k^{-1}} h(r)\,dF_{k+1}^1(r) = \int_k^{\bar{Y}} h(u^1(y))\,dy =$$

$$= \int_0^{\bar{Y}} h(u^2(y))\,dy = \int_0^{p_k^{-1}} h(r)\,dF_{k+1}^2(r),$$

so that $F_{k+1}^1(r) = F_{k+1}^2(r)$ when $r \leq p_k^{-1}$; Q.E.D.

4. We recall that $Y^i(p) = \Phi(p) + \Psi^i(p)$, i = 1, 2, where $\Psi^i(p)$ are analytic in (p_{k+1}, p_k). That is, for coincidence of $Y^1(p)$ and $Y^2(p)$ in (p_{k+1}, p_k) it is sufficient that the subset (p_{k+1}, p_k), where $Y^1(p) = Y^2(p)$, has a limit point within the interval (p_{k+1}, p_k).

5. In conclusion, let us formulate another property of the pair of solutions $u^1(y)$ and $u^2(y)$, which follows from the continuity of the difference $Y^1(p) - Y^2(p)$ in (p_{k+1}, p_k) and consists of the following:

Let $p^0 \in (p_{k+1}, p_k)$, and let a set of level $u^1(y) = 1/p^0$ contain a segment of length λ situated between the k-th and (k−1)-th waveguides ($u^1(y) = 1/p^0$ when $y \in [y^1, y^1+\lambda] \in [\bar{y}_k^1, y_{k+1}^1)$). Then the same is true of a set of level $u^2(y) = 1/p^0$: $u^2(y) = 1/p^0$ when $y \in [y^2, y^2+\lambda] \in [\bar{y}_k^2, y_{k+1}^2)$.

For the numbers $p_1, \ldots, p_n$ a similar assertion has the following form. The measure of subset j_k^1, where $u^1(y) = p_k^{-1}$, is equal to the measure of subset j_k^2, in which $u^2(y) = p_k^{-1}$. This property was already formulated in a somewhat different form in Section 13:

$$\lambda = \Phi(p^0 - 0) - \Phi(p^0 + 0),$$

$$\text{mes}\{y,\ y \in j_k^i,\ u^i(y) = p_k^{-1}\} = \Phi(p_k - 0) - \Phi(p_k + 0).$$

With this we conclude our study of the problem of determining u(y) from X(p) and T(p).

In the next chapter, we shall determine X(p) and T(p) from a travel-time curve Γ. We shall assume at first that the travel-time curve Γ corresponds to a velocity distribution u(y) on which there is an additional restriction. No pencil of rays propagated with velocity u(y) is focused on emergence at the surface at one point. In other words, the function X(p) is not constant in any interval. The general case (without this additional restriction) will be considered in Chapter V.

Chapter IV

GEOMETRIC PROPERTIES OF TRAVEL-TIME CURVE PARAMETRIZATION OF TRAVEL-TIME CURVE

Above, we assumed that the functions X(p) and T(p) were unknown. In this chapter, by studying the geometric properties of the travel-time curve, we shall show that from a given travel-time curve Γ these functions can, in fact, be uniquely determined for all p(0, 1) (except, perhaps, at a finite number of points). In other words, it will be shown that among all possible parametrizations of the curve

$$\Gamma\{x = 2X(p),\quad t = 2T(p)\},\quad p \in (0,\ 1)$$

one can be found in which the parameter is the ray parameter p.

The proof will use the above assumption that the function X(p), which is unknown to start with, is not constant in any interval. This restriction is not connected with the method of proof but with the essence of the matter: an example will be constructed below that shows that without this assumption the assertion ceases to be valid.

Let us turn to the proof. We recall, first of all, some properties of the functions X(p) and T(p) that were shown in Chapter II, and introduce a number of symbols.

22. Properties of X(p) and T(p). Symbols

Each point of the interval (0, 1) on the p axis will be related to one of four disjoint sets:

$$(0,\ 1) = \tilde{\Pi} \cup \Pi' \cup \Pi_0 \cup \Pi^*.$$

To fix $\tilde{\Pi}$ we relate a finite number of points: points $f(y-0)$ and $f(y+0)$ if $f(y-0) > f(y+0)$ and points $f(y^0)$ if $f(y)$ is continuous but $f'(y)$ or $f''(y)$ either does not exist or is discontinuous when $y = y^0$. The set $\tilde{\Pi}$ is finite, in view of the assumption that the function u(y) is piecewise twice continuously differentiable and in view of the relation $f(y) = [\sup u(y^0),\ y^0 \leq y]^{-1}$.

To fix Π' we relate to points p ∈ (0, 1) at which $f'(Y(p)) = 0$, but which do not belong to $\tilde{\Pi}$. When p ∈ Π'

$$X(p) = X(p \pm 0) = T(p) = T(p \pm 0) = +\infty.$$

The measure of Π' is zero.

The set of points p ∈ (0, 1) that are not contained in $\tilde{\Pi} \cup \Pi'$ we shall denote by Π. When p ∈ Π, the functions X(p) and T(p) are continuously differentiable and the equality T'(p) = pX'(p) is satisfied. Set Π is obviously open.

Let Π_0 be a subset of Π in which X'(p) = 0. Since X'(p) is continuous when p ∈ Π, Π_0 is closed in Π. Since X(p) is not constant in any interval, Π_0 is nowhere dense in Π.

Finally, $\Pi^* = \Pi \setminus \Pi_0$. If $p \in \Pi^*$, T'(p)/X'(p) = p. Set Π* is open and, therefore, is the sum of a finite or denumerable number of nonoverlapping intervals:

$$\Pi^* = \bigcup_n I_n; \quad I_k \cap I_l = 0 \quad \text{for} \quad k \neq l.$$

The transformation x = 2X(p), t = 2T(p) maps the interval (0, 1) of the p-axis onto the x, t plane. This transformation generates on the x, t plane the images of sets $\tilde{\Pi}$, Π', Π_0, Π*, and I_n, which will be denoted by $\tilde{M}$, M', M_0, M*, and γ_n, correspondingly.

23. Construction of Sets M_0 and γ_n

Lemma 1. Each of the projections of set M_0 onto the coordinate axes has measure zero.

Proof. Being open, set Π is the sum of a finite or denumerable set of intervals. Let us consider one of these intervals I. When p ∈ I, the function X(p) is continuously differentiable. At points $p \in I_0 = I \cap \Pi_0$, the derivative X'(p) = 0.

We shall use the following theorem from analysis: the set of values taken by a continuously differentiable function at points where its derivative is zero is of zero measure.

According to this theorem, the set of values taken by the function X(p) when $p \in I_0$ is of measure zero. Thus, the projection of M_0 onto the x axis is of measure zero. Since when $p \in I_0$ the derivative T'(p) is also zero (T'(p) = pX'(p) = 0), the projection of M_0 onto the t axis is also of measure zero.

Lemma 2. For any n, γ_n is a smooth arc that is uniquely projected onto each of the coordinate axes. Let $p^* \in I_n$ and let x* = 2X(p*) and t* = 2T(p*). Then the tangent of the angle between the x axis and the tangent to the arc γ_n at the point (x*, t*) is equal to p*. When $k \neq l$, arcs γ_k and γ_l either do not intersect or they have one common point.

Proof. When $p \in I_n$, X'(p) and T'(p) are continuous and do not vanish. Therefore, γ_n is a smooth arc.

Since X'(p) and T'(p) retain their sign when $p \in I_n$, then X(p) and T(p) vary monotonically in the interval I_n. Therefore, γ_n is uniquely projected onto each of the coordinate axes. Thus, γ_n is a graph of the smooth function $t = t_n(x)$ (this function is defined in the interval of the x axis that is the transform of the interval I_n when x = 2X(p) is mapped).

It is clear that $t_n'(x^*) = T'(p^*)/X'(p^*) = p^*$, i.e., the tangent of the angle between the x axis and the tangent to the arc γ_n at the point (x*, t*) is equal to p*.

The last assertion of lemma 2 remains to be verified. We shall prove it by contradiction. Let us assume that when $k \neq l$ arcs γ_k and γ_l have two common points (x_1, t_1) and (x_2, t_2), where $x_1 < x_2$. Let us consider the function $r(x) = t_k(x) - t_l(x)$ in the interval (x_1, x_2). Since $r(x_1) = r(x_2)$, then at some point $c \in (x_1, x_2)$ the derivative r'(c) = 0, i.e., $t_k'(c) - t_l'(c) = 0$. The latter equation is impossible, however, since the intervals I_k and I_l do not overlap.

Thus, lemma 2 is proved.

24. Geometry of Travel-Time Curve Γ

We can describe the geometric properties of the travel-time curve Γ using the results demonstrated in Sections 22 and 23. Curve Γ is a curve in the plane x, t that is constructed as follows.

There is a finite or denumerable set of open smooth arcs $\gamma_1, \gamma_2, \ldots$. Let M^* be the sum of all these arcs. In addition, there is a set M_0 (as we denote the sum $\widetilde{M} \cup M_0$) which is projected onto each of the coordinate axes into a set of measure zero. Curve Γ is the sum of the sets $\Gamma = M^* \cup M^0$.

Thus, "most" of the points of Γ belong to the arcs γ_n. The arcs γ_n either do not intersect at all or intersect pointwise.

Each of the arcs γ_n serves as a graph of the monotone increasing function $t = t_n(x)$, which has the monotonic derivative $t_n'(x)$ (see Fig. 18). Here, the relation $t_n'(2X(p)) = p$ is satisfied for any point $(2X(p), 2T(p)) \in \gamma_n$. It is easy to verify this.

Lemma 3. Let μ be any smooth arc that belongs to Γ. Then a point $(x, t) \in \mu$ is found that possesses the following property: there exists an arc γ_n that contains a point (x, t) and is such that the tangents to γ_n and μ coincide at the point (x, t).

Proof. We shall represent μ as the sum

$$\mu = (\mu \cap M^*) \cup (\mu \cap M^0).$$

The projections of the set $\mu \cap M^0$ onto the coordinate axes are each of measure zero. Therefore, the set $\mu \cap M^*$ is nondenumerable. Since $M^* = \bigcup_n \gamma_n$, an arc γ_n is found such that the set $\mu \cap \gamma_n$ is nondenumerable. Let (x, t) be any point of γ_n that is a limit point for the set $\mu \cap \gamma_n$. The point (x, t) and the arc γ_n obviously satisfy the requirements of the lemma.

This lemma has an obvious corollary.

Corollary. On an arbitrary smooth arc $\mu \in \Gamma$ a point $(2X(p), 2T(p))$ is found such that p is the tangent of the angle between the x axis and the tangent to μ at that point.

Lemma 4. Let $(x_0, t_0) \in \Gamma$, and let Γ contain a smooth arc μ that passes through (x_0, t_0) at an angle φ_0 to the x axis. Let $p^0 = \tan \varphi_0$. Let $p^0 \neq 0$, $p^0 \neq 1$, and $p^0 \overline{\in} \widetilde{\Pi}$. Then $2X(p^0) = x_0$ and $2T(p^0) = t_0$.

Proof. According to the corollary of lemma 3, a sequence of points $(2X(p^n), 2T(p^n))$ is found that converges on (x_0, t_0) and such that p^n is the tangent of the angle between the x axis and the tangent to μ at the point $(2X(p^n), 2T(p^n))$. Since the arc μ is smooth, $\lim_{n\to\infty} p^n = p^0$. Therefore, $p^0 \in [0, 1]$. Since $p^0 \neq 0$, $p^0 \neq 1$, and $p^0 \overline{\in} \widetilde{\Pi}$, then $p^0 \in \Pi \cup \Pi'$, and this means that $X(p^n) \to X(p^0)$, $T(p^n) \to T(p^0)$ when $n \to \infty$. Thus, $2X(p^0) = x^0$ and $2T(p^0) = t_0$, Q.E.D.

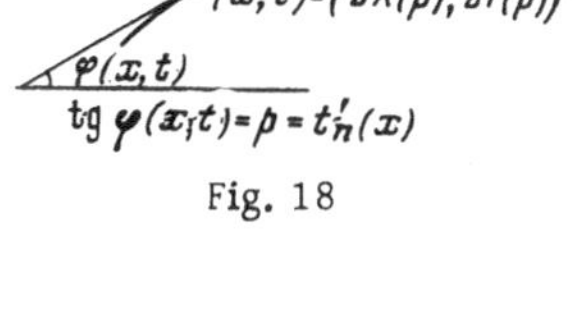

Fig. 18

We shall return to our study of the geometric properties of the travel-time curve in a paper on determining Γ from $\widetilde{\Gamma}$. There we shall be concerned with the question of what properties a curve in the plane x, t must have in order to be a travel-time curve Γ for some velocity distribution $u(y)$. Now we shall solve the main problem of this chapter – finding $X(p)$ and $T(p)$ from the travel-time Γ.

25. Determination of Functions X(p) and T(p) from the Travel-Time Curve Γ

Let the travel-time curve Γ be given as a curve in the plane x, t. Let Γ^* be a set of points of the travel-time curve Γ that possess the following property: if $(x_0, t_0) \in \Gamma^*$, then Γ contains a smooth arc μ that passes through (x_0, t_0).

Let us consider some point $(x_0, t_0) \in \Gamma^*$ and an arc $\mu \in \Gamma$ that contains that point. Let p^0 be the tangent of the angle between the x axis and the arc μ at the point (x_0, t_0).

Let $X(p^0) = x^0/2$ and $T(p^0) = t_0/2$.

We shall define $X(p)$ and $T(p)$ as functions in some subset of the interval $(0, 1)$. This subset automatically includes Π^*, for $\Gamma^* \supset M^*$. It possibly also contains points that do not belong to Π^*. Here, lemma 4 of the preceding section guarantees that we will determine incorrectly $X(p)$ and $T(p)$ only at a finite number of points.

At points of the interval $(0, 1)$ where $X(p)$ and $T(p)$ are still not defined, we can determine them by continuity. Here, errors or multivaluedness can arise only at a finite number of points (at points of $\widetilde{\Pi}$), since $X(p)$ and $T(p)$ are continuous in the set $\Pi \cup \Pi'$ (at points of Π they are continuous in the usual sense; at points $p \in \Pi'$, $X(p) = X(p \pm 0) = T(p) = T(p \pm 0) = +\infty$). Moreover, since the set $\Pi_0 = \Pi \setminus \Pi^*$ is nowhere dense and Π' is of measure zero, $X(p)$ and $T(p)$ are defined over the entire interval. At points where multivaluedness occurs, we select one value arbitrarily.

Thus, assuming that $X(p)$ is not constant in any interval, the problem of parametrization of the travel-time curve Γ is completely solved: $X(p)$ and $T(p)$ are uniquely determined from Γ everywhere, except perhaps for a finite number of points.

If we compare the latter result with the result of Chapter III, we obtain a solution of the problem of finding the velocity distribution $u(y)$ from the travel-time curve Γ.

If X(p) takes a constant value over an entire interval, then multivaluedness can arise in determining X(p) and T(p) from Γ.

26. Example of a Travel-Time Curve from which X(p) and T(p) Are Not Determined Uniquely

Let velocity u(y) be a monotone increasing function (see Fig. 19) such that

$$Y(p)=\begin{cases}\operatorname{arch} p^{-1}, & 1<p^{-1}\leqslant 2,\\ \operatorname{arch} p^{-1}, +\operatorname{arch} p^{-1}/2, & 2\leqslant p^{-1}<\infty .\end{cases}$$

We write X(p) as

$$X(p)=\int_0^{Y(p)}\frac{u(y)\,dy}{\sqrt{p^{-2}-u^2(y)}}=\int_1^{p^{-1}}\frac{u\,dY(u^{-1})}{\sqrt{p^{-2}-u^2}} .$$

When $1 < p^{-1} < 2$

$$X(p)=\int_1^{p^{-1}}\frac{u\,d(\operatorname{arch} u)}{\sqrt{p^{-2}-u^2}}=\int_1^{p^{-1}}\frac{u\,du}{\sqrt{(p^{-2}-u^2)(u^2-1)}}=\frac{\pi}{2}.$$

When $2 < p^{-1} < \infty$

$$X(p)=\frac{\pi}{2}+\int_2^{p^{-1}}\frac{u\,d(\operatorname{arch} u/2)}{\sqrt{p^{-2}-u^2}}=$$

$$=\frac{\pi}{2}+\frac{1}{2}\int_2^{p^{-1}}\frac{u\,du}{\sqrt{(p^{-2}-u^2)(u^2/4-1)}}=\pi .$$

Thus the graph of X(p) has the form in Fig. 20.

Therefore sets Π' and Π* are empty, set $\tilde{\Pi}$ consists of one point $p = \frac{1}{2}$, and set Π^0 coincides with Π and is the sum of the intervals $(0, \frac{1}{2})$ and $(\frac{1}{2}, 1)$.

It is easy to verify that

$$T(p)=\begin{cases}\pi/2 & \text{for } p\in(1/2,\ 1)\\ 3\pi/4 & \text{for } p\in(0,\ 1/2).\end{cases}$$

The travel-time curve Γ degenerates into two points in the plane x, t (see Fig. 21).

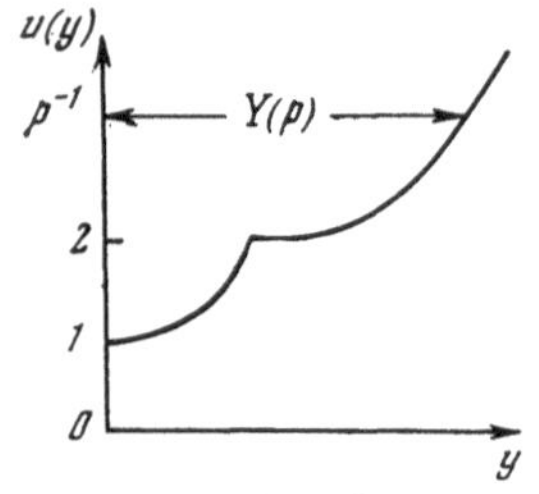

Fig. 19

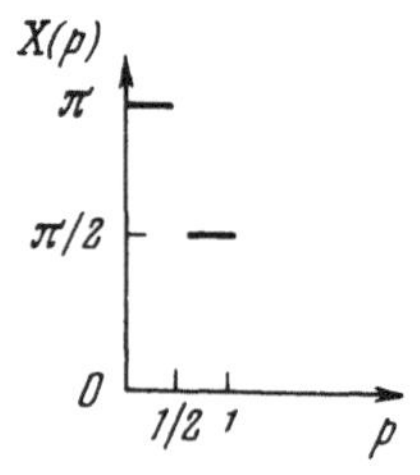

Fig. 20

Precisely the same travel-time curve would be obtained when

$$X(p)=\begin{cases}\pi/2, & 1>p>p^0\\ \pi, & p^0>p>0\end{cases};\quad T(p)=\begin{cases}\pi/2, & 1>p>p^0\\ \frac{3\pi}{4}, & p^0>p>0.\end{cases}$$

For any $p^0 \in (0, \frac{1}{2})$, we can determine from these functions a set of solutions $\{u_\nu(y)\}$ by operating according to the rules described in Chapter III. Thus, in this case X(p) and T(p) are not determined uniquely from the plotted travel-time curve Γ.

Remark. It is not difficult to verify that a function X(p) that takes two values cannot have a positive discontinuity: this readily follows from the monotonicity of Φ(q). Thus, we find all pairs of the functions X(p) and T(p) that correspond to the "two-point" travel-time curve Γ.

Since all these pairs are "equally justified," some additional information is required to determine the number p^0. For this purpose we can use the travel-time curves from deep sources.

Chapter V

USE OF TRAVEL-TIME CURVES FROM DEEP SOURCES

So far we have assumed that the source O from which seismic rays emerged was at the free surface. Now we assume that along with the travel-time curve from the surface source O, we also know the travel-time curve from a source O_1 that is at some unknown depth y = d.

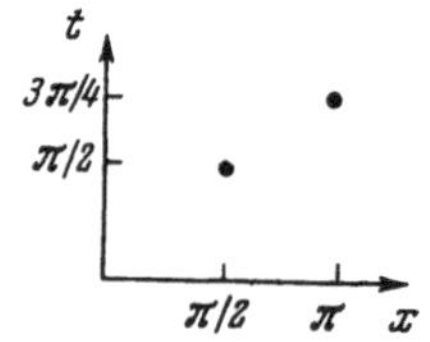

Fig. 21

27. Rays Emerging from a Deep Source

Let us consider the propagation of rays from the source O_1 (Fig. 22). From symmetry, it is sufficient to consider the rays that emerge to the left upward and to the right downward from the source.

Each ray that departs to the right downward from the source will be joined to the ray departing to the left upward that has the same ray parameter (we recall that the ray parameter of a seismic ray is a number equal to $\sin \alpha(y)/u(y)$, where $\alpha(y)$ is the angle between the ray and the axis of the ordinates at point y; this number does not vary along the ray). Rays with parameters $p \in (0, f(d+0))$ are joined in pairs. If $f(d+0) < f(d-0)$, some of the rays that depart to the left upward will remain unpaired. If $f(d+0) = f(d-0)$, rays that have a common tangent at O_1 will be paired: for example, rays O_1A and O_1B in Fig. 22. The total p value of two paired rays will be called the parameter of the curve obtained when the rays are joined. If we let $f(d+0) = p_d$, then $p \in (0, p_d)$.

Each curve with a parameter $p \in (0, p_d)$ coincides with a ray that emerges from a surface source with the same parameter p, except for a translation along the x axis. Therefore, each such curve is symmetric relative to a vertical line that passes through its deepest point.

Each curve with a parameter $p \in (0, p_d)$ has at least one point of emergence at the surface. If there are two such points, they are situated to the left and to the right of the epicenter O, the point on the surface directly above the source O_1.

28. Functions $X_1(p)$, $X_2(p)$, $T_1(p)$, and $T_2(p)$

Let $X_1(p)$ and $T_1(p)$ be the distance from the epicenter to the emergence point at the surface, and the emergence time of a pulse that moves with velocity u(y) along a ray with parameter p when the emergence point is to the left of O. Let $X_2(p)$ and $T_2(p)$ be the same values for rays that emerge at the surface to the right of O. We recall that $X_1(p)$

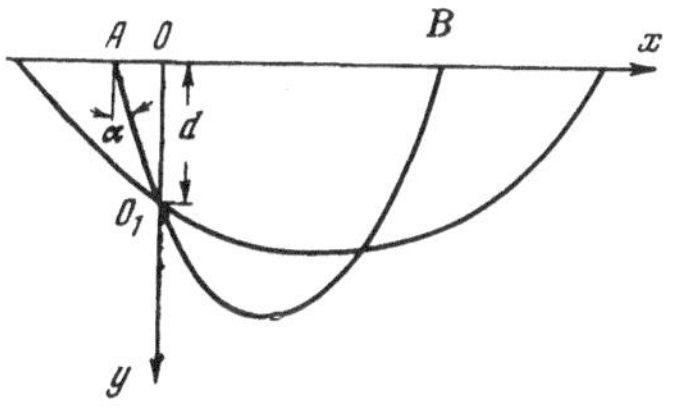

Fig. 22

and $X_2(p)$ are known from mod $2\pi R/v(R)$. Then in the plane (x, t) we have given the two curves $\tilde{\Gamma}_i\{\tilde{X}_i(p), T_i(p)\}$, $p \in (0, p_d)$, where

$$0 < \tilde{X}_i(p) \leqslant \frac{2\pi R}{v(R)}, \quad X_i(p) \equiv X(p) \left(\operatorname{mod} \frac{2\pi R}{v(R)}\right), \quad i = 1, 2.$$

We must explain how these curves can be used in determining u(y).

Just as in our study of surface-source travel-time curves, we shall assume that not only the curves $\tilde{\Gamma}_i$ but also the curves $\Gamma_i\{X_i(p), T_i(p)\}$, $p \in (0; p_j)$, i = 1, 2, are known.

The formulas for $X_i(p)$ and $T_i(p)$, i = 1, 2, are obtained exactly as those for X(p) and T(p) in Chapter II:

$$X_1(p) = \int_0^d \frac{pu(y)\,dy}{\sqrt{1 - p^2u^2(y)}},$$

$$X_2(p) = 2\int_d^{Y(p)} \frac{pu(y)\,dy}{\sqrt{1 - p^2u^2(y)}} + X_1(p),$$

$$T_1(p) = \int_0^d \frac{dy}{u(y)\sqrt{1 - p^2u^2(y)}},$$

$$T_2(p) = 2\int_0^{Y(p)} \frac{dy}{u(y)\sqrt{1 - p^2u^2(y)}} + T_1(p),$$

where

$$Y(p) = \inf\{y, pu(y) \geqslant 1\}, \quad p \in (0, p_d).$$

Remark. Everywhere below, when curve Γ_1 is not considered jointly with curve Γ_2 (Sections 29 and 32), it must be assumed that $p_d = f(d-0)$ (but not $f(d+0)$; see Section 27).

29. Properties of the Curve $\Gamma_1\{X_1(p), T_1(p)\}$

The functions $X_1(p)$ and $T_1(p)$ have the following properties.

Lemma. $X_1(p)$ and $T_1(p)$ are differentiable monotone increasing functions, $X_1'(p) > 0$, $p \in [0, p_d)$, and the function $X_1(p)$ vanishes when p = 0.

Proof. $X_1(p)$ and $T_1(p)$ are proper integrals of the functions, which are differentiable with respect to p and increase as p increases. Hence the first assertion follows. The fact that $X_1'(p) > 0$, $X_1(0) = 0$, is verified directly.

Corollary 1. The value of the parameter p can be determined at each point on the curve Γ.

Just as in Section 10 (Chapter II), it is verified that $T_1'(p) - pX_1'(p) = 0$. According to the lemma,

$X_1'(p) \neq 0$, $p \in [0, p_d)$; therefore,

$$p = \frac{dT_1}{dX_1}.$$

Corollary 2. If the curve $\tilde{\Gamma}_1$ is known, then the curve Γ_1 is uniquely determined.

The assertion follows from the fact that from the curve $\tilde{\Gamma}_1$ we can uniquely find a Γ_1 that is continuous and such that $X_1(0) = 0$.

Corollary 3. We can uniquely determine $X_1(p)$ and $T_1(p)$ from $\tilde{\Gamma}_1$.

The assertion follows from corollaries 1 and 2.

30. Joint Parametrization of Γ_1, Γ_2, and Γ

We shall show how joint examination of curves Γ_1 and Γ_2 and the travel-time curve Γ aids in the parametrization of Γ_2 and Γ.

In Chapter IV, the problem of parametrization of the travel-time curve was solved assuming that $X(p)$ was not constant in any interval. We shall show that if, along with curve Γ, curves Γ_1 and Γ_2 are known, this assumption can be disregarded.

The following theorem is valid.

Theorem. Let curves Γ_1, Γ_2, and Γ be known. Then we can determine $X_1(p)$, $X_2(p)$, $X(p)$, $T_1(p)$, $T_2(p)$, and $T(p)$ for all $p \in (0, p_d)$, except, perhaps, at a finite number of points.

Proof. According to the preceding section, $X_1(p)$ and $T_1(p)$, $p \in [0, p_d)$, can be considered known.

It is easy to see that when $p \in (0, p_d)$,

$$X_2(p) = 2X(p) - X_1(p), \quad T_2(p) = 2T(p) - T_1(p).$$

Thus if the value of one of the functions $X(p)$ or $X_2(p)$ at some point $p \in (0, p_d)$ is known, the value of the other function at the same point is known. This is also true of $T(p)$ and $T_2(p)$.

According to Section 25 (Chapter IV), we can define $X(p)$ and $T(p)$ in some subset of the interval $(0, 1)$ that necessarily includes Π^*, wherein errors can arise only at a finite number of points. Here the assumption that $X(p)$ is not constant in any interval is not used.

In an entirely analogous way it is proved that $X_2(p)$ and $T_2(p)$ are determined from Γ_2 in some subset of $(0, p_d)$ that includes Π_d^* (denoting thus the subset $(0, p_d)$ where $X_2'(p) \neq 0$).

Since $X_1'(p)$ is strictly greater than zero, $X_2'(p)$ and $X'(p)$ do not vanish simultaneously. Therefore,

$$\Pi_d^* \supset (0, p_d) \cap \Pi_0.$$

(We recall that Π_0 is the set of those points $p \in (0, 1)$ where $X'(p) = 0$.)

Thus $X(p)$, $X_2(p)$, $T(p)$, and $T_2(p)$ are defined at all points $p \in (0, p_d) \cap \Pi$.

At points $p \in (0, p_d) \cap \Pi'$, the following relations are obviously valid:

$$X_2(p) = X_2(p \pm 0) = T_2(p) = T_2(p \pm 0) = +\infty.$$

Thus, $X(p)$, $X_2(p)$, $T(p)$, and $T_2(p)$ are defined everywhere in $(0, p_d)$, except perhaps at a finite number of points.

31. Determination of Source Depth

In this section we shall show that if $X_1(p)$ is known when $p \in [0, p_d)$, we can determine the source depth d.

Let

$$F_d(r) = \operatorname{mes}\{y,\ y \leqslant d,\ u(y) \leqslant r\}.$$

This function can be defined for $r \in (0, p_d^{-1})$ if $X_1(p)$ is known. The proof of this is essentially contained in Paragraph 3 of Section 21 (Chapter III), where it is shown that from the known function

$$\int_0^{\bar{y}_k^1} \frac{u^1(y)\,dy}{\sqrt{1-[pu^1(y)]^2}} \qquad p \in (p_{k+1},\ p_k)$$

we can uniquely determine for all $r \leq p_k^{-1}$ the measure

$$F_{k+1}^1(r) = \operatorname{mes}\{y,\ y \leqslant y_{k+1}^1,\ u(y) \leqslant r\}.$$

Since $u(y) \leq p_d^{-1}$ when $y \leq d$, then $F_d(p_d^{-1}) = d$.

32. Travel-Time Curve for Reflected Waves

When there is a boundary at depth d from which reflected waves arrive, we can consider the travel-time curve of these waves.

Let $T_0(p)$ be the total travel-time along a ray with parameter p that goes from a surface source O to the reflecting boundary and back up to the surface. Let $X_0(p)$ be the distance along the x axis between the source and the emergence point. Comparison of Figs. 22 and 23 shows that $T_0(p) = 2T_1(p)$ and $X_0(p) = 2X_1(p)$, where $X_1(p)$ and $T_1(p)$ are functions defined earlier for a source at depth d. Thus, the travel-time curve for the reflected wave, $\{x = X_0(p),\ t = T_0(p)\}$, after twofold uniform compression gives the curve $\Gamma_1\{X_1(p),\ T_1(p)\}$, $p \in [0, p_d)$, which

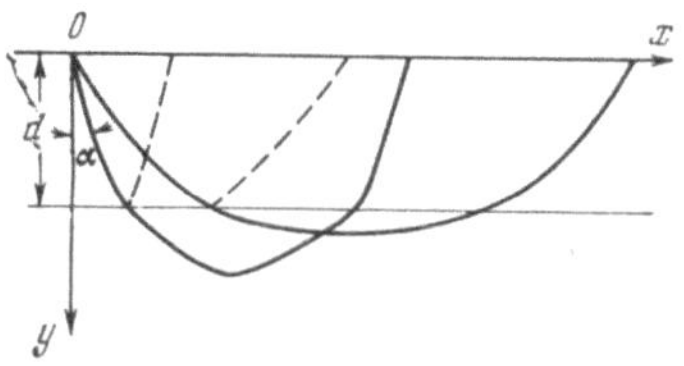

Fig. 23

we could obtain from a deep source at the depth of the reflecting boundary.

Below, we shall discuss the information that can be extracted by joint examination of the curves $\Gamma\{X(p), T(p)\}$, $p \in (0, 1)$, and $\Gamma_1\{X_1(p), T_1(p)\}$, $p \in [0, p_d)$. We do not care whether Γ_1 is obtained from the reflected-wave travel-time curve or is a part of the deep-source travel-time curve.

Remark. We should point out one difference between the travel-time curve for a source at depth d and the travel-time curve of waves reflected from a source at the same depth: since in the case of a reflecting boundary, the curve Γ_2 does not correspond to curve Γ_1, the reflected-wave travel-time curves do not facilitate parametrization of a surface-source travel-time curve.

33. Determination of u(y) from X(p), $X_1(p)$, T(p), and $T_1(p)$

Let us consider the extent to which the ambiguity in determining u(y) is reduced when, along with X(p) and T(p), the functions $X_1(p)$ and $T_1(p)$ are known.

1. We introduce the functions

$$X_d(p) = X(p) - X_1(p), \quad T_d(p) = T(p) - T_1(p).$$

If rays emerging from a source at depth d and propagated in the half-plane $y > d$ were recorded at the same depth d (see Fig. 24), the following travel-time curve would be obtained:

$$\Gamma_d\{x = 2X_d(p),\ t = 2T_d(p)\}, \quad p \in (0, p_d).$$

Therefore the problem of determining u(y) for $y > d$ from $X_d(p)$ and $T_d(p)$ can be solved in the same way as the problem of determining u(y) for $y > 0$ from

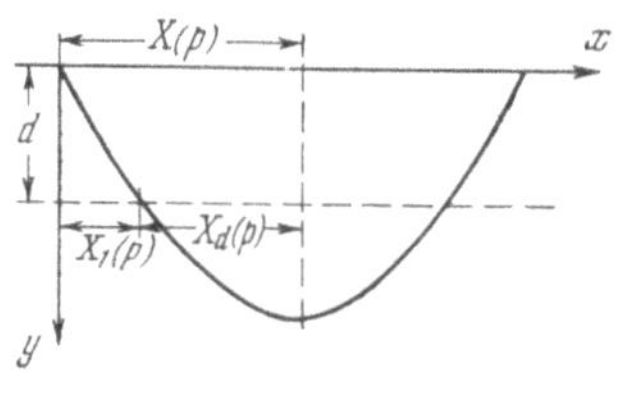

Fig. 24

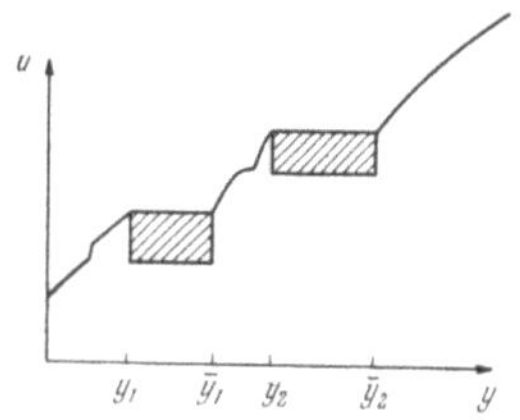

Fig. 25

X(p) and T(p). Thus, from $X_d(p)$ and $T_d(p)$ a set G_d is constructed that is populated by the graphs of the solutions for $y > d$.

2. There is a complete analogy with Chapter III when $p_d \in (p_{k+1}, p_k)$, $1 \le k \le n$. If $p_d = p_k$, the structure of set G_d is somewhat different from that of set G. In considering the case of a surface source, we assumed that a waveguide did not begin immediately at the surface. If this requirement is not met, the structure of set G is entirely the same except that the interval $(0, y_1)$ in which u(y) was uniquely determined disappears. If $p_d = p_k$, $k = 1, \ldots, n$, then G_d does not contain an interval in which velocity is uniquely determined.

3. Thus, if we know X(p), T(p), $X_1(p)$, and $T_1(p)$, we can determine the following:

a) The depth d of a deep source (or reflecting boundary);

b) The function $F_d(r) = \text{mes}\,\{y,\ y \le d,\ u(y) \le r\}$ for $r \in (0, p_d^{-1})$;

c) The set G_d in which are situated the graphs of the solutions $u_\nu(y)$ for $y > d$.

4. We emphasize that the numbers $\bar{y}_k$ and y_{k+1} and velocity distribution u(y) for $y \in (\bar{y}_k, y_{k+1})$ are uniquely determined when $p_d \in (p_{k+1}, p_k)$.

5. In particular, if the travel-time curves of a surface source and several deep sources are known, and there is a source between any two adjacent waveguides, then (see Fig. 25):

a) the positions of the waveguides j_k (the numbers $y_1, \bar{y}_1, \ldots, y_n, \bar{y}_n$) are exactly determined;

b) the velocity u(y) between the waveguides (when $y \in (y_k, \bar{y}_{k+1})$, $0 \le k \le n$) is uniquely determined; and

c) although u(y) cannot be exactly determined in the waveguides, the functions

$$F_k(r) = \text{mes}\,\{y,\ y \in j_k,\ u(y) \leqslant r\}, \quad 1 \leqslant k \leqslant n,$$

are known, and $\inf\limits_{y \in j_k} u(y)$ is found specifically for each j_k.

The authors sincerely thank V. A. Gerver, E. G. Dirdovskaya, and L. N. Pozharskaya, who provided in a very short time a great deal of painstaking technical assistance in preparation of the article for printing.

PROPERTIES OF SURFACE-SOURCE TRAVEL-TIME CURVES†

M. L. Gerver and V. M. Markushevich

INTRODUCTION

This is a continuation of the preceding paper [1]. The problem treated in [1] may be formulated as follows.

Problem of Rays in Circle

Let K be a circle of radius R with center C, and let A be a point on its circumference (Fig. 1). We introduce the polar coordinates r, θ, and take C as the pole and CA as the polar axis.

A pulsed perturbation occurs at point A at time t = 0 and is propagated in K along rays, according to the laws of geometric optics, with velocity v(r), $r \in [0, R]$. The arrival time t is recorded at each point B at which the pulse again emerges at the circumference.

Let α, $\alpha \in (0, \pi/2)$, be the angle between the ray AB and radius AC. The angle ACB (angular epicentral distance) we shall call $\tilde{\theta}$, $0 \leq \tilde{\theta} < 2\pi$.

Then in the plane (θ, t) we obtain the curve

$$\tilde{\Gamma}_0\{\theta = \tilde{\theta}(\alpha),\ t = t(\alpha)\}.$$

It is required to determine v(r) from $\tilde{\Gamma}_0$.

The Curve Γ_0

Assuming that we knew not only the angle $\tilde{\theta}$ but also the number k of revolutions of the ray about the center of the circle, we examined in [1] the curve

$$\Gamma_0\{\theta = \theta(\alpha),\ t = t(\alpha)\},$$

where

$$\theta(\alpha) = \tilde{\theta}(\alpha) + 2\pi k.$$

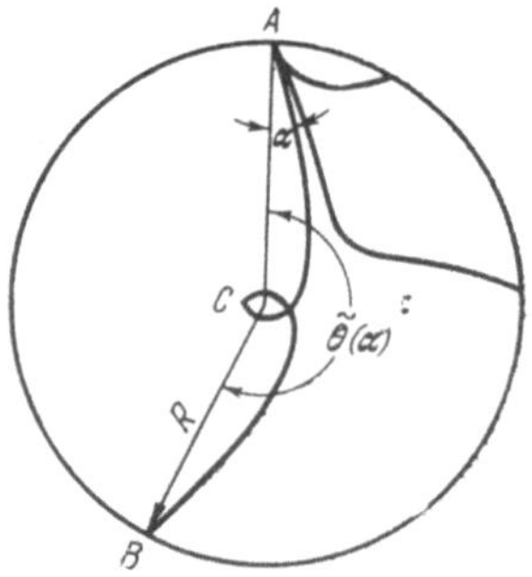

Fig. 1

The problem of rays in a circle was divided into two parts: 1) determination of Γ_0 from $\tilde{\Gamma}_0$ and 2) determination of v(r) from Γ_0. The first problem was not considered in [1], but it will be examined in Chapter III of the present paper. By conformally mapping the circle K onto the half-plane (x, y), $y \geq 0$:

$$x = \frac{R\theta}{v(R)}, \quad y = \frac{R}{v(R)} \ln \frac{R}{r}$$

and by introducing the function

$$u(y) = \frac{Rv(r)}{v(R)r} = \frac{v(Re^{-yv(R)/R})}{v(R)e^{-yv(R)/R}}$$

the second problem was reduced to the following problem.

Problem of Rays in Half-Plane

Rays emerge into the lower (y > 0) half-plane from point O of the plane (x, y). Let L be one of

† Translated from Vychislitel'naya Seismologiya, No. 4, Moscow, Nauka (1968), pp. 15-63.

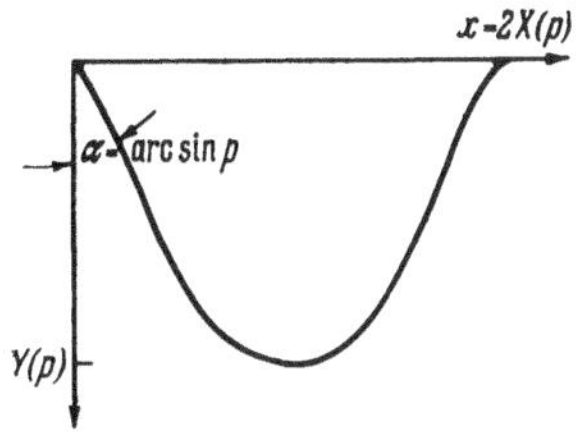

Fig. 2

them, which forms the angle α with the y axis, $0 < \alpha < \pi/2$ (Fig. 2).

The number $p = \sin\alpha \in (0, 1)$ is called the ray parameter. A pulse that arises at point O at $t = 0$ moves along L with velocity u(y), where we have scaled the velocities so that $u(0) = 1$. X(p) and Y(p) are the coordinates of the deepest point of L, and T(p) is the travel-time from 0 to this point. The curve $\Gamma\{x = 2X(p),\ t = 2T(p)\}$ (the travel-time curve) is known. It is necessary to find u(y).

Determination of Γ_0 from $\tilde{\Gamma}_0$, and the Curve $\tilde{\Gamma}$

The travel-time curve Γ can be considered known if curve Γ_0 is given. If only $\tilde{\Gamma}_0$ is known, then instead of Γ we must examine the curve

$$\tilde{\Gamma}\{x = 2\tilde{X}(p),\quad t = 2T(p)\},$$

where

$$\tilde{X}(p) \equiv X(p)\left(\mathrm{mod}\,\frac{\pi R}{v(R)}\right),\quad 0 \leqslant \tilde{X}(p) < \frac{\pi R}{v(R)}.$$

Thus, determination of Γ_0 from $\tilde{\Gamma}_0$ comes down to determining Γ from $\tilde{\Gamma}$. This problem will be considered in Chapter III.

Parametrization of the Travel-Time Curve Γ

In the problem of rays in a half-plane, the travel-time curve Γ is given as a curve in the plane (x, y). In [1] we studied in detail the question of parametrization of Γ, i.e., of finding the functions X(p) and T(p) from Γ. As a rule, Γ is parametrized uniquely. Sometimes, determination of X(p) and T(p) requires additional information. Deep-source travel-time curves were used in [1] for this information.

Determination of u(y) from X(p) and T(p)

Let X(p) and T(p) be defined by Γ. The problem arises of determining u(y) from them. The central result of [1] was discovery of the fact that this problem does not have a unique solution, and a complete study of this ambiguity.

In this paper we shall treat questions of existence in determining u(y) from X(p) and T(p).

Functions X(p) and T(p) are expressed in terms of u(y) as follows (see [1]):

$$X(p) = \int\limits_0^{Y(p)} \frac{pu(y)\,dy}{\sqrt{1 - p^2u^2(y)}},\quad T(p) = \int\limits_0^{Y(p)} \frac{dy}{u(y)\sqrt{1 - p^2u^2(y)}},$$

$$p \in (0, 1), \tag{1}$$

where

$$Y(p) = \inf\{y,\ pu(y) \geqslant 1\}.$$

When X(p) and T(p) are known, (1) can be considered a system of equations in u(y). We already know that this system, generally speaking, does not have a unique solution. But it may not have any solutions at all. As in [1], we shall call u(y) a solution of equations (1) if u(y) satisfies them for all $p \in (0, 1)$, except perhaps for a finite number of points.

In Chapter I we shall find necessary and sufficient conditions for the existence of at least one solution of system (1). In other words, we shall ascertain which curves in the plane (x, y) are travel-time curves.

In Chapter II we shall study these conditions in greater detail and draw from them some inferences about the properties of travel-time curves. These properties will be needed in Chapter III. They will probably also be useful in the practical construction of a travel-time curve for the earth.

Chapter I
CHARACTERISTICS OF TRAVEL-TIME CURVES

§1. Allowable Velocity Profiles

1. As in [1], we shall assume that velocity u(y) is a positive function, piecewise twice continuously differentiable, that is bounded on any interval of the semiaxis $y \geq 0$ and unbounded on the entire semiaxis; and scaled so that $u(0) = 1$. Moreover, as in [1], we shall assume that u(y) has only a finite number of waveguides (see paragraph 2, below) for exact definition).

2. Let (Fig. 3a):

$$f(y) = (\sup\{u(y^0),\ y^0 \in [0, y]\})^{-1}.$$

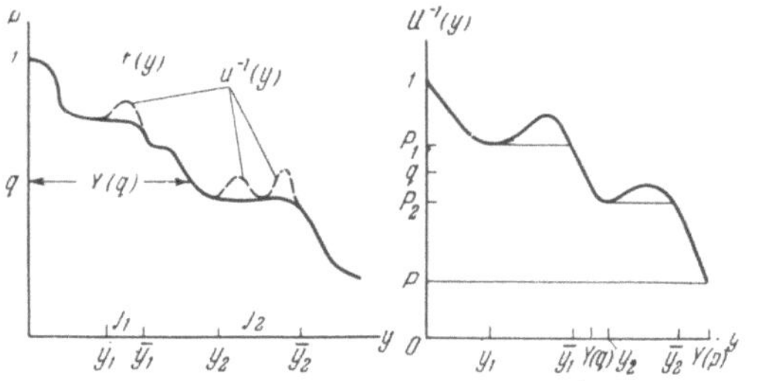

a b

Fig. 3

By definition, waveguides $j_1, \ldots, j_n$ are intervals of the y axis with the following properties:

1) $f(y)$ is constant in each of them;

2) each of them contains at least one interval where $u^{-1}(y) > f(y)$;

3) outside these intervals, $u^{-1}(y) = f(y)$.

We shall denote the ends of the waveguide j_k by y_k and $\bar{y}_k$ and the values of $f(y)$ taken in the interval j_k by p_k, $1 \le k \le n$:

$$y_1 < \bar{y}_1 < y_2 < \bar{y}_2 < \ldots < y_n < \bar{y}_n;$$

$$p_1 > p_2 > \ldots > p_n.$$

For definiteness, we shall assume that $y_1 > 0$, but all reasoning extends to the case of $y_1 = 0$ with slight changes.

3. The velocity profiles u(y) that satisfy the formulated requirements will be called "allowable." Solutions of equations (1) (see Introduction) will be sought in the class of allowable profiles.

§2. Existence Theorem

The following conditions are necessary and sufficient for a curve $\Gamma\{2X(p), 2T(p)\}$, $p \in (0, 1)$, to be a travel-time curve for some allowable velocity profile u(y).

A. Functions X(p) and T(p) have the following properties:

1) they are positive;

2) they are differentiable almost everywhere;

3) at points p where X(p) and T(p) are differentiable, $T'(p) - pX'(p) = 0$;

4) at points p where X(p) and T(p) are not differentiable (except, perhaps, at a finite number of points),

$$X(p \pm 0) = X(p) = T(p \pm 0) = T(p) = \infty.$$

B. The function $\tau(p) = T(p) - pX(p)$† has the following properties:

1) it is decreasing monotonically

2) $\tau(1 - 0) = 0$;

3) it is continuous everywhere except at a finite number of points p_k, $p_1 > p_2 > \ldots > p_n$, where it has discontinuities $\sigma_k = \tau(p_k - 0) - \tau(p_k + 0)$.

C. The function $\Phi(q) = \frac{2}{\pi}\int_q^1 \frac{X(p)\,dp}{\sqrt{p^2 - q^2}}$

1) is finite for all $q \in (0, 1)$;

2) is nonincreasing;

3) $\Phi(+0) = +\infty$;

4) there exists a $C > 0$ such that everywhere on (p_{k+1}, p_k), where $\Phi'(q)$ is finite,

$$\Phi'(q) < -Cq/\sqrt{p_k^2 - q^2}, \quad 1 \leqslant k \leqslant n, \quad p_{n+1} = 0;$$

5) the function g(y), which is the inverse of $\Phi(q)$, is piecewise twice continuously differentiable.

D. The function $\tau(p) + \int_p^1 \sqrt{q^2 - p^2}\,d\Phi(q)$ is absolutely continuous for $p \in (p_{k-1}, p_k)$, $0 \le k \le n$, $p_0 = 1$, $p_{n+1} = 0$.‡

† See Note 1.

‡ In the system of sufficient conditions, condition D can be replaced by condition D_0: The function $\tau(p)$ is absolutely continuous when

$$p \in (p_{k+1}, p_k), 0 \leqslant k \leqslant n, \quad p_0 = 1, \quad p_{n+1} = 0.$$

It was unclear, however, whether or not condition D_0 is necessary (see Section 5).

Recently (when the paper had already been sent to press), E. L. Reznikov and M. L. Gerver found the following simple proof of the necessity of condition D_0.

We subtract from function $\tau(p)$ the step function

$$\tilde{\tau}(p) = \tau(p) - \sum_{p_i > p} \sigma_i \equiv \begin{cases} \tau(p), & p \in (p_1, 1), \\ \tau(p) - \sigma_1, & p \in (p_2, p_1), \\ \vdots & \\ \tau(p) - \sum_{i=1}^{n} \sigma_i, & p \in (0, p_n). \end{cases}$$

To prove the absolute continuity of $\tilde{\tau}(p)$ (which is equivalent to condition D_0), it is sufficient to verify that

$$\tilde{\tau}(p) = \int_p^1 X(q)\,dq.$$

The following calculation demonstrates the validity of the latter formula (see Fig. 3b):

$$\int_p^1 X(q)\,dq =$$

$$= \int_p^1 \left(\int_0^{Y(q)} \frac{q\,dy}{\sqrt{u^{-2}(y) - q^2}} \right) dq = \int_0^{Y(p)} \left(\int_p^{f(y)} \frac{q\,dq}{\sqrt{u^{-2}(y) - q^2}} \right) dy =$$

$$= \int_0^{Y(p)} \left[\sqrt{u^{-2}(y) - q^2} \,\Big|_{f(y)}^{p} \right] dy = \int_0^{Y(p)} \sqrt{u^{-2}(y) - p^2}\,dy -$$

$$- \sum_{p_i > p} \int_{y_i}^{\bar{y}_i} \sqrt{u^{-2}(y) - p_i^2}\,dy = \tau(p) - \sum_{p_i > p} \sigma_i = \tilde{\tau}(p).$$

Note 1. In [1], we introduced $\tau(p)$ by the formula

$$\tau(p)=\int_0^{Y(p)}\sqrt{u^{-2}(y)-p^2}\,dy,\quad p\in(0,1),$$
$$Y(p)=\inf\{y,\ pu(y)\geqslant 1\}.$$

If u(y) is an allowable velocity profile, the function thus defined is continuous everywhere on (0, 1), except at a finite number of points, and satisfies the relation $\tau(p) = T(p) - pX(p)$ where the right side is finite.

In this paper, we shall write

$$\tau(p)=T(p)-pX(p),$$

without mentioning that $\tau(p)$ is defined only at points p where X(p) and T(p) are finite (according to A2, this set is dense on (0, 1)) and at the other points $p \in (0, 1)$, $p \neq p_1, \ldots, p_n$, the definition of $\tau(p)$ is extended by continuity.

Note 2. The number of discontinuities of $\tau(p)$ equals the number of waveguides of u(y). If the first waveguide begins immediately at the surface ($y_1 = 0$), then $p_1 = 1$. In this case, condition B2 becomes condition B2':

$$\tau(1-0)=\sigma_1>0.$$

Note 3. Condition B1 (as a sufficient condition) is used only once (Section 4, paragraph 1.1) and then only to deduce that $\sigma_k = \tau(p_k - 0) - \tau(p_k + 0) > 0$. Therefore in the system of sufficient conditions it can be replaced by the requirement that the numbers $\sigma_1, \ldots, \sigma_n$ be positive.

§3. Necessity of Conditions A, B, C, and D

The necessity of most of the conditions formulated in Section 2 was proved in [1]. The exceptions are conditions C3, C4, C5, and D.

We shall show that if $\Gamma\{2X(p), 2T(p)\}$ is a travel-time curve, then these conditions are satisfied.

Condition C3

A main formula linking Y(p), X(p), and u(y) was proved in [1]: when $p \in (p_{k+1}, p_k)$, $1 \le k \le n$, $Y(p) = \Phi(p) + \Psi(p)$, where $\Phi(p)$ is the function of condition C and

$$\Psi(p)=\frac{2}{\pi}\sum_{i=1}^{k}\int_{y_i}^{\bar y_i}\operatorname{arctg}\sqrt{\frac{u^{-2}(y)-p_i^2}{p_i^2-p^2}}\,dy.$$

Since $Y(+0) = +\infty$, but $\Psi(+0)$, then $\Phi(+0) = +\infty$.

Condition C4

Let us differentiate $\Psi(p)$:

$$\Psi'(p)=\frac{2p}{\pi}\sum_{i=1}^{k}\frac{1}{\sqrt{p_i^2-p^2}}\int_{y_i}^{\bar y_i}\frac{\sqrt{u^{-2}(y)-p_i^2}}{u^{-2}(y)-p^2}\,dy.$$

It is obvious that

$$\Psi'(p)\geqslant\frac{2p}{\pi\sqrt{p_k^2-p^2}}\int_{y_k}^{\bar y_k}\frac{\sqrt{u^{-2}(y)-p_k^2}}{u^{-2}(y)-p^2}\,dy>$$

$$>\frac{p}{\sqrt{p_k^2-p^2}}\,\frac{2}{\pi}\int_{y_k}^{\bar y_k}\frac{\sqrt{u^{-2}(y)-p_k^2}}{u^{-2}(y)}\,dy=$$

$$=\frac{C_kp}{\sqrt{p_k^2-p^2}},\quad C_k>0,\ p\in(p_{k+1},p_k),\ 1\leqslant k\leqslant n.$$

If we let $C=\min\limits_{1\leqslant k\leqslant n} C_k$, we obtain finally

$$\Psi'(q)>\frac{Cq}{\sqrt{p_k^2-p^2}},\quad q\in(p_{k+1},p_k),\ 1\leqslant k\leqslant n.$$

Since $Y(p) = \inf\{y, pu(y) \ge 1\}$ is a nonincreasing function, then everywhere on (p_{k+1}, p_k) where $\Phi'(q)$ is finite, $\Phi'(q) < -Cq/\sqrt{p_k^2 - q^2}$, $1 \le k \le n$.

Condition C5

1. First of all, let us recall some properties of the function $f(y)$, which was defined in Section 1; these properties were demonstrated in [1], but they are easily verified directly.

Function $f(y)$ (see Fig. 3a) does not increase in the interval $0 < f(y) \le 1$.

We add to the graph of $f(y)$ segments that join the points $(y, f(y - 0))$ and $(y, f(y + 0))$, if $f(y - 0) > f(y + 0)$. We treat the graph of Y(p) similarly: we add to it segments with ends $(Y(p-0), p)$ and $(Y(p + 0), p)$ if $Y(p - 0) > Y(p + 0)$. Then the graphs of $f(y)$ and Y(p) coincide as curves in the plane (y, p). In this sense $f(y)$ and Y(p) are inverse functions; g(y) and $\Phi(p)$ are inverse in the same sense.

Since

$$f(y)=\begin{cases}u^{-1}(y), & y\in\bigcup\limits_{l=1}^{k} j_k,\\ p_k, & y\in j_k,\quad 1\leqslant k\leqslant n,\end{cases}$$

$f(y)$ is a piecewise twice continuously differentiable function.

2. Since, according to [1], $Y(p) = \Phi(p) + \Psi(p)$ for $p \in (p_1, 1)$, $\Psi(p)$ is equal to zero by definition), then (Fig. 4)

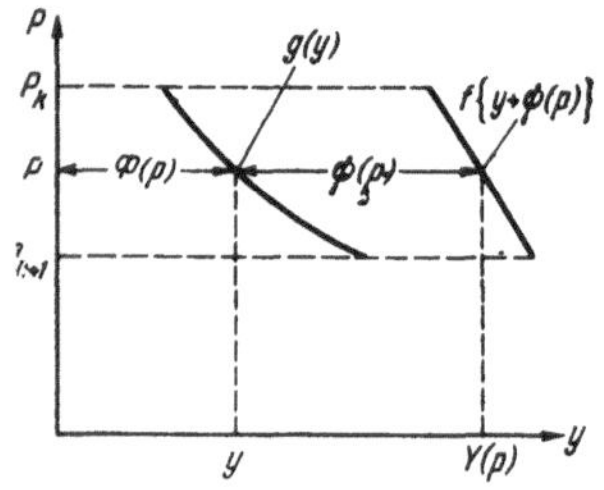

Fig. 4

$$g(y) = f\{y + \Psi[g(y)]\}.$$

In particular, $g(y) = f(y)$ when $y \in (0, y_1)$.

We shall use the implicit function theorem to prove that $g(y)$ is a piecewise twice continuously differentiable function.

Let

$$\varphi(y, p) = f[y + \Psi(p)] - p.$$

The function $\varphi(y, p)$ is chosen such that $\varphi[y, g(y)] = 0$.

Since $f(y)$ and $\Psi(p)$ are piecewise twice continuously differentiable functions (and moreover, $\Psi(p)$ is piecewise analytic) and since $f' \leq 0$ but $\Psi' > 0$ everywhere that f' and Ψ' exist, then $\varphi(y, p)$ has continuous second partial derivatives at the point $[y, g(y)]$ for all y except a finite number of points:

$$\frac{\partial \varphi}{\partial p} = f'[y + \Psi(p)] \cdot \Psi'(p) - 1 \leqslant -1.$$

Hence it follows that $g(y)$ is a piecewise twice continuously differentiable function.

Condition D

We transform the function

$$\tau(p) + \int_p^1 \sqrt{q^2 - p^2}\, d\Phi(q),$$

using the formula

$$\tau(p) = \int_0^{Y(p)} \sqrt{u^{-2}(y) - p^2}\, dy$$

(by examining expressions (1) for X(p) and T(p) and the relation $\tau(p) = T(p) - pX(p)$, it is easy to see that this formula is valid).

We shall assume that $y_0 = \bar{y}_0 = 0$. Then, for $p \in (p_{k+1}, p_k)$, $0 \leq k \leq n$,

$$\tau(p) = \int_0^{Y(p)} \sqrt{f^2(y) - p^2}\, dy + \sum_{i=0}^{k} \int_{y_i}^{\bar{y}_i} \left(\sqrt{u^{-2}(y) - p^2} - \sqrt{p_i^2 - p^2}\right) dy.$$

We write the first term as a Stieltjes integral:

$$\int_0^{Y(p)} \sqrt{f^2(y) - p^2}\, dy = \int_1^p \sqrt{q^2 - p^2}\, dY(q) = \int_1^p \sqrt{q^2 - p^2}\, d\Phi(q) + \int_1^p \sqrt{q^2 - p^2}\, d\Psi(q).$$

Thus, for $p \in (p_{k+1}, p_k)$, $0 \leq k \leq n$,

$$\tau(p) + \int_p^1 \sqrt{q^2 - p^2}\, d\Phi(q) = \int_1^p \sqrt{q^2 - p^2}\, d\Psi(q) + \sum_{i=0}^{k} \int_{y_i}^{\bar{y}_i} \left(\sqrt{u^{-2}(y) - p^2} - \sqrt{p_i^2 - p^2}\right) dy.$$

This function is obviously absolutely continuous on each of the intervals (p_{k+1}, p_k), $0 \leq k \leq n$.

§4. Sufficiency of Conditions A, B, C, and D

We shall show that if conditions A, B, C, and D of Section 2 are satisfied, then equations (1) (see Introduction) have a solution in the class of allowable velocity profiles.

According to [1], instead of this it is sufficient to show the following.

Assertion. There exists a function $u^*(y)$ that is an allowable velocity profile such that when $Y^*(p) = \inf\{y, pu^*(y) \geq 1\}$ the following identity holds:

$$\int_0^{Y^*(p)} \sqrt{[u^*(y)]^{-2} - p^2}\, dy \equiv \tau(p),$$

$$p \in (0, 1), \quad p \neq p_1, \ldots, p_n. \tag{2}$$

The proof of this assertion repeats, in many respects, Sections 16 and 17 of [1]. It is rather lengthy, and we shall divide it into three parts:

1) plot $u^*(y)$,

2) verify that it is an allowable velocity profile, and

3) show that $u^*(y)$ satisfies identity (2).

Then (repeating, for convenience in reading, the reasoning of [1]) we explain how it follows from the proved assertion that $u^*(y)$ satisfies equations (1).

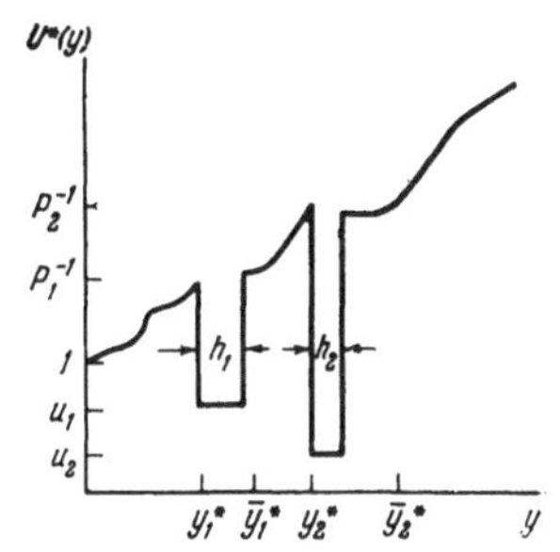

Fig. 5

1. Plotting the Function u*(y)

1.1. The function u*(y) will have the following form (Fig. 5). We take the numbers p_k, σ_k, and C from conditions B3 and C4 and define the positive numbers u_k and h_k $(1 \leq k \leq n)$ such that

$$\frac{\sigma_k}{u_k^{-2} - p_k^2} \leqslant \frac{C\pi}{2n}, \tag{3}$$

$$h_k \sqrt{u_k^{-2} - p_k^2} = \sigma_k. \tag{4}$$

This definition is correct, for according to condition B1, the numbers σ_k are positive.

The intervals $j_k^* = (y_k^*, \bar{y}_k^*)$ will be selected in a special way (to be explained below); the inequalities $y_k^* + h_k \leq \bar{y}_k^*$, $1 \leq k \leq n$, will be valid.

The function u*(y) will be a step function on each interval j_k^*

$$u^*(y) = \begin{cases} u_k, & y_k^* < y < y_k^* + h_k, \\ p_k^{-1}, & y_k^* + h_k \leqslant y \leqslant \bar{y}_k^*. \end{cases}$$

In order to select the intervals j_k^* and define u*(y) outside of these intervals, we construct two auxiliary functions Ψ*(q) and Y*(q).

1.2. The function Ψ*(q), which in general is related to u*(y) by the formula

$$\Psi^*(q) = \frac{2}{\pi} \sum_{i=1}^{k} \int_{y_i^*}^{\bar{y}_i^*} \operatorname{arctg} \sqrt{\frac{[u^*(y)]^{-2} - p_i^2}{p_i^2 - q^2}}\, dy,$$

$$q \in (p_{k+1}, p_k), 1 \leqslant k \leqslant n,$$

in our case takes the form

$$\Psi^*(q) = \frac{2}{\pi} \sum_{i=1}^{k} h_i \operatorname{arctg} \sqrt{\frac{u_i^{-2} - p_i^2}{p_i^2 - q^2}}.$$

Note that it does not depend on the arrangement of intervals j_i^* (that is, on y_i^* and $\bar{y}_i^*$).

1.3. The derivative of Ψ*(q) on the interval (p_{k+1}, p_k), $1 \leq k \leq n$, is

$$\Psi^{*\prime}(q) = \frac{2q}{\pi} \sum_{i=1}^{k} \frac{1}{\sqrt{p_i^2 - q^2}} \frac{h_i \sqrt{u_i^{-2} - p_i^2}}{u_i^{-2} - q^2},$$

or, using equation (4),

$$\Psi^{*\prime}(q) = \frac{2q}{\pi} \sum_{i=1}^{k} \frac{1}{\sqrt{p_i^2 - q^2}} \frac{\sigma_i}{u_i^{-2} - q^2}.$$

When $q \in (p_{k+1}, p_k)$, the relations

$$\frac{1}{\sqrt{p_i^2 - q^2}} \leqslant \frac{1}{\sqrt{p_k^2 - q^2}}, \quad \frac{\sigma_i}{u_i^{-2} - q^2} < \frac{\sigma_i}{u_i^{-2} - p_k^2},$$

$$1 \leqslant i \leqslant k.$$

are obviously valid.

Therefore, using equation (3), when $q \in (p_{k+1}, p_k)$

$$\Psi^{*\prime}(q) < \frac{2q}{\pi} \frac{k}{\sqrt{p_k^2 - q^2}} \frac{C\pi}{2n} < \frac{Cq}{\sqrt{p_k^2 - q^2}}, \quad 1 \leqslant k \leqslant n.$$

1.4. We let $\Psi^*(q) \equiv 0$ when $q \in (p_1, 1)$ and give Y*(q) by the formula

$$Y^*(q) = \Phi(q) + \Psi^*(q), \quad q \in (0, 1).$$

We shall show that Y*(q) does not increase.

For the interval $(p_1, 1)$, this is ensured by condition C2.

Now let us consider the interval (p_{k+1}, p_k). According to condition C5, everywhere on it (except, perhaps, at a finite number of points) there exists a derivative Φ'(q) that is finite or equal to −∞ (for the discontinuity points, there are one-sided backward and forward derivatives). Where Φ'(q) is finite, we have

$$\Phi'(q) < \frac{-Cq}{\sqrt{p_k^2 - q^2}}$$

(see condition C4). According to the preceding paragraph,

$$\Psi^{*\prime}(q) < \frac{Cq}{\sqrt{p_k^2 - q^2}}.$$

That is, $Y^{*\prime}(q) < 0$ where Φ'(q) is finite and $Y^{*\prime}(q) = -\infty$ where Φ'(q) is infinite.

The function Ψ*(q) is continuous on (p_{k+1}, p_k). According to condition C2, at the discontinuity points of Φ(q), $\Phi(q - 0) > \Phi(q + 0)$; thus, $Y^*(q-0) > Y^*(q+0)$ at these points. That is, Y*(q) does not increase on (p_{k+1}, p_k). To verify that Y*(q) does not increase on the entire interval (0, 1), it remains to prove the inequalities

$$Y^*(p_k - 0) - Y^*(p_k + 0) \geqslant 0, \quad 1 \leqslant k \leqslant n.$$

It is easy to see that they are satisfied. In fact, according to condition C2, $\Phi(p_k - 0) - \Phi(p_k + 0) \geq 0$ for any $k = 1, \ldots, n$. The difference $\Psi^*(p_k - 0) - \Psi^*(p_k + 0)$ is equal to h_k, since

$$\lim_{q \to p_k - 0} \frac{2}{\pi} h_k \operatorname{arctg} \sqrt{\frac{u_k^{-2} - p_k^2}{p_k^2 - q^2}} = h_k.$$

Thus, even the stronger inequalities

$$Y^*(p_k - 0) - Y^*(p_k + 0) \geqslant h_k, \quad 1 \leqslant k \leqslant n.$$

are valid,

1.5. Now we can conclude the plotting of $u^*(y)$. Let $y_k^* = Y^*(p_k + 0)$ and $\bar{y}_k^* = Y^*(p_k - 0)$. Just as we had intended, on the intervals $j_k^*(y_k^*, \bar{y}_k^*)$ we determine $u^*(y)$ by the formula

$$u^*(y) = \begin{cases} u_k, & y_k^* < y < y_k^* + h_k, \\ p_k^{-1}, & y_k^* + h_k \leqslant y \leqslant \bar{y}_k^*. \end{cases}$$

This is possible, since $\bar{y}_k^* - y_k^* \geq h_k$.

Outside of the intervals j_k^* we define $u^*(y)$ such that it does not decrease outside of $\bigcup_{k=1}^{n} j_k^*$ and satisfies the relation $Y^*(p) = \inf\{y, pu^*(y) \geq 1\}$. This is possible, since $Y^*(p)$ does not increase.

2. Proof That $u^*(y)$ Is an Allowable Profile

2.1. We shall show that $u^*(y)$ is a piecewise twice continuously differentiable function. Obviously it is sufficient to verify that

$$f^*(y) = \begin{cases} p_k, & y \in j_k^*, \quad 1 \leqslant k \leqslant n, \\ [u^*(y)]^{-1}, & y \in \bigcup_{k=1}^{n} j_k^*. \end{cases}$$

has this property.

The function $f^*(y)$ is the inverse of $Y^*(p)$ and, therefore, is related to $g(y)$ as follows (compare with the proof of condition C5 of Section 3):

$$f^*(y) = g\{y - \Psi^*[f^*(y)]\}.$$

We introduce the notation $\varphi^*(y, p) = p - g[y - \Psi^*(p)]$. Then $f^*(y)$ satisfies the equation $\varphi^*[y, f^*(y)] = 0$.

If we exclude a finite number of points y, then at the point $(y, p = f^*(y))$ the function $\varphi^*(y, p)$ is twice continuously differentiable and we have

$$\frac{\partial \varphi^*}{\partial p} = 1 + g'[y - \Psi^*(p)] \cdot \Psi^{*\prime}(p).$$

Let us verify that $\partial\varphi^*/\partial p \neq 0$ at each such point.

If $g'[y - \Psi^*(p)] = 0$, then $\partial\varphi^*/\partial p = 1$.

If $g'[y - \Psi^*(p)] \neq 0$,(and, therefore, $g'[y - \Psi^*(p)] < 0$), then $g'[y - \Psi^*(p)]\,\Phi'(p) = 1$. Since $\Psi^*(p) < -\Phi'(p)$ (this inequality was proved when $u^*(y)$ was plotted in 1.4), then in this case $0 > g'[y - \Psi^*(p)] \cdot \Psi^{*\prime}(p) > -1$, therefore, $\partial\varphi^*/\partial p > 0$.

According to the implicit function theorem, $f^*(y)$ is piecewise twice continuously differentiable. Thus, the piecewise double continuous differentiability of $u^*(y)$ is proved.

2.2. It is clear that $u^*(0) = 1$. According to conditions C3 and C1, $u^*(y)$ is bounded on each finite interval of the y axis and unbounded on the entire axis. The number of waveguides of $u^*(y)$ is equal to n.

Therefore, $u^*(y)$ is an allowable profile.

3. Proof that $u^*(y)$ Satisfies Inequality (2)

3.1. Let

$$\tau^*(p) = \int_0^{Y^*(p)} \sqrt{[u^*(y)]^{-2} - p^2}\, dy, \quad p \neq p_k;$$

$$\tau^*(p_k) = \tau(p_k), \quad 1 \leqslant k \leqslant n.$$

We must establish that $\tau^*(p) \equiv \tau(p)$, $p \in (0, 1)$.

3.2. Let

$$X^*(p) = \int_0^{Y^*(p)} \frac{pu^*(y)\, dy}{\sqrt{1 - p^2[u^*(y)]^2}},$$

$$\Phi^*(q) = \frac{2}{\pi} \int_q^1 \frac{X^*(p)\, dp}{\sqrt{p^2 - q^2}}; \quad p, q \in (0, 1).$$

According to the main formula (which is valid, since $u^*(y)$ is an allowable profile), $Y^*(q) = \Phi^*(q) + \Psi^*(q)$. If we compare this formula with the definition of $Y^*(q)$ (given in 1.4), according to which $Y^*(q) = \Phi(q) + \Psi^*(q)$, we find $\Phi^*(q) \equiv \Phi(q)$, $q \in (0, 1)$.

3.3. According to condition B3, $\tau(p)$ is continuous everywhere, except at the points $p = p_1, \ldots, p_n$, where there are discontinuities $\sigma_1, \ldots, \sigma_n$.

Since $u^*(y)$ is an allowable profile with n waveguides, then according to notes 1 and 2 in Section 2, $\tau^*(p)$ has exactly n discontinuities. According to formula (4), at points $p_1, \ldots, p_n$ it has discontinuities $\sigma_1, \ldots, \sigma_n$.

That is, the difference $\tau^*(p) - \tau(p)$ is continuous on (0, 1).

3.4. We shall show that this difference is absolutely continuous. According to condition D, the functions

$$\chi^*(p) = \tau^*(p) + \int_p^1 \sqrt{q^2 - p^2}\, d\Phi^*(q)$$

and

$$\chi(p) = \tau(p) + \int_p^1 \sqrt{q^2 - p^2}\, d\Phi(q)$$

are absolutely continuous when $p \in (p_{k+1}, p_k)$. Here condition D is used as a necessary condition for $X^*(p)$ and as a sufficient condition for $X(p)$.

Since $\Phi^*(q) \equiv \Phi(q)$, $q \in (0, 1)$ (see 3.2), then $\tau^*(p) - \tau(p) \equiv \chi^*(p) - \chi(p)$. That is, the difference $\tau^*(p) - \tau(p)$ is absolutely continuous on (p_{k+1}, p_k) and therefore, according to 3.3, over the entire interval $(0, 1)$ as well.

3.5. According to conditions A2 and A3,

$$\tau'(p) = -X(p), \quad \tau^{*\prime}(p) = -X^*(p)$$

almost everywhere on $(0, 1)$. In fact, if $T'(p)$ and $X'(p)$ exist, then $\tau'(p) = [T(p) - pX(p)]' = T'(p) - pX'(p) - X(p) = -X(p)$. The proof is the same for $\tau^{*\prime}(p)$.

According to condition B2, $\tau(1 - 0) = \tau^*(1 - 0) = 0$.

Therefore, using 3.4,

$$\tau^*(p) - \tau(p) = \int_p^1 [X^*(q) - X(q)]\, dq.$$

3.6. Let us verify that the latter integral is zero for $p \in (0, 1)$.

It is easy to see that

$$\frac{\pi}{2} = \int_r^p \frac{q\,dq}{\sqrt{(p^2 - q^2)(q^2 - r^2)}}.$$

Therefore,

$$\frac{\pi}{2}\int_r^1 X(p)\,dp = \int_r^1 X(p)\left[\int_r^p \frac{q\,dq}{\sqrt{(p^2 - q^2)(q^2 - r^2)}}\right] dp.$$

In the triangle $\Delta_r \{r < p < q < 1\}$ on the plane (p, q) (Fig. 6) the function $qX(p)/\sqrt{(p^2 - q^2)(q^2 - r^2)}$

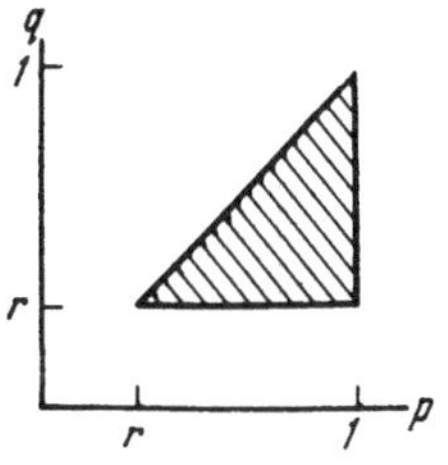

Fig. 6

is measurable and positive (using condition A1), and one of the iterated integrals exists. Therefore, this function is integrable in Δ_r. Using the Fubini theorem and changing the order of integration we obtain

$$\frac{\pi}{2}\int_r^1 X(p)\,dp = \int_r^1 \frac{q\int_q^1 \frac{X(p)\,dp}{\sqrt{p^2 - q^2}}}{\sqrt{q^2 - p^2}}\,dq = \frac{\pi}{2}\int_r^1 \frac{q\Phi(q)\,dq}{\sqrt{q^2 - r^2}}.$$

Quite similarly it is proved that

$$\int_r^1 X^*(p)\,dp = \int_r^1 \frac{q\Phi^*(q)\,dq}{\sqrt{q^2 - r^2}}.$$

Since according to 3.2, $\Phi^*(q) \equiv \Phi(q)$, $q \in (0, 1)$, then

$$\int_r^1 [X^*(p) - X(p)]\,dp \equiv 0.$$

Thus, $\tau^*(p) \equiv \tau(p)$, and the assertion is proved.

4. Proof That $u^*(p)$ Satisfies System (1)

4.1. We shall show that $X(p)$ and $T(p)$ are defined from $\tau(p)$ everywhere on $(0, 1)$, except perhaps at a finite number of points.

Let Π be the set of those p for which $X(p)$ and $T(p)$ are differentiable, Π' the set of those p for which $X(p) = X(p \pm 0) = T(p) = T(p \pm 0) = \infty$, and let $\widetilde{\Pi}$ be a subset of $(0, 1)$ that does not belong to $\Pi \cup \Pi'$. According to condition A4, set $\widetilde{\Pi}$ consists of a finite number of points.

It was shown in **3.5** that $\tau'(p) = -X(p)$ when $p \in \Pi$.

Let us consider the set $\hat{\Pi}$, where $\tau'(p)$ exists and is finite and continuous. According to what we have already proved, $\hat{\Pi} \supseteq \Pi$. It is easy to see that $\hat{\Pi} \cap \Pi' = 0$. In fact, let $\hat{p} \in \hat{\Pi}$: according to condition A2, set Π is of full measure; therefore, there exists a sequence $q_n \in \Pi$ such that $q_n \to \hat{p}$. Since $\tau'(q_n) = -X(q_n)$ and since $\tau'(p)$ is continuous at point $\hat{p}$, then $\lim_{n\to\infty} X(q_n) = -\tau'(\hat{p})$. Since $\tau'(\hat{p}) < \infty$, this limit is finite. That is, $\hat{p} \overline{\in} \Pi'$, i.e., $\hat{\Pi} \cap \Pi' = 0$. Thus, $\Pi \subseteq \hat{\Pi} \subseteq \Pi \cup \widetilde{\Pi}$.

Let

$$X(p) = \begin{cases} -\tau'(p), & p \in \hat{\Pi}, \\ +\infty, & p \overline{\in} \hat{\Pi}; \end{cases} \qquad T(p) = \tau(p) + pX(p).$$

In this, incorrect values can occur only at a finite number of points (when $p \in \widetilde{\Pi}$).

4.2. Similarly, from $\tau^*(p)$ we can determine $X^*(p)$ and $T^*(p)$, where

$$T^*(p) = \int_0^{Y^*(p)} \frac{dy}{u^*(y)\sqrt{1-p^2[u^*(y)]^2}}.$$

According to 3.6, $\tau(p) \equiv \tau^*(p)$. Therefore $X(p) = X^*(p)$ and $T(p) = T^*(p)$ everywhere on $(0, 1)$, except perhaps at a finite number of points. But this does imply that $u^*(y)$ is a solution of system (1).

The sufficiency of conditions A, B, C, and D has been proved.

Note. Since formula (3) in 1.1 is not an exact equality, the numbers h_k (and also u_k) are not determined uniquely. Different sets $h_1, \ldots, h_k$ correspond to different functions $u^*(y)$. Thus, having proved the existence of a solution $u^*(y)$, we have simultaneously proved its nonuniqueness.

Chapter II

PROPERTIES OF TRAVEL-TIME CURVES

§5. Condition D'

The complicated condition D can be replaced (in the system of sufficient conditions) by the condition:

D'. $X(p) = \infty$ in not more than a denumerable set of values $p \in (0, 1)$.

Condition D' is not a necessary condition. But satisfaction of conditions A, B, C, and D' is sufficient for $\Gamma\{2X(p), 2T(p)\}$ to be a travel-time curve. Let us prove this.

Condition D was used only to prove the absolute continuity of the difference $\tau^*(p) - \tau(p)$ (see **3.4** in Section 4). If condition D' is fulfilled, then each of these functions (both $\tau(p)$ and $\tau^*(p)$) is absolutely continuous when $p \in (p_{k+1}, p_k)$, $0 \le k \le n$.

Proof. We shall use the following theorem.

If the derivative of a monotone decreasing continuous function goes to $-\infty$ only on a denumerable set of values, then this function is absolutely continuous.

Therefore $\tau(p)$, $p \in (p_{k+1}, p_k)$, is absolutely continuous, since according to **3.5** of Section 4, condition A4, and condition D', outside the denumerable set of points $\tau'(p) = -X(p) > -\infty$; and since, according to condition B1 and B3, $\tau(p)$ is monotonic and continuous at $p \ne p_1, \ldots, p_n$.

To prove the absolute continuity of $\tau^*(p)$ it is sufficient to verify that $X^*(p) = \infty$ on not more than a denumerable set of points p.

Since by definition

$$\tau^*_{(p)} = \int_0^{Y^*(p)} \sqrt{[u^*(y)]^{-2} - p^2}\, dy,$$

this is a monotonic function that is bounded on the interval $(r, 1)$ for any $r > 0$. Since, in addition, almost everywhere $\tau^{*\prime}(p) = -X^*(p)$ (see **3.5** of Section 4), then the integral

$$\int_r^1 X^*(p)\,dp$$

exists for any positive r.

Using this, we can, just as in **3.6**, establish the equality

$$\int_r^1 [X^*(p) - X(p)]\,dp = 0.$$

Thus, taking condition A2 into account, $X^*(p) - X(p) = 0$ almost everywhere.

According to conditions A3 and A4, we can exclude from $(0, 1)$ a finite number of points such that at the remaining points

$$X(p) = X(p \pm 0),\ X^*(p) = X^*(p \pm 0).$$

If $X(p^0)$ were finite at some point p^0 that did not belong to the excluded set, while $X^*(p^0) = \infty$, then the integral of the difference $X^*(p) - X(p)$ in a small vicinity of p^0 would be positive, which would be inconsistent with the coincidence of $X^*(p)$ and $X(p)$ almost everywhere. That is, $X^*(p) = \infty$ also on not more than a denumerable set of points p. Therefore, $\tau^*(p)$ is absolutely continuous on (p_{k+1}, p_k), $0 \le k \le n$; Q.E.D.

§6. Points of Continuity of X(p) and T(p)

In this section we shall show that $X(p)$ and $T(p)$ are continuous at the same points $p \in (0, 1)$ and that this set of points of continuity is open.

1. We shall use necessary condition B3. Let p be a point of continuity of $\tau(p)$, i.e., $p \ne p_1, \ldots, p_n$. Since $\tau(p) = T(p) - pX(p)$, functions $X(p)$ and $T(p)$ are either both discontinuous or both continuous at point p. At points p_k, $1 \le k \le n$, $X(p)$ and $T(p)$ are obviously discontinuous. This follows immediately from relations (1):

$$X(p) = \int_0^{Y(p)} \frac{p\,dy}{\sqrt{u^{-2}(y) - p^2}},\quad T(p) = \int_0^{Y(p)} \frac{dy}{u^2(y)\sqrt{u^{-2}(y) - p^2}}.$$

Refinement. We decided to regard u(y) as a solution of equations (1) (see Introduction) if u(y) satisfied these equations for all $p \in (0, 1)$, except perhaps at a finite number of points.

Thus, if a curve $\{2X(p), 2T(p)\}$ is a travel-time curve and if functions $\hat{X}(p)$ and $\hat{T}(p)$ differ from X(p) and T(p) only at a finite number of points, then $\{2\hat{X}(p), 2\hat{T}(p)\}$ is also a travel-time curve.

Therefore, the proved assertion is formally simply invalid: for example, we can change X(p) at some point of continuity p^0, and then X(p) will be discontinuous and T(p) will remain continuous at $p = p^0$.

The exact formulation of the above assertion must therefore be: "Let $\{2X(p), 2T(p)\}$ be a travel-time curve. Then X(p) and T(p) are continuous at the same points $p \in (0, 1)$ or, in any case, become such if they are changed at a finite number of points."

Some assertions that we shall prove later will require similar refinements (Lemma 7.3, Section 7; theorem of Section 10). At the corresponding places, we shall limit ourselves to references to the refinement just made.

All these assertions are valid without any reservations if X(p) and T(p) are understood not as one of the parametrizations of the travel-time curve but as functions occurring on the right side of equations (1) when u(y) is an allowable profile.

2. The set for which X(p) and T(p) are continuous we shall call N. The intersection of N with the closure $\overline{\Pi'}$ offset Π', where $X(p) = X(p \pm 0) = T(p) = T(p \pm 0) = +\infty$, is obviously empty. Set Π, where X(p) and T(p) are differentiable, obviously belongs to N. Thus, according to condition A4, the complement of N on the interval (0, 1) either coincides with $\overline{\Pi'}$ or is obtained by adding to $\overline{\Pi'}$ a finite number of points. In both cases, the complement of N is closed while N itself is open.

Note. The behavior of functions X(p) and T(p) is related in the following way to the properties of the function $f(y)$, which was defined in Section 1.

We note a finite number of points $p \in (0, 1)$: points $p = f(y \pm 0)$ if $f(y - 0) > f(y + 0)$ and points $p = f(y^0)$ if when $y = y^0$ the function $f(y)$ itself is continuous but $f'(y)$ or $f''(y)$ is either discontinuous or does not exist.

We divide the remaining points $p \in (0, 1)$ into two sets. The first contains those p for which $f'(Y(p)) \neq 0$ and the second those p for which $f'(Y(p)) = 0$.

Then at the points of the first set X(p) and T(p) are continuously differentiable and their derivatives are related as $T'(p) = pX'(p)$, while at points of the second set $X(p) = X(p \pm 0) = T(p) = T(p \pm 0) = +\infty$. The first set is open; the second set is closed on (0, 1) and is of measure zero.

This assertion was proved in [1]. We shall use it below.

§7. Conditions Related to the Monotonicity of $\Phi(q)$

According to condition C2, the function $\Phi(q)$ does not increase on (0, 1). If condition C4 is used, it is apparent that it even strictly decreases on the interval $(0, p_1)$.

When we wish to determine whether or not a given curve $\Gamma\{x = 2X(p),\ t = 2T(p)\}$ is a travel time curve, verification of the monotonicity of $\Phi(q)$ can present certain difficulties. It is useful to have some simple conditions for X(p) whose satisfaction ensures the monotonicity of $\Phi(q)$, and also conditions whose violation necessarily disrupts the monotonicity of $\Phi(q)$.

Lemma 7.1. Let $q_0 \geq 0$. If X(p) does not increase when $p \in (q_0, 1)$, then $\Phi(q)$ decreases when $q \in (q_0, 1)$.

Proof. Let $q_0 < q < Q < 1$. We shall denote $(Q - q)$ by δ. Since in the expression for $\Phi(q)$ the integrand is positive,

$$\frac{\pi}{2}\Phi(q) = \int_q^1 \frac{X(p)\,dp}{\sqrt{p^2 - q^2}} > \int_q^{1-\delta} \frac{X(p)\,dp}{\sqrt{p^2 - q^2}}.$$

We make the substitution $r = p + Q - q = p + \delta$. Then $p - q = r - Q$ and $p + q = r + Q - 2\delta$, so that

$$\frac{\pi}{2}\Phi(q) > \int_Q^1 \frac{X(r-\delta)\,dr}{\sqrt{(r-Q)(r+Q-2\delta)}} > \int_Q^1 \frac{X(p-\delta)dp}{\sqrt{p^2 - Q^2}}.$$

By convention

$$X(p - \delta) \geqslant X(p).$$

That is,

$$\frac{\pi}{2}\Phi(q) > \int_Q^1 \frac{X(p)\,dp}{\sqrt{p^2 - Q^2}} = \frac{\pi}{2}\Phi(Q),$$

Q.E.D.

The condition of Lemma 7.1 is, of course, not necessary for the monotonicity of $\Phi(q)$. This is shown by the following example.

Example. Let the function u(y), the velocity profile corresponding to travel-time curve Γ, be discontinuous at some point Y that does not belong to the waveguides $j_1, \ldots, j_n$ (Fig. 7):

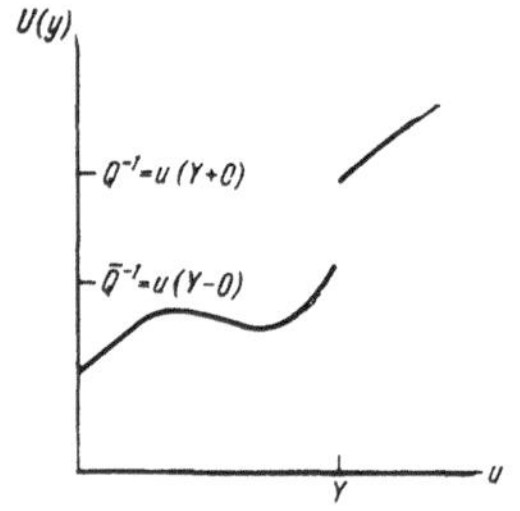

Fig. 7

$$u(Y-0) < u(Y+0).$$

Then the function $Y(p)$ is constant on the interval $Q = u^{-1}(Y+0) < p < \bar{Q} = u^{-1}(Y-0)$

$$Y(p) = Y,\quad p \in (Q, \bar{Q}).$$

Thus when $p \in (Q, \bar{Q})$, the function

$$X(p) = \int_0^Y \frac{pu(y)\,dy}{\sqrt{1 - p^2u^2(y)}}$$

increases monotonically.

When the inequality that we shall prove in the following lemma is not satisfied, the monotonicity of $\Phi(q)$ is necessarily disrupted.

Lemma 7.2. For any $q \in (0, 1)$,

$$\overline{\lim_{p\to q-0}}\, X(p) \geqslant \underline{\lim_{p\to q+0}}\, X(p).$$

Proof. We shall assume the converse: for some $q_0 \in (0, 1)$ we shall let

$$\overline{\lim_{p\to q_0-0}}\, X(p) < \underline{\lim_{p\to q_0+0}}\, X(p),$$

and show that then

$$\Phi(q) < \Phi(q_0)$$

for q less than q_0 and sufficiently close to q_0.

According to our assumption, we can find a and b, $0 < a < b$, and q_1 and q_2, $0 < q_1 < q_0 < q_2 < 1$, such that

$$X(p) < a \quad \text{when} \quad p \in (q_1, q_0),$$
$$X(p) > b \quad \text{when} \quad p \in (q_0, q_2)$$

(Fig. 8).

When $q \in (q_1, q_0)$

$$\frac{\pi}{2}\Phi(q) = \int_q^{q_0} \frac{X(p)\,dp}{\sqrt{p^2-q^2}} + \int_{q_0}^{q_2} \frac{X(p)\,dp}{\sqrt{p^2-q^2}} + \int_{q_2}^{1} \frac{X(p)\,dp}{\sqrt{p^2-q^2}},$$

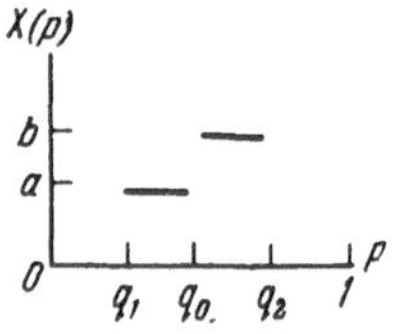

Fig. 8

$$\frac{\pi}{2}\Phi(q_0) = \int_{q_0}^{q_2} \frac{X(p)\,dp}{\sqrt{p^2-q_0^2}} + \int_{q_2}^{1} \frac{X(p)\,dp}{\sqrt{p^2-q_0^2}}.$$

Therefore, when $q \in (q_1, q_0)$

$$\frac{\pi}{2}[\Phi(q_0) - \Phi(q)] >$$

$$> \int_{q_0}^{q_2} X(p)\left(\frac{1}{\sqrt{p^2-q_0^2}} - \frac{1}{\sqrt{p^2-q^2}}\right)dp -$$

$$- \int_q^{q_0} \frac{X(p)\,dp}{\sqrt{p^2-q^2}} >$$

$$> b\int_{q_0}^{q_2}\left(\frac{1}{\sqrt{p^2-q_0^2}} - \frac{1}{\sqrt{p^2-q^2}}\right)dp - a\int_q^{q_0}\frac{dp}{\sqrt{p^2-q^2}}.$$

Integrating, we obtain

$$\frac{\pi}{2}[\Phi(q_0) - \Phi(q)] > b\left(\operatorname{arch}\frac{p}{q_0} - \operatorname{arch}\frac{p}{q}\right)\Bigg|_{p=q_0}^{p=q_2} -$$

$$- a\operatorname{arch}\frac{p}{q}\Bigg|_{p=q}^{p=q_0} = b\left(\operatorname{arch}\frac{q_2}{q_0} - \operatorname{arch}\frac{q_2}{q}\right) + (b-a)\operatorname{arch}\frac{q_0}{q}.$$

The function on the right side of the latter inequality we shall denote by $\varphi(q)$. Let us take its derivative:

$$\varphi'(q) = \frac{bq_2}{q\sqrt{q_2^2 - q^2}} - \frac{(b-a)q_0}{q\sqrt{q_0^2-q^2}}.$$

For q less than q_0 and sufficiently close to q_0, obviously,

$$\varphi'(q) < 0.$$

Moreover, since $\varphi(q_0) = 0$, then near point q_0, and at the left of it,

$$\varphi(q) > 0$$

and thus,

$$\Phi(q) < \Phi(q_0).$$

We have arrived at a contradiction of the condition that the function $\Phi(q)$ does not increase. Thus the lemma is proved.

Fig. 9

According to Lemma 7.2, the curve in Fig. 9 is clearly not a travel-time curve.

Lemma 7.2 describes the behavior of X(p) in the vicinity of point q, but it says nothing about the value of this function at the point q itself. The following lemma (note that its formulation requires a refinement similar to the one made in Section 6) indicates how the value of X(q) is related to the values of X(p) when $p < q$.

Lemma 7.3. For any $q \in (0, 1)$

$$\lim_{p \to q-0} X(p) \geqslant X(q).$$

Proof. Let us assume the converse: for some $q \in (0, 1)$ and for some $J > 0$ let

$$\lim_{p \to q-0} X(p) < J < X(q).$$

We select a monotonic sequence $p_n \to q - 0$ such that for any n

$$X(p_n) = \int_0^{Y(p_n)} \frac{u(y)\,dy}{\sqrt{p_n^{-2} - u^2(y)}} < J.$$

Since $p_n < q$, then $Y(p_n) \geq Y(q)$, we only strengthen the latter inequality by writing

$$\int_0^{Y(q)} \frac{u(y)\,dy}{\sqrt{p_n^{-2} - u^2(y)}} < J.$$

When the numbers p_n, increasing monotonically, approach q, the functions $u(y)/\sqrt{p_n^{-2} - u^2(y)}$, also increasing monotonically, approach $u(y)/\sqrt{q^{-2} - u^2(y)}$. According to the theorem of B. Levi, the limit function is integrable, and

$$\int_0^{Y(q)} \frac{u(y)\,dy}{\sqrt{p_n^{-2} - u^2(y)}} \to \int_0^{Y(q)} \frac{u(y)\,dy}{\sqrt{q^2 - u^2(y)}} = X(q).$$

Moving to the limit in the latter of the above inequalities, we obtain

$$X(q) \leqslant J.$$

The obtained contradiction proves the lemma.

§8. Allowable Variations of Travel-Time Curves

Let $\Gamma = \{2X(p), 2T(p)\}$ be a travel-time curve. Let us vary the functions X(p) and T(p). If the changed functions $\hat{X}(p)$ and $\hat{T}(p)$ also imply a travel-time curve, then this variation will be called allowable.

Instead of specifying Γ by the two functions X(p) and T(p), we can indicate (see 4.1 of Section 4) only one function $\tau(p)$. If some other function $\hat{\tau}(p)$ also gives a travel-time curve Γ, then transition from $\tau(p)$ to $\hat{\tau}(p)$ is also allowable.

Lemma 8.1. For some $\hat{q} \in (0, 1)$ let

$$\tau(\hat{q} - 0) - \tau(\hat{q} + 0) \geqslant K\hat{q},$$

where K is a positive constant.

Then transition to the function

$$\hat{\tau}(p) = \begin{cases} \tau(p) - Kp, & 0 < p < \hat{q}, \\ \tau(p), & \hat{q} \leqslant p < 1, \end{cases}$$

is allowable.

Proof. Transition from $\tau(p)$ to $\hat{\tau}(p)$ is the same as leaving the function T(p) unchanged and replacing X(p) by

$$\hat{X}(p) = \begin{cases} X(p) - Kp, & 0 < p < \hat{q}, \\ X(p), & \hat{q} \leqslant p < 1. \end{cases}$$

It must be shown that the curve $\hat{\Gamma} = \{2\hat{X}(p), 2T(p)\}$ is a travel-time curve if $\Gamma = \{2X(p), 2T(p)\}$ was a travel-time curve.

Conditions A and B are obviously satisfied for $\hat{\Gamma}$. Since

$$\hat{\Phi}(p) = \begin{cases} \Phi(p) + K \operatorname{arch} \frac{\hat{q}}{p}, & 0 < p < \hat{q}, \\ \Phi(p), & \hat{q} \leqslant p < 1, \end{cases}$$

then condition C is satisfied.

Since $\tau(p)$ and $\hat{\tau}(p)$, just as $\Phi(p)$ and $\hat{\Phi}(p)$, coincide on $(\hat{q}, 1)$ and differ by an analytic function on $(0, \hat{q})$, then condition D is satisfied, and the lemma is proved.

Lemma 8.2 is proved similarly. Let

$$X^*(p) = \begin{cases} X(p) + K_1, & 0 < p < \hat{p}, \\ X(p), & \hat{p} \leqslant p < 1, \end{cases}$$

$$T^*(p) = \begin{cases} T(p) + K_2, & 0 < p < \hat{p}, \\ T(p), & \hat{p} \leqslant p < 1. \end{cases}$$

If $K_2 \geq K_1\hat{p} - \tau(\hat{p} - 0) + \tau(\hat{p} + 0)$, then such a variation is allowable.

Lemma 8.3. Everywhere on (0, 1), except perhaps at a finite number of points (in particular, at all continuity points of X(p)),

$$X(p) = \int_0^{\Phi(p)} \frac{p\,dy}{\sqrt{g^{-2}(y) - p^2}}.$$

Proof. In 3.6 of Section 4 we proved the formula

$$\int_r^1 X(p)\,dp = \int_r^1 \frac{q\Phi(q)\,dq}{\sqrt{q^2 - r^2}}.$$

Therefore, X(p) is uniquely defined by Φ(p) at the continuity points, which means (according to A4), everywhere on (0, 1) as well, except perhaps at a finite number of points.

For velocity profiles u(y) and $\hat{u}(y) = g(y)$, obviously,

$$\Phi(p) = \hat{\Phi}(p),$$

so that (except perhaps at a finite number of points)

$$X(p) = \hat{X}(p).$$

Since

$$\hat{Y}(p) = \hat{\Phi}(p) = \Phi(p),$$

then

$$\hat{X}(p) = \int_0^{\hat{Y}(p)} \frac{p\,dy}{\sqrt{\hat{u}^{-2}(y) - p^2}} = \int_0^{\Phi(p)} \frac{p\,dy}{\sqrt{g^{-2}(y) - p^2}},$$

Q.E.D.

Lemma 8.4. Let $\hat{q}$ be a continuity point of X(p). We shall vary the functions X(p) and T(p) on the interval $(0, \hat{q})$, letting $\hat{X}(p) = X(\hat{q}) = \hat{X}$, $\hat{T}(p) = T(\hat{q}) = \hat{T}$ when $p \in (0, \hat{q})$; when $p \geq \hat{q}$, we let $\hat{X}(p) = X(p)$, $\hat{T}(p) = T(p)$. Such a variation is allowable.

Before proving Lemma 8.4, let us explain its geometric meaning.

Geometric Meaning of Lemma 8.4. Let $\hat{q}$ be a continuity point of X(p). We erase the part of the travel-time curve Γ formed by the rays with parameters $p < \hat{q}$. We call the remaining curve $\hat{\Gamma}$. Then $\hat{\Gamma}$ is itself a travel-time curve, and the rays with parameters $p < \hat{q}$ for this new travel-time curve are focused at one point $(X(\hat{q}), T(\hat{q}))$.

Proof.

1. Conditions A, B, and D are obviously satisfied for the curve $\hat{\Gamma} = \{2\hat{X}(p), 2\hat{T}(p)\}$. Since $\hat{\Phi}(q) = \Phi(q)$ when $q \in [\hat{q}, 1)$, then conditions C are satisfied for these q values. It remains to verify conditions C for $q \in (0, \hat{q})$.

2. When $q \in (0, \hat{q})$, obviously,

$$\frac{\pi}{2}\hat{\Phi}(q) = \hat{X} \operatorname{arch} \frac{\hat{q}}{q} + \int_{\hat{q}}^1 \frac{X(p)\,dp}{\sqrt{p^2 - q^2}}. \quad (5)$$

Therefore, $\hat{\Phi}(+0) = +\infty$ – condition C3 is satisfied. It also follows from formula (5) that condition C1 is satisfied: $\hat{\Phi}(q) < \infty$ when $q \in (0, 1)$.

Let us use the formula

$$X(p) = \int_0^{\Phi(p)} \frac{p\,dy}{\sqrt{g^{-2}(y) - p^2}}$$

(see Lemma 8.3) to transform formula (5). We have

$$\int_{\hat{q}}^1 \frac{X(p)\,dp}{\sqrt{p^2 - q^2}} = \int_{\hat{q}}^1 \int_0^{\Phi(p)} \frac{p\,dy\,dp}{\sqrt{[g^{-2}(y) - p^2](p^2 - q^2)}}.$$

By changing the order of integration (Fig. 10), we establish that the latter expression equals

$$\int_0^{\Phi(\hat{q})} \left\{ \int_{\hat{q}}^{g^{-1}(y)} \frac{p\,dp}{\sqrt{[g^{-2}(y) - p^2](p^2 - q^2)}} \right\} dy =$$

$$= \int_0^{\Phi(\hat{q})} \operatorname{arctg} \sqrt{\frac{g^{-2}(y) - \hat{q}^2}{\hat{q}^2 - q^2}}\, dy$$

(see 3.6 of Section 4, where in a similar situation the possibility of changing the order of integration was demonstrated).

Since (according to Lemma 8.3)

$$\hat{X} \operatorname{arch} \frac{\hat{q}}{q} = \int_0^{\Phi(\hat{q})} \frac{\hat{q}\,dy}{\sqrt{g^{-2}(y) - \hat{q}^2}} \operatorname{arch} \frac{\hat{q}}{q},$$

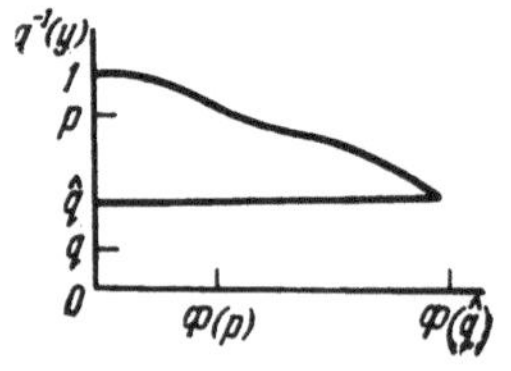

Fig. 10.

then (see formula (5)) we finally obtain: when $q \in (0, \hat{q})$

$$\frac{\pi}{2}\hat{\Phi}(q) = \int_0^{\Phi(\hat{q})} \left[\operatorname{arctg}\sqrt{\frac{g^{-2}(y)-\hat{q}^2}{\hat{q}^2-q^2}} + \frac{\hat{q}}{\sqrt{g^{-2}(y)-\hat{q}^2}}\operatorname{arch}\frac{\hat{q}}{q}\right]dy.$$

In the latter formula, for any fixed $y \in (0, \Phi(q))$ the integrand decreases monotonically as q increases since its partial derivative with respect to q is

$$\frac{q}{g^{-2}(y)-q^2}\sqrt{\frac{g^{-2}(y)-\hat{q}^2}{\hat{q}^2-q^2}} - \frac{\hat{q}}{\sqrt{g^{-2}(y)-\hat{q}^2}}\,\frac{\hat{q}}{q\sqrt{\hat{q}^2-q^2}} =$$

$$= -\frac{g^{-2}(y)\sqrt{\hat{q}^2-q^2}}{[g^{-2}(y)-q^2]\,q\sqrt{g^{-2}(y)-\hat{q}^2}} < 0.$$

Thus, $\hat{\Phi}(q)$ decreases on the interval $(0, \hat{q})$.

3. At point $\hat{q}$ the function $\hat{\Phi}(q)$ is continuous to the left. In fact,

$$\frac{\pi}{2}\Phi(\hat{q}) = \int_{\hat{q}}^{1} \frac{X(p)\,dp}{\sqrt{p^2-\hat{q}^2}}.$$

Further,

$$\lim_{q\to\hat{q}-0}\left(\hat{X}\operatorname{arch}\frac{\hat{q}}{q}\right) = 0,$$

and, finally (according to the theorem of B. Levi),

$$\int_{\hat{q}}^{1}\frac{X(p)\,dp}{\sqrt{p^2-q^2}} \underset{q\to\hat{q}-0}{\longrightarrow} \int_{\hat{q}}^{1}\frac{X(p)\,dp}{\sqrt{p^2-\hat{q}^2}}.$$

Thus, according to formula (5), $\hat{\Phi}(\hat{q}-0) = \hat{\Phi}(\hat{q})$. Thus, $\hat{\Phi}(q)$ does not increase on (0, 1); condition C2 is verified.

4. Let us verify condition C4.

Let the discontinuity points of $\tau(p)$ be $p_1 > p_2 > \ldots > p_n$. That for which

$$\hat{q} \in (p_{m+1}, p_m)$$

we shall denote by p_m ($\hat{q}$ does not coincide with any of the points $p_1, \ldots, p_n$, since $\hat{q}$ is a continuity point of X(p), while at the points $p_1, \ldots, p_n$ the function X(p) is discontinuous).

Then the function $\hat{\tau}(p)$ (which is linear when $0 < p < \hat{q}$) will have discontinuities only at the points $p_1, \ldots, p_m$. To ensure that $\hat{\Phi}(q)$ satisfies condition C4, therefore, it is necessary only to verify that for some $\hat{C} > 0$

$$\hat{\Phi}'(q) < \frac{-\hat{C}q}{\sqrt{p_m^2-q^2}}, \quad q \in (0, \hat{q}).$$

According to paragraph 2,

$$\frac{\pi}{2}\hat{\Phi}'(q) = -\int_0^{\Phi(\hat{q})} \frac{g^{-2}(y)\sqrt{\hat{q}^2-q^2}\,dy}{[g^{-2}(y)-q^2]\,q\sqrt{g^{-2}(y)-\hat{q}^2}}, \quad q \in (0, \hat{q}).$$

Let us write this integral as a Stieltjes integral (see Fig. 10):

$$\frac{\pi}{2}\hat{\Phi}'(q) = \int_{\hat{q}}^{1} \frac{p^2\sqrt{\hat{q}^2-q^2}\,d\Phi(p)}{(p^2-q^2)\,q\sqrt{p^2-\hat{q}^2}}.$$

Since according to condition C4, which was used as a necessary condition for $\Phi(p)$, when $p \in (p_{m+1}, p_m)$

$$\Phi'(p) < \frac{-Cp}{\sqrt{p_m^2-p^2}},$$

then

$$\frac{\pi}{2}\hat{\Phi}'(q) < -\int_{\hat{q}}^{p_m} \frac{p^2\sqrt{\hat{q}^2-q^2}\,Cp\,dp}{(p^2-q^2)\,q\sqrt{p^2-\hat{q}^2}\sqrt{p_m^2-p^2}},$$

$$q \in (0, \hat{q}).$$

Denoting by J(q) the integral

$$\int_{\hat{q}}^{p_m} \frac{\sqrt{\hat{q}^2-q^2}\,p\,dp}{(p^2-q^2)\sqrt{p^2-\hat{q}^2}}, \quad q \in (0, \hat{q}),$$

and using the inequality $q < p$, we obtain: when $q < \hat{q}$

$$\frac{\pi}{2}\hat{\Phi}'(q) < -\frac{Cq^2J(q)}{q\sqrt{p_m^2-q^2}} = -\frac{CJ(q)\,q}{\sqrt{p_m^2-q^2}}.$$

Since

$$J(q) = \operatorname{arctg}\sqrt{\frac{p_m^2-\hat{q}^2}{\hat{q}^2-q^2}} > J(0),$$

then, assuming

$$\hat{C} = \frac{2}{\pi}CJ(0),$$

we have finally

$$\hat{\Phi}'(q) < \frac{-\hat{C}q}{\sqrt{p_m^2-q^2}}, \quad q \in (0, \hat{q}).$$

Condition C4 is satisfied.

5. Since according to formula (5), $\hat{\Phi}(q)$ is an analytic function and since $\hat{\Phi}'(q) \neq 0$ when $q \in (0, \hat{q})$, the function g(y), which is the inverse of $\hat{\Phi}(q)$, satisfies condition C5. Lemma 8.4 is proved.

Lemma 8.5. Let the function X(p) take two values on the interval $(q_1, q_2) \subset (0, 1)$:

$$X(p) = \begin{cases} X_2, & q_1 < p < r, \\ X_1, & r \leqslant p < q_2, \end{cases} \quad (X_2 - X_1 = R > 0).$$

Here obviously,

$$T(p) = \begin{cases} T_2 = \text{const}, & q_1 < p < r, \\ T_1 = \text{const}, & r \leqslant p < q_2. \end{cases}$$

We take an arbitrary point $\hat{r} \in (q_1, r)$ and let

$$\hat{X}(q) = \begin{cases} X_2, & q_1 < p < \hat{r}, \\ X_1, & \hat{r} \leqslant p < q_2; \end{cases} \quad T(p) = \begin{cases} T_2, & q_1 < p < \hat{r}, \\ T_1, & \hat{r} < p < q_2. \end{cases}$$

Outside the interval (q_1, q_2) we let $\hat{X}(p) = X(p)$, $\hat{T}(p) = T(p)$. This variation is allowable.

Proof. We varied X(p) and T(p) only on the interval $(\hat{r}, r)$. When $q \in (\hat{r}, r)$, X(p) and T(p) are equal to X_2 and T_2, respectively. The new functions $\hat{X}(p)$ and $\hat{T}(p)$ are equal on this interval to X_1 and T_1, respectively.

In order to establish that a curve $\Gamma = \{2\hat{X}(p), 2\hat{T}(p)\}$ is a travel-time curve, it is obviously sufficient to verify that condition C is satisfied (conditions A, B, and D are automatically satisfied).

Let us verify it separately on each of the following three intervals of variation of q: $(0, \hat{r}]$, $(\hat{r}, r)$, and $[r, 1)$.

When $q \in [r, 1)$, $\hat{\Phi}(q) = \Phi(q)$, and condition C is satisfied here.

For $q \in (\hat{r}, r)$, $\hat{\Phi}(q)$ would not be changed if we let $\hat{X}(p)$ equal X_1 not only for $p \in (\hat{r}, r)$ but also for all $p \in (0, r)$. That is, condition C is satisfied here (according to Lemma 8.4).

When $q \in (0, \hat{r}]$

$$\hat{\Phi}(q) = \Phi(q) + \frac{2}{\pi} \int_{\hat{r}}^{r} \frac{(X_1 - X_2)\, dp}{\sqrt{p^2 - q^2}} =$$

$$= \Phi(q) + \frac{2R}{\pi} \left(\operatorname{arch} \frac{\hat{r}}{q} - \operatorname{arch} \frac{r}{q} \right)$$

so that

$$\hat{\Phi}'(q) = \Phi'(q) + \frac{2R}{\pi q} \left(\frac{r}{\sqrt{r^2 - q^2}} - \frac{\hat{r}}{\sqrt{\hat{r}^2 - q^2}} \right).$$

In other words, when $q \leq \hat{r}$, $\hat{\Phi}(q)$ is obtained by adding to $\Phi(q)$ a decreasing function whose derivative is less than

$$\frac{-\hat{C}q}{\sqrt{\hat{r}^2 - q^2}}; \quad \hat{C} > 0.$$

Thus, condition C is also satisfied on (0, r).

Since when $q \in (\hat{r}, r)$

$$\hat{\Phi}(q) = \Phi(q) - \frac{2R}{\pi} \operatorname{arch} \frac{r}{q},$$

then

$$\hat{\Phi}(\hat{r} - 0) - \hat{\Phi}(\hat{r} + 0) = \Phi(\hat{r} - 0) - \Phi(\hat{r} + 0)$$

and

$$\hat{\Phi}(r - 0) - \hat{\Phi}(r + 0) = \Phi(r - 0) - \Phi(r + 0).$$

That is, condition C is satisfied on the entire interval $0 < q < 1$. The lemma is proved.

§9. Parametrization of Travel-Time Curves

Parametrization Theorem. Let Γ be any travel-time curve. Then, on the p axis a finite number of intervals ω_i, $i = 1, \ldots, l$, can be selected whose closures are disjoint, so that the following requirements are satisfied:

I. Any two parametrizations of a travel-time curve Γ

$$\{2X(p), 2T(p)\} \; u \; \{2X_1(p), 2T_1(p)\}$$

coincide everywhere on (0, 1) outside of $\omega_1, \ldots, \omega_l$ (except perhaps at a finite number of points).

II. For any parametrization of a travel-time curve Γ

$$\{2X_1(p), 2T_1(p)\}$$

and for any $i = 1, \ldots, l$, when $p \in \omega_i$ the functions $X_1(p)$ and $T_1(p)$ are nonincreasing step functions (or in any case become such if they are varied at a finite number of points).

Proof.

1. We specify some parametrization of a travel-time curve Γ

$$\{2X(p), 2T(p)\}$$

and use it to construct intervals $\omega_1, \ldots, \omega_l$. Then we show that the intervals constructed satisfy the requirements of the theorem.

2. From conditions A3 and A4 it follows that the equality $X(p) = X(p \pm 0)$ can be violated only at a finite number of points. Some of them can be adjoined by intervals on which $X(p)$ is constant. For definiteness, we shall assume that such intervals actually exist. We shall denote them by $\Delta_1, \ldots, \Delta_N$. It can happen that $X(p)$ is constant on an interval that adjoins 0 or 1; such intervals are also included in $\Delta_1, \ldots, \Delta_N$. Intervals of the greatest possible length adjacent to (0, 1) are taken: Δ_i is not contained on any interval on which $X(p) = \text{const}$.

If among the intervals thus constructed, two or more adjoin one another, we join this group of intervals (and their common ends) into one interval. These new intervals are also called $\omega_1, \ldots, \omega_l$. The closures of intervals ω_i are obviously disjoint.

3. Our proof that the intervals $\omega_1, \ldots, \omega_l$ satisfy the requirements of the theorem will be based on the following facts.

a) Just as in Section 25 of [1], Γ^* will denote a set of points of a travel-time curve Γ that have the following property: if $(x_0, t_0) \in \Gamma^*$, then Γ contains a smooth arc μ that passes through (x_0, t_0).

Let p^0 be the tangent of the angle between the x axis and the arc μ at a point $(x_0, t_0) \in \Gamma^*$; we then relate p^0 to set B.

According to Section 24 of [1], for any parametrization of a travel-time curve Γ

$$\{2X_1(p),\ 2T_1(p)\}$$

we can exclude from Γ^* a finite number of points (x_0, t_0), and from B a finite number of points p^0, so that for the remaining points

$$2X_1(p^0) = x_0, \quad 2T_1(p^0) = t_0.$$

In other words, no two parametrizations of Γ can differ on an infinite subset of set B.

b) According to the note in Section 6, the derivatives $X'(p)$ and $T'(p)$ exist, are continuous, and are related as $T'(p) = pX'(p)$ on an open set of full measure. We divide this set into two parts, relating to Π_0 those points p where $X'(p) = 0$ and to Π^* those points p where $X'(p) \neq 0$. Set Π^* is obviously open.

We shall call Π' the set of points at which $X(p) = X(p \pm 0) = +\infty$. According to the note in Section 6, the complement of $\Pi' \cup \Pi_0 \cup \Pi^*$ on (0, 1) – we call it $\widetilde{\Pi}$ – consists of a finite number of points.

Sets Π_1', Π_1^*, Π_{01}, and $\widetilde{\Pi}_1$, which are defined just as Π', Π^*, Π_0, and $\widetilde{\Pi}$ but with using functions $X_1(p)$ and $T_1(p)$, have similar properties.

c) According to Section 23 of [1], $\Pi^* \cup \Pi_1^* \subseteq$ B. This is easily verified directly.

In fact, let $p^0 \in \Pi^*$ and let $2X(p^0) = x_0$ and $2T(p^0) = t_0$. Since Π^* is an open set and the relation $T'(p)/X'(p) = p$ is satisfied when $p \in \Pi^*$, then Γ contains a smooth arc μ that passes through (x_0, t_0) at an angle to the x axis whose tangent is p^0; that is, $\Pi^* \subseteq$ B.

d) According to a) and c), the parametrizations of Γ

$$\{2X(p),\ 2T(p)\} \text{ and } \{2X_1(p),\ 2T_1(p)\}$$

cannot differ on an infinite subset of set $\Pi^* \cup \Pi_1^*$.

4. Let us verify that $\omega_1, \ldots, \omega_l$ satisfy the first requirement of the theorem.

We mark all points $p \in (0, 1)$ that lie outside of $\omega_1, \ldots, \omega_l$ at which at least one of the equalities $X(p) = X(p \pm 0)$ and $X_1(p) = X_1(p \pm 0)$ is violated. We shall call the marked points $\mathscr{P}_1, \ldots, \mathscr{P}_r$.

We shall show that when $p \neq \mathscr{P}_1, \ldots, \mathscr{P}_r$, outside of $\omega_1, \ldots, \omega_l$,

$$X(p) = X_1(p).$$

Let us assume the converse: let $X(p^0) \neq X_1(p^0)$ at some point $p^0 \in (0, 1)$ that lies outside of $\omega_1, \ldots, \omega_l$ and differs from $\mathscr{P}_1, \ldots, \mathscr{P}_r$. Since $X(p^0) = X(p^0 \pm 0)$ and $X_1(p^0) = X_1(p^0 \pm 0)$, the inequality $X(p) \neq X_1(p)$ is satisfied on the entire interval containing p^0. We take the maximum interval $\Delta \ni p^0$ on which $X(p)$ and $X_1(p)$ differ everywhere, except perhaps at a finite number of points.

According to 3d), interval Δ cannot contain infinitely many points of $\Pi^* \cup \Pi_1^*$.

Any point of Π' is obviously a limit point for Π^*. That is, if Δ overlaps Π', then Δ overlaps Π^*.

Set Π^* is closed. That is, if Δ contains at least one point of Π^*, then Δ contains infinitely many such points (complete interval).

The same is true if Π' is replaced by Π_1' and Π^* by Π_1^*.

Thus Δ does not overlap either $\Pi' \cup \Pi^*$ or $\Pi_1' \cup \Pi_1^*$. In other words, $\Delta \subseteq \Pi_0 \cup \widetilde{\Pi}$ and $\Delta \subseteq \Pi_{01} \cup \widetilde{\Pi}_1$.

Since $\Delta \subseteq \Pi_0 \cup \widetilde{\Pi}$ and since $X(p^0) = X(p^0 \pm 0)$, then $p^0 \in \Pi_0$; that is, $X(p)$ is constant and equal to $X(p^0)$ in some vicinity of p^0. We shall denote the ends of interval Δ by $\mathscr{P}$ and Q: $\mathscr{P} < p^0 < Q$. Since p^0 lies outside of $\omega_1, \ldots, \omega_l$, then

1) $X(p)$ is constant on the entire interval Δ,

2) $0 < \mathscr{P}$, $Q < 1$, and

3) $X(\mathscr{P} - 0) = X(Q + 0) = X(p^0)$.

From the imbedding $\Delta \subseteq \Pi_{01} \cup \widetilde{\Pi}_1$ it follows that $X_1(p)$ is a step function when $p \in \Delta$. Since Δ is the maximum interval on which $X(p) \neq X_1(p)$, then

$$X(\mathscr{P}-0) \geqslant \varliminf_{p \to \mathscr{P}-0} X_1(p), \quad \varlimsup_{p \to Q+0} X_1(p) \geqslant X(Q+0).$$

Therefore,

$$\varlimsup_{p \to Q+0} X_1(p) \geqslant X(p^0) \geqslant \varliminf_{p \to \mathscr{P}-0} X_1(p).$$

According to Lemma 7.2, from the latter inequality and the fact that $X_1(p)$ is a step function it follows that

$$X_1(p) \equiv X(p^0) \text{ on } \Delta \text{ when } p \neq \mathscr{P}_1, \ldots, \mathscr{P}_r,$$

in particular, $X_1(p^0) = X(p^0)$. We have arrived at the contradiction.

It is shown similarly that outside of $\omega_1, \ldots, \omega_l$ when $p \neq \mathscr{P}_1, \ldots, \mathscr{P}_r$

$$T_1(p) = T(p).$$

5. The second assertion of the theorem is proved quite simply.

If for some parametrization of Γ

$$\{2X_1(p), 2T_1(p)\}$$

$X_1(p)$ and $T_1(p)$ were not step functions on the interval ω_i $(1 \le i \le l)$, then this interval would contain points of $\Pi_1^! \cup \Pi_1^*$. Thus (see 4), it would contain infinitely many points of Π_1^*. That is (see 3d), on the entire interval belonging to ω_i the function $X(p)$ would not be a step function. But by definition, $X(p)$ is a step function on ω_i.

Thus, $X_1(p)$ and $T_1(p)$ are step functions on ω_i $(1 \le i \le l)$. According to Lemma 7.2, they do not increase (or, in any case, they become nondecreasing if they are varied at a finite number of points).

The theorem is proved.

§10. Travel-Time Curves Consisting of a Finite Number of Points

An example of a travel-time curve consisting of two points was constructed in Section 26 of [1].

Below, we shall see which finite sets in the plane (x, t) are travel-time curves. First, let us make the following stipulation.

If a travel-time curve $\{2X(p), 2T(p)\}$ consists of a finite number of points, $X(p)$ takes only a finite number of values. We shall agree to consider only those functions $X(p)$ and $T(p)$ that at each point on the interval (0, 1) are continuous either to the right or to the left (see refinement in Section 6).

Theorem. We consider an arbitrary finite set Γ that is located in the first quadrant in plane

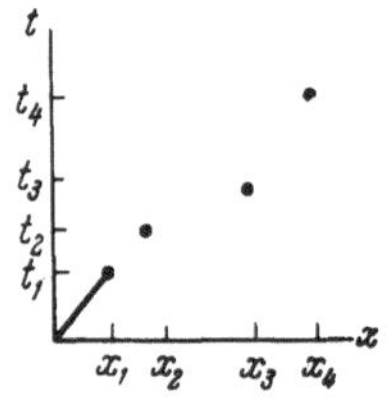

Fig. 11

(x, t) and is uniquely projected onto the axis of the abscissas x. We number the points belonging to Γ such that their abscissas increase monotonically:

$$(x_1, t_1), (x_2, t_2), \ldots, (x_r, t_r);$$
$$x_1 < x_2 < \ldots < x_r.$$

In order that set Γ be a travel-time curve, it is necessary and sufficient that

$$x_1 \leqslant t_1 < t_2 < \ldots < t_r$$

(the first point must lie on the bisectrix of the coordinate angle or above it; the ordinates of the points must increase monotonically).

Proof of Necessity. Let Γ be a travel-time curve. Let us consider any of its parametrizations $\{2X(p), 2T(p)\}$. In accordance with Lemma 7.2 and the stipulation made at the beginning of this section, $X(p)$ is a nonincreasing step function. According to Section 6, the discontinuity points of $T(p)$ and $X(p)$ coincide; we shall number them: $1 = q_0 > q_1 > q_2 \ldots > q_{r-1} > q_r = 0$. Let $X_i = x_i/2$, $T_i = t_i/2$, $i = 1, \ldots, r$. Then $\tau(p) = T_i - pX_i$ when $p \in (q_i, q_{i-1})$. Since $\tau(p)$ does not increase, $T_i - q_iX_i \ge T_{i-1} - q_{i-1}X_{i-1}$, i.e., $T_i - T_{i-1} \ge q_{i-1}(X_i - X_{i-1}) > 0$, or $t_i > t_{i-1}$, $i \ge 2$. Since $\tau(1-0) \ge 0$ (see note 2 of Section 2), $T_1 - X_1 \ge 0$, or $t_1 \ge x_1$.

Thus, $x_1 \le t_1 < t_2 < \ldots < t_r$.

Note. The conditions formulated in describing Γ (set Γ is located in the first quadrant and is uniquely projected onto the x axis) are, of course, necessary for Γ to be a travel-time curve. This follows from the positiveness of $X(p)$ and $T(p)$ and the coincidence of their discontinuity points.

Proof of Sufficiency. We mark on the p axis the points

$$1 = q_0 > q_1 > q_2 > \ldots > q_{r-1} > q_r = 0$$

such that

$$t_i - t_{i-1}/x_i - x_{i-1} \geqslant q_{i-1}, \quad 2 \leqslant i \leqslant r.$$

For this, it is sufficient, for example, to take

$$q_1 = \min_{2 \leqslant i \leqslant r} \frac{t_i - t_{i-1}}{x_i - x_{i-1}}.$$

Let

$$X(p) = x_i/2, \quad T(p) = t_i/2, \quad p \in [q_i, q_{i-1}), \quad 1 \leqslant i \leqslant r.$$

It is easy to see that X(p) and T(p) satisfy conditions A, B, C, and D', i.e., Γ is a travel-time curve.

Chapter III

DETERMINATION OF TRAVEL-TIME CURVE Γ FROM CURVE $\tilde{\Gamma}$

§11. Substance of the Chapter

1. As was indicated in the introduction, each travel-time curve $\Gamma\{x = 2X(p),\ t = 2T(p)\}$ is related to a curve $\tilde{\Gamma}\{x = 2\tilde{X}(p),\ t = 2T(p)\}$, such that

$$\tilde{X}(p) \equiv X(p) \left(\mathrm{mod}\ \frac{\pi R}{v(R)}\right), \quad 0 \leqslant \tilde{X}(p) < \frac{\pi R}{v(R)}.$$

We shall denote $\pi R/v(R)$ by $\mathfrak{M}$. Then we obtain

$$\tilde{X}(p) \equiv X(p) \quad (\mathrm{mod}\ \mathfrak{M}), \quad 0 \leqslant \tilde{X}(p) < \mathfrak{M}.$$

In this chapter we shall consider the problem of determining Γ from $\tilde{\Gamma}$.

2. The mapping $\tilde{\Gamma} \rightarrow \Gamma$ is, generally speaking, not unique. It can happen that different travel-time curves

$$\Gamma\{x = 2X(p), t = 2\ T(p)\} \text{ and } \Gamma^*\{x = 2X^*(p), t = 2\ T(p)\}$$

correspond to the same curve

$$\tilde{\Gamma}\ \{x = 2\tilde{X}(p),\ t = 2T(p)\}.$$

Below, for brevity, we shall call these equivalent travel-time curves.

Note. In defining equivalent travel-time curves, we assumed that Γ and Γ* were parametrized. Let us refine the definition for the case when they are not parametrized uniquely (see Section 9).

We shall call Γ and Γ* equivalent if there exist parametrizations

$$\Gamma = \{2X(p),\ \ 2T(p)\}, \quad \Gamma^* = \{2X^*(p),\ \ 2T^*(p)\},$$

such that

$$T(p) = T^*(p), \quad X(p) \equiv X^*(p) \quad (\mathrm{mod}\ \mathfrak{M}).$$

Such parametrizations of equivalent travel-time curves we shall call "matched."

3. In Section 12, we shall introduce the idea of the "travel-time curve component" and use it to study the relative positions of equivalent travel-time curves.

In Section 13, we shall construct examples of travel-time curves Γ that have an infinite number of equivalent travel-time curves Γ*.

In Section 14, we shall prove an important theorem about the farthest travel-time curve on the right, among the equivalent ones.

In Section 15, the travel-time curves in question have certain constraints imposed on them and a class of "simple travel-time curves" is isolated. It is shown that the number of travel-time curves that are equivalent to a simple one is finite. It is explained how all travel-time curves that are equivalent to a simple travel-time curve Γ are found from a curve $\tilde{\Gamma}$ that corresponds to Γ.

§12. Components of Travel-Time Curves

Let $\Gamma = \{2X(p),\ 2T(p)\}$ be any travel-time curve.

Functions X(p) and T(p) are continuous at the same points $p \in (0, 1)$ (see paragraph 1 of Section 6). The set of such points is open (see paragraph 2 of Section 6). Therefore, it consists of a finite or countable number of nonoverlapping intervals. The images of these intervals in the map $x = 2X(p)$, $t = 2T(p)$ of the p axis on the plane (x, t) we shall call components of travel-time curve Γ.

Lemma. Let Γ* be a travel-time curve equivalent to Γ and let γ* be any component of Γ*. Then we find a component γ of travel-time curve Γ and an integer s such that γ* is obtained from γ by a shift of $2s\mathfrak{M}$ along the x axis.

Proof. Let (see paragraph 2 of Section 11)

$$\{2X(p),\ 2\ T(p)\} \text{ and } \{2X^*(p),\ 2T^*(p)\}$$

be matched parametrizations of travel-time curves Γ and Γ*.

Let Δ be an interval of the p axis that is the original of γ* in the map $x = 2X^*(p)$, $t = 2T^*(p)$. Functions X*(p) and T*(p) are continuous when $p \in \Delta$, wherein Δ is not contained in any larger interval on which these functions would also be continuous. Functions T(p) and T*(p) are equal to one another. Functions X(p) and T(p) are continuous at the same points p. Therefore, the image of Δ in the map $x = 2X(p)$, $t = 2T(p)$ is a component of Γ. We shall denote it by γ.

For any $p \in (0, 1)$, the difference $X^*(p) - X(p)$ is an integral multiple of $\mathfrak{M}$. When $p \in \Delta$, this difference is continuous. Therefore when $p \in \Delta$,

$$\frac{X^*_{(p)} - X_{(p)}}{\mathfrak{M}} = \text{const}.$$

We shall denote this constant by s.

It is clear that γ^* is obtained by shifting γ by $2s\mathfrak{M}$ along the x axis. The lemma is proved.

Classification of Components

Let γ be a component of Γ that is the image of the interval $\Delta \subseteq (0, 1)$ in the map $x = 2X(p)$, $t = 2T(p)$ of the p axis on the plane (x, t).

According to conditions A3 and A4, X(p) and T(p) are either both constant or both not constant on Δ. In the former case, we shall call γ a one-point component, and in the latter case, an extended component.

Let q and $\bar{q}$ be the ends of the interval Δ: $0 \le q < \bar{q} \le 1$. According to condition A4, for all components γ (except perhaps a finite number of them) $X(q + 0) = T(q + 0) = X(\bar{q} - 0) = T(\bar{q} - 0) = +\infty$. In particular, any travel-time curve Γ contains only a finite number of one-point components.

Let γ be an extended component. Let Q be the lower bound and $\bar{Q}$ the upper bound of all $p \in \Delta$, where $X'(p) \neq 0$. Only for a finite number of components can the inequalities $q < Q$ or $\bar{Q} < \bar{q}$ be satisfied; if X(p) and T(p) tend to $+\infty$ when the ends of interval Δ are approached, then $q = Q$ and $\bar{q} = \bar{Q}$.

All one-point components of Γ and those extended components for which at least one of the inequalities $q < Q$ or $\bar{Q} < \bar{q}$ is satisfied we shall call "singular."

Each travel-time curve contains only a finite number of singular components. If a travel-time curve does not contain any singular components, then (according to Section 9) it has a unique parametrization $\{2X(p), 2T(p)\}$.

§13. Examples of Travel-Time Curves with Infinitely Many Equivalents

Example 1. Let Γ be any travel-time curve consisting of two points (x_1, t_1) and (x_2, t_2), $x_1 = t_1$, $x_1 < x_2$, $t_1 < t_2$.

Let Γ^*_s be a set consisting of the points (x_1, t_1) and $(x_2 + 2s\mathfrak{M}, t_2)$. It is obvious that Γ^*_s, $s = 1, 2, \ldots$, is a travel-time curve (see Section 10) that is equivalent to Γ.

Note 1. A similar construction is possible when Γ consists of $n > 2$ points.

A more general assertion is valid.

Let Γ contain a one-point component (x_0, t_0) on which are focused all rays with sufficiently small ray parameters p (Fig. 13). Then there exists a travel-time curve Γ^* that contains the same one-point component (x_0, t_0), differs as little as desired from Γ, and has infinitely many equivalents. These equivalent travel-time curves differ from Γ^* only in that instead of (x_0, t_0), they contain the component $(x_0 + 2s\mathfrak{M}, t_0)$, $s = 1, 2, \ldots$.

We shall prove this assertion, having constructed Γ^*.

Consider the graphs of X(p) and T(p).

When $p < p_0$, $X(p) = X_0 = x_0/2$ and $T(p) = T_0 = t_0/2$.

On the interval $(p_0, 1)$ in an arbitrarily small vicinity of p_0 we find the continuity point of X(p). We shall call it $\hat{q}$.

Let

$$\hat{X}(p) = \begin{cases} X(\hat{q}) = \hat{X}, & p < \hat{q} \\ X(p), & p \geqslant \hat{q} \end{cases}$$

and

$$\hat{T}(p) = \begin{cases} T(\hat{q}) = \hat{T}, & p < \hat{q}, \\ T(p), & p \geqslant \hat{q}. \end{cases}$$

According to Lemma 8.4, $\{2\hat{X}(p), 2\hat{T}(p)\}$ is a travel-time curve.

We fix the integer $s \ge 0$ and let

$$X_0 - \hat{X} + s\,\mathfrak{M} = K_{1s}, \quad T_0 - \hat{T} = K_2, \quad K_2/K_{1s} = \hat{p}_s.$$

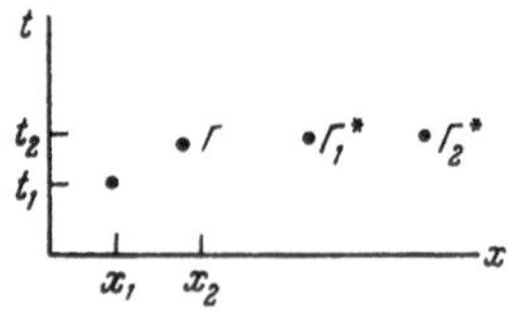

Fig. 12

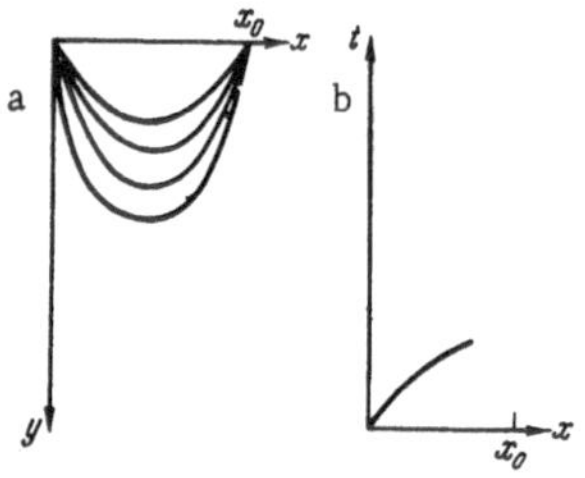

Fig. 13

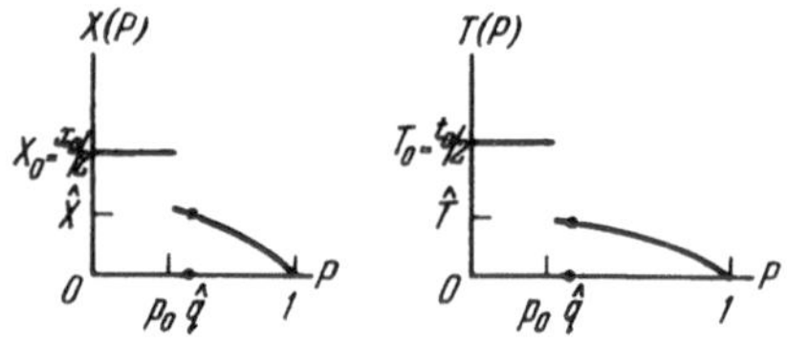

Fig. 14

Then we let

$$X_s^*(p) = \begin{cases} X_0 + s\mathfrak{M} = \hat{X} + K_{1s}, & p < \hat{p}_s, \\ \hat{X}(p), & p \geqslant \hat{p}_s, \end{cases}$$

$$T_s^*(p) = \begin{cases} T_0 = \hat{T} + K_2, & p < \hat{p}_s, \\ \hat{T}(p), & p \geqslant \hat{p}_s. \end{cases}$$

According to Lemma 8.2, $\Gamma_s^* = \{2X_s^*(p),\ 2T_s^*(p)\}$ is a travel-time curve. According to Lemma 8.5, for different s travel-time curves Γ_s^* can be (pairwise) consistently parametrized. The travel-time curve $\Gamma^* = \Gamma_0^*$ is obviously the desired one.

Note 2. It can be shown that not only the presence in Γ of a one-point component (on which are focused all rays with sufficiently small parameters), but in general the presence in Γ of a singular component (see end of Section 12) causes Γ to have infinitely many equivalent travel-time curves.

The corresponding examples, just as example 1, are based on the fact that Γ admits infinitely many different parametrizations (compare with Section 9).

The following example is based on another idea.

Example 2. We construct a sequence of velocity profiles $u_n(y)$ such that the travel-time curves corresponding to them will be equivalent. Each of these travel-time curves will contain infinitely many components.

Auxiliary Functions U(y) and F(r)

1. We select U(y) such that it has the following two properties (Fig. 15):

1) U(y) is an allowable velocity profile that does not have waveguides;

2) the numbers Y_n, increasing monotonically, tend to ∞; $U'(Y_n) = 0$; if $y \neq Y_n$, then $U'(y) > 0$.

Let $U(Y_n) = r_n$; we shall denote the inverse of U(y) by E(r).

2. We introduce the function F(r) (Fig. 16). When $r \in [1, r_1]$, $F(r) = E(r)$. When $r \in [r_n, r_{n+1}]$

$$F(r) = E(r) + \sum_{i=1}^{n} \left(\frac{2\mathfrak{M}}{\pi} \operatorname{arch} \frac{r}{r_i} + 1 - \frac{2}{\pi} \operatorname{arctg} \frac{\mathfrak{M}}{\sqrt{1 - r_i^2 r^{-2}}} \right).$$

Velocity Profile u(y)

3. The inverse of F(r) we shall call u(y). The numbers of $F(r_n)$ we shall denote by y_n.

Figure 17 shows the relative position of the graphs of U(y) and u(y).

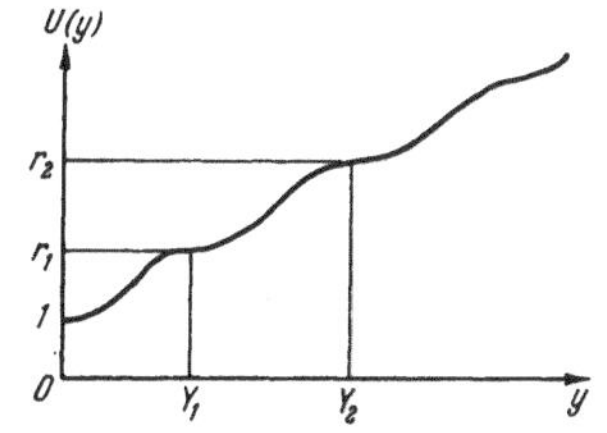

Fig. 15

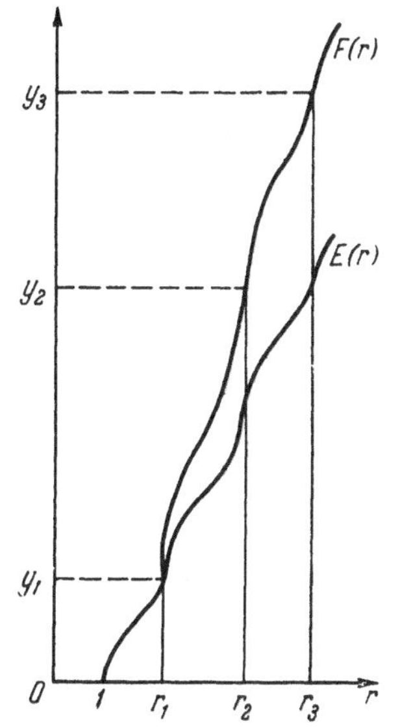

Fig. 16

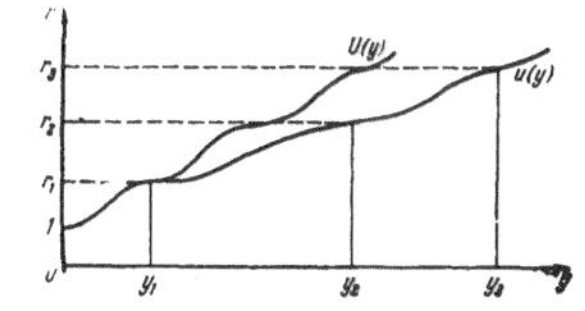

Fig. 17

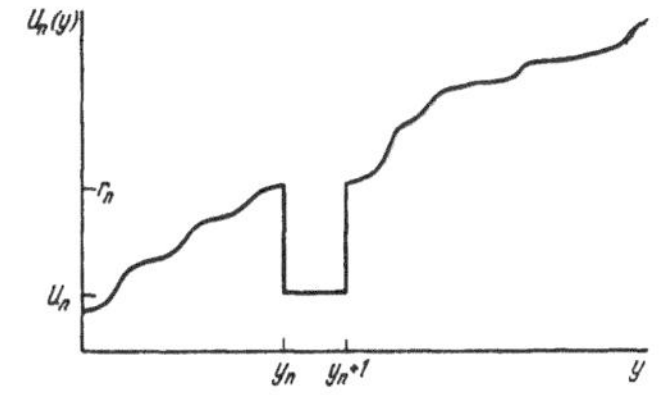

Fig. 18

It is clear that u(y) is an allowable velocity profile; $u'(y_n) = 0$; if $y \neq y_n$, then $u'(y) > 0$.

Velocity Profiles $u_n(y)$

4. We arbitrarily fix the number $n \geq 1$. The velocity profile $u_n(y)$ is shown in Fig. 18.

When $y \in [0, y_n]$, $u_n(y) = u(y)$. The interval $(y_n; y_n+1)$ is a waveguide. When $y \in (y_n, y_n+1)$, $u_n(y) = u_n = \text{const}$; the value $u_n > 0$ is chosen such that $\sqrt{u_n^{-2} - r_n^{-2}} = \mathfrak{M}/r_n$, i.e.,

$$u_n = r_n / \sqrt{\mathfrak{M}^2 + 1}.$$

When $y \geq y_n+1$, $u_n(y)$ is the inverse of the function

$$F_n(r) = F(r) - \frac{2\mathfrak{M}}{\pi} \operatorname{arch} \frac{r}{r_n} + \frac{2}{\pi} \operatorname{arctg} \frac{\mathfrak{M}}{\sqrt{1-r_n^2 r^{-2}}},$$

$$r \geqslant r_n.$$

This definition is correct, since by virtue of the selection of $F(r)$: 1) $F_n(r)$ increases monotonically, and 2) $F_n(r_n) = y_n + 1$.

Equivalence of Travel-Time Curves Γ_n, $n = 1, 2, \ldots$

5. We shall denote the travel-time curves that correspond to $u(y)$ and $u_n(y)$ by Γ and Γ_n. We shall show that Γ and Γ_n are equivalent.

As usual, the functions related to Γ and $u(y)$ we shall denote by $X(p)$, $T(p)$, $\tau(p)$, and $\Phi(p)$. The corresponding functions related to Γ_n and $u_n(y)$ we shall call $X_n(p)$, $T_n(p)$, $\tau_n(p)$, and $\Phi_n(p)$.

Let $r_n^{-1} = q_n$.

To establish the equivalence of Γ and Γ_n it is sufficient to show that

$$T_n(p) = T(p), \text{ but } X_n(p) = \begin{cases} X(p) - \mathfrak{M}, & p < q_n, \\ X(p), & p \geqslant q_n. \end{cases}$$

The equality $X_n(p) = X(p)$, $p \in (q_n, 1)$, is obvious (since $u_n(y) = u(y)$ when $0 \leq y \leq y_n$).

The equality $X_n(p) = X(p) - \mathfrak{M}$, $p \in (0, q_n)$, is equivalent to the relation

$$\Phi_n(p) = \Phi(p) - \frac{2\mathfrak{M}}{\pi} \operatorname{arch} \frac{q_n}{p}, \quad p < q_n,$$

which, in turn, follows from the easily verified formulas

$$\Phi(p) = F(p^{-1}), \quad \Phi_n(p) =$$

$$= F_n(p^{-1}) - \frac{2}{\pi} \operatorname{arctg} \sqrt{\frac{u_n^{-2} - q_n^2}{q_n^2 - p^2}},$$

$$p < q_n.$$

In our example, $X(p)$ and $T(p)$ have a countable number of discontinuities, since $u'(y)$ vanishes only when $y = y_k$, $k = 1, 2, \ldots$ (see note in Section 6). Therefore, $\tau(p)$ and $\tau_n(p)$ are absolutely continuous. Since, moreover, $\tau(p)$ is continuous while $\tau_n(p)$ has a single jump

$$\tau_n(q_n - 0) - \tau_n(q_n + 0) = \sqrt{u_n^{-2} - q_n^2} = \mathfrak{M} q_n,$$

then

$$\tau(p) = \int_p^1 X(q)\, dq, \quad \tau_n(p) - \int_p^1 X_n(q)\, dq = \begin{cases} \mathfrak{M} q_n, & p < q_n, \\ 0, & p \geqslant q_n. \end{cases}$$

Considering the relationship between $X_n(p)$ and $X(p)$, we find:

$$\tau_n(p) = \begin{cases} \tau(p) + p\mathfrak{M}, & p < q_n, \\ \tau(p), & p \geqslant q_n. \end{cases}$$

Since

$$T(p) = \tau(p) + pX(p), \quad T_n(p) = \tau_n(p) + pX_n(p),$$

then, thereby,

$$T_n(p) = T(p).$$

The equivalence of Γ and Γ_n (for any $n \geq 1$) is established.

Note on the Number of Waveguides. We know that all velocity profiles that correspond to one (parametrized) travel-time curve have the same number of waveguides: it is equal to the number of discontinuities of $\tau(p)$. Profiles that correspond to equivalent travel-time curves can, conversely, have different numbers of waveguides: in the example in question, $u(y)$ has no waveguides and $u_n(y)$ has one waveguide. The travel-time curves Γ_n constructed above do not exhaust the entire set of travel-time curves equivalent to Γ. We shall now show that for any k there exists a travel-time curve equivalent to Γ and such that each velocity profile corresponding to it has k waveguides.

We arbitrarily fix k integers:

$$n_1 < n_2 < \ldots < n_k.$$

We introduce the function

$$F_{n_1, \ldots, n_k}(r) = \begin{cases} F(r), \quad 1 \leqslant r < r_{n_1}, \\ F(r) + \sum_{i=1}^{l} \frac{2}{i\pi} \left(\operatorname{arctg} \frac{\mathfrak{M}}{\sqrt{1 - r_{n_i}^2 r^{-2}}} - \mathfrak{M} \operatorname{arch} \frac{r}{r_{n_i}} \right), \\ \qquad r_{n_l} \leqslant r < r_{n_{l+1}}, \; 1 \leqslant l < k, \\ F(r) + \sum_{i=1}^{k} \frac{2}{\pi} \left(\operatorname{arctg} \frac{\mathfrak{M}}{\sqrt{1 - r_{n_i}^2 r^{-2}}} - \mathfrak{M} \operatorname{arch} \frac{r}{r_{n_i}} \right), \\ \qquad r_{n_k} \leqslant r. \end{cases}$$

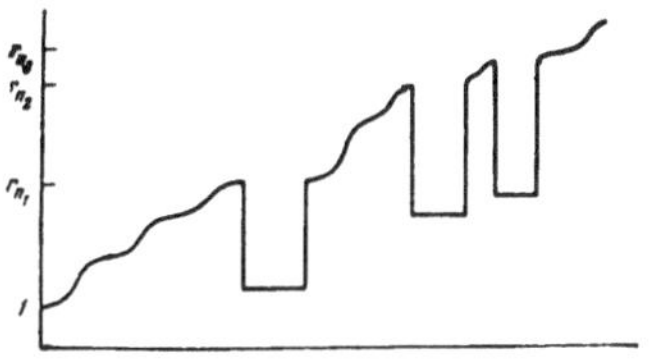

Fig. 19

Let

$$F_{n_1,\dots,n_k}(r_{n_i}-0)=y_{n_1,\dots,n_k;\,i},\quad i=1,\dots,k.$$

Outside the intervals $(y_{n_1,\dots,n_k;\,i},\ y_{n_1,\dots,n_k;\,i+1})$ we define $u_{n_1,\dots,n_k}(y)$ as the inverse of the function $F_{n_1,\dots,n_k}(r)$, and on the interval $(y_{n_1,\dots,n_k;\,i},\ y_{n_1,\dots,n_k;\,i+1})$ we let $u_{n_1,\dots,n_k}(y)=r_{n_i}/\sqrt{\mathfrak{M}^2+1}$ (see Fig. 19).

It is easy to verify that $u_{n_1,\dots,n_k}(y)$ is an allowable profile and that the travel-time curve corresponding to it $\Gamma_{n_1,\dots,n_k}$ is equivalent to Γ. This profile obviously has k waveguides.

§ 14. Rightmost Travel-Time Curve

Definition of Travel-Time Curve That Is Shifted to the Right. Let a travel-time curve Γ have the following property.

There exists a parametrization of Γ such that for the corresponding function $\tau(p)$ and for some point $\hat{q}\in(0,1)$

$$\tau(\hat{q}-0)-\tau(\hat{q}+0)\geqslant\mathfrak{M}\hat{q}.$$

According to Lemma 8.1, the function

$$\hat{\tau}(p)=\begin{cases}\tau(p)-\mathfrak{M}p, & p<\hat{q},\\ \tau(p), & p\geqslant\hat{q}\end{cases}$$

is given by travel-time curve $\hat{\Gamma}$. This travel-time curve (as a curve on the plane (x, t)) is obtained from Γ by shifting to the right by $2\mathfrak{M}$, part of the components of Γ. (The components of Γ that are the image of the interval $0<p<\hat{q}$ are shifted.)

In accordance with this, we shall call travel-time curve Γ right-shifted.

Travel-time curve Γ is also so-called when for some parametrization,

$$\tau(1-0)\geqslant\mathfrak{M}.$$

In this case the function

$$\hat{\tau}(p)=\tau(p)-\mathfrak{M}p,\quad p\in(0,1),$$

obviously gives the travel-time curve $\hat{\Gamma}$ which is obtained by shifting the entire curve Γ to the right by $2\mathfrak{M}$.

Let us consider the class of equivalent travel-time curves.

As example 1 of Section 13 shows, each travel-time curve can be shifted to the right. The following uniqueness theorem is valid, however.

Theorem. If in the class of equivalent travel-time curves there exists a travel-time curve Γ_r that is not shiftable to the right, there is exactly one such travel-time curve.

Proof. Let Γ be any travel-time curve that is equivalent to Γ_r.

We take matched parametrizations of Γ and Γ_r:

$$\Gamma=\{2X(p),\ 2T(p)\},\quad \Gamma_r=\{2X_r(p),\ 2T(p)\}.$$

Consider the corresponding functions $\tau_r(p)$ and $\tau(p)$.

We shall show that $\tau_r(p)\le\tau(p)$ for any $p\in(0,1)$ (it is obvious that the uniqueness of Γ_r will follow from this).

We shall assume the converse: for any $q\in(0,1)$ let

$$\tau_r(q)>\tau(q).$$

The difference $\tau_r(q)-\tau(p)$ is equal to $q\,[X(q)-X_r(q)]$. Functions $X(p)$ and $X_r(p)$ are comparable mod $\mathfrak{M}$. Therefore,

$$\tau_r(q)\geqslant\tau(q)+\mathfrak{M}q.$$

Let $\hat{q}$ be the upper bound of those q for which the latter inequality is valid.

Since $\tau(1-0)\ge0$ and $\tau_r(1-0)\le\mathfrak{M}$, then $\hat{q}<1$.

It is easy to see that:

1) $\tau_r(\hat{q}-0)\ge\tau(\hat{q}-0)+\mathfrak{M}\hat{q}$,

2) $\tau_r(\hat{q}+0)\le\tau(\hat{q}+0)\le\tau(\hat{q}-0)$.

Thus,

$$\hat{\tau}_r(\hat{q}-0)\geqslant\tau_r(\hat{q}+0)+\mathfrak{M}\hat{q},$$

i.e., Γ_r is shifted to the right. This contradiction proves the theorem.

Definition. If among the travel-time curves that are equivalent to Γ there exists a travel-time curve Γ_r that is not shiftable to the right, we shall call it (in accordance with the theorem just proved) the rightmost travel-time curve among all those equivalent to Γ.

§ 15. Simple Travel-Time Curves

The examples constructed in Section 13 show that the existence of an infinite set of travel-time curves that are equivalent to a given travel-time curve Γ can be due to the fact that Γ contains infinitely many components, or to the fact that Γ contains singular components.

The set of equivalent travel-time curves can be infinite in no other way.

Definition. We shall call a travel-time curve Γ *simple* if it contains a finite number of components, and none of them is singular.

Note. Each simple travel-time curve allows a unique parametrization $\{2X(p), 2T(p)\}$ (see end of Sections 12 and 9). Any travel-time curve that is equivalent to a simple travel-time curve is obviously simple (see lemma in Section 12).

Theorem. The number of travel-time curves that are equivalent to a simple travel-time curve is finite.

Proof.

1. Let $\Gamma\{2X(p), 2T(p)\}$ be a simple travel-time curve and let $\Gamma^*\{2X^*(p), 2T(p)\}$ be any travel-time curve that is equivalent to Γ. We shall denote the components of Γ and Γ^* (respectively) by

$$\gamma_1, \ldots, \gamma_r \text{ and } \gamma_1^*, \ldots, \gamma_r^*.$$

The intervals of the p axis whose images are these components we shall call

$$(q_1, \bar{q}_1), \ldots, (q_r, \bar{q}_r).$$

2. Let component γ_i^* be obtained from γ_i by shifting by $2s_i\mathfrak{M}$ to the left along the x axis (see lemma in Section 12). Then when $p \in (q_i, \bar{q}_i)$,

$$X^*(p) = X(p) - s_i\mathfrak{M}.$$

We shall let X_i be the lower bound of X(p) when $p \in (q_i, \bar{q}_i)$. Since $X^*(p) > 0$, then in the case in question $s_i \leq X_i/\mathfrak{M}$.

3. If component γ_i^* is obtained from γ_i by shifting by $2s_i\mathfrak{M}$ to the right along the x axis, then the following equality obviously holds for the corresponding functions $\tau^*(p)$ and $\tau(p)$:

$$\tau^*(p) = \tau(p) - s_i p\mathfrak{M}, \quad p \in (q_i, \bar{q}_i).$$

Let

$$\tau_i = \tau(q_i - 0).$$

Since $\tau^*(q_i - 0) \geq 0$, in the case in question

$$s_i \leqslant \tau_i / q_i\mathfrak{M}.$$

4. These estimates show that the number of travel-time curves that are equivalent to a simple one is finite.

Corollary. Among the travel-time curves that are equivalent to a simple travel-time curve there is a rightmost travel-time curve (see Section 14).

Finding Travel-Time Curves Equivalent to a Given Simple Travel-Time Curve Γ. If we are given a simple travel-time curve Γ, by enumeration we can easily find all travel-time curves that are equivalent to it. Let us explain this in more detail.

Using the symbols introduced in the proof of the theorem on the number of travel-time curves that are equivalent to a simple travel-time curve, we assert the following. We shift each component γ_i, $1 \leq i \leq r$, along the x axis by an integral multiple of $2\mathfrak{M}$ either to the left (by not more than X_i) or to the right (by not more than τ_i/q_i). We obtain some curve Γ^*. If Γ^* is a travel-time curve, then it is a travel-time curve that is equivalent to Γ. All travel-time curves that are equivalent to Γ can be thus obtained.

In checking whether or not Γ^* is a travel-time curve, we should first of all see whether the discontinuities of $\tau^*(p)$ have the required sign and whether the inequality of Lemma 7.2 is violated for $X^*(p)$. This allows some curves Γ^* that are not travel-time curves to be rejected immediately and reduces sorting time.

Finding a Simple Travel-Time Curve Γ from Curve $\tilde{\Gamma}$.

1. Let a curve $\tilde{\Gamma} = \{2\tilde{X}(p), 2T(p)\}$ that corresponds to a simple travel-time curve Γ be given. We already know that the number of travel-time curves that are equivalent to Γ is finite, and we are able to find all the others from one of these equivalents by sorting.

Thus our problem consists in finding from $\tilde{\Gamma}$ one of the travel-time curves that is equivalent to Γ. We find the travel-time curve Γ_r – the rightmost of all the equivalents of Γ.

2. First, let us construct $\tilde{X}(p)$ and T(p) from curve $\tilde{\Gamma}$. This construction is done in the same way as parametrization of a travel-time curve.

Let $\tilde{\Gamma}^*$ be a subset of $\tilde{\Gamma}$ that has the following property: if $(x_0, t_0) \in \tilde{\Gamma}^*$, then $\tilde{\Gamma}$ contains a smooth arc μ that passes through (x_0, t_0). The tangent of the angle between the x axis and the arc μ at a point $(x_0, t_0) \in \tilde{\Gamma}^*$ we shall call p^0. For each of the points of $\tilde{\Gamma}^*$, let

$$\tilde{X}(p^0) = x_0/2, \quad T(p^0) = t_0/2.$$

After this, generally speaking, points where $\tilde{X}(p)$ and T(p) are still not defined will remain on (0, 1); the set where they are already defined we shall call

Π_1. We shall denote the closure of Π_1 by Π_2: $\Pi_2 = \bar{\Pi}_1$. For all points $p \in \Pi_2 \setminus \Pi_1$ (except perhaps a finite number of them) there exists a limit $T(p^0)$ when $p^0 \to p$, $p^0 \in \Pi_1$; we shall extend the definition of T(p) at these points by continuity. Let Π_3 be the open complement of Π_2 on the interval (0, 1). Set Π_3 is not, generally speaking, empty. Let (a, b) be one of its component intervals. Since, by definition, $\tilde{\Gamma}$ corresponds to a simple travel-time curve, then T(p) is already defined at points a and b, and necessarily $T(a) = T(b)$. We extend the definition of T(p) for $p \in (a, b)$ by letting $T(p) = T(a) = T(b)$. Similarly, we define T(p) on the remaining component intervals of set Π_3. After this, T(p) is defined everywhere on (0, 1), except a finite number of points, which are its discontinuity points.

We leave $\tilde{X}(p)$ defined only in set Π_1.

3. Let $\Delta_1, \ldots, \Delta_r$ be intervals on which T(p) is continuous. Intervals Δ_i do not overlap; their closures give the entire interval [0, 1]. The intervals are numbered such that they follow one another on the p axis from right to left.

On each of the intervals Δ_i we fix an arbitrary point Q_i at which $\tilde{X}(p)$ is defined. We denote by $X_0(p)$ a function that is continuous on Δ_i, equal to $\tilde{X}(p)$ when $p = Q_i$, constant on each of the component intervals of set Π_3, and such that

$$X_0(p) \equiv \tilde{X}(p) (\mathrm{mod}\, \mathfrak{M}), \quad p \in \Delta_i \cap \Pi_1, \quad 1 \leqslant i \leqslant r.$$

Such a function obviously exists and is unique.

4. Using $X_0(p)$, let us construct the following function $X_r(p)$. When $p \in \Delta_i$, the difference $X_r(p) - X_0(p)$ is an integral multiple of $\mathfrak{M}$:

$$X_r(p) - X_0(p) = n_i \mathfrak{M}.$$

The integers n_i, $1 \le i \le r$, are selected such that for the function

$$\tau_r(p) = T(p) - pX_r(p)$$

the relations

$$0 \leqslant \tau_r(1-0) < \mathfrak{M}; \quad 0 \leqslant \tau_r(\mathscr{P}_i - 0) - \tau_r(\mathscr{P}_i + 0) < \mathfrak{M}\mathscr{P}_i$$

are satisfied ($\mathscr{P}_i$ is the common end of intervals Δ_i and Δ_{i+1}).

Curve $\Gamma_r\{2X_r(p), 2T(p)\}$ is obviously a travel-time curve. It is the rightmost travel-time curve among all those that are equivalent to Γ.

LITERATURE CITED

1. Gerver, M. L., and V. M. Markushevich, "Determining seismic-wave velocity from travel-time curves," this volume, p. 123.

CHARACTERISTIC PROPERTIES OF DEEP-FOCUS TRAVEL-TIME CURVES†

V. M. Markushevich

1. Formulation of Problem

1. In this paper we discuss the problem of determining velocity structure from a travel-time curve for a source at a certain depth in the following formulation [1].

The source is located in a half-plane x, y, $y \geq 0$, at a point (0, d) (Fig. 1). At time t = 0 a pulse is generated and propagated along the rays with velocity u(y), a function only of the depth y which is scaled so that u(0) = 1. The arrival time is recorded at each emergence point on the surface. From these data we must determine u(y).

2. In this paper we are concerned with the rays that depart to the right upward from the source. On the plane x, t, they form a curve Γ_1, such as that shown in Fig. 2. The broken line in this figure corresponds to those rays directed to the right downward at the source, which are not considered here. The arc IJ appears only when the source is not in a waveguide and $u(d+0) > u(d-0)$. The point I may be at infinity.

3. We assume that u(y) is positive, bounded, and measurable.

We denote by p the ray parameter, which is equal to the sine of the angle between the ray and the vertical at the point of emergence of the ray at the surface (Fig. 1).

We introduce the function $f(y) = (\sup \{u(y^0), 0 \leq y^0 \leq y\})^{-1}$ and let $f(d-0) = \mathscr{P}$, $f(d+0) = Q$. It

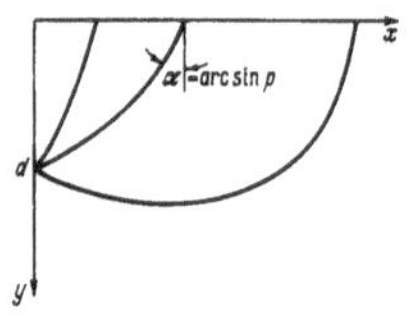

Fig. 1

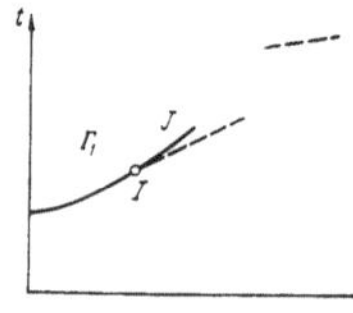

Fig. 2

is clear that $\mathscr{P} \geq Q$. The parametric equations of the curve $\Gamma_1\{X_1(p), T_1(p)\}$ are [1]:

$$X_1(p) = \int_0^d \frac{pu(y)\,dy}{\sqrt{1-p^2u^2(y)}}, \quad T_1(p) = \int_0^d \frac{dy}{u(y)\sqrt{1-p^2u^2(y)}},$$

$$p \in [0, \mathscr{P}).$$

4. Using the function $H(r) = \text{mes}\,\{y,\ y \leq d,\ u(y) \leq r\}$, these equations can be written as:

$$X_1(p) = \int_0^{\mathscr{P}^{-1}} \frac{pr\,dH(r)}{\sqrt{1-p^2r^2}}, \quad T_1(p) = \int_0^{\mathscr{P}^{-1}} \frac{dH(r)}{r\sqrt{1-p^2r^2}},$$

$$p \in [0, \mathscr{P}).$$

Hence it is already clear that u(y) is not uniquely defined by Γ_1, except when u(y) is monotonic for $y \in [0, d]$, since a single curve Γ_1, may correspond to different functions u(y) that have the same H(r).

On the other hand, it was shown in [1] that H(r) was defined by any of the functions $X_1(p)$ and $T_1(p)$ if it was taken into account that H(0) = 0 (otherwise, $T_1(p) = \infty$ for all p).

5. We shall consider (1) as a system of equations in H(r). From the definition it follows that

† Translated from Vychislitel'naya Seismologiya, No. 4, Moscow, Nauka (1968), pp. 64-77.

H(r) is a monotone nondecreasing function and that H(0) = 0.

The question arises as to which properties the functions $X_1(p)$ and $T_1(p)$ must possess in order that some nondecreasing function H(r) may satisfy Eqs. (1).

In other words, along with the problem of the uniqueness of a solution of Eqs. (1), we have the problem of the conditions for the existence of a solution of these equations in the class of nondecreasing functions. It is obvious that this is equivalent to the problem of what properties $X_1(p)$ and $T_1(p)$ must have in order that the curve Γ_1 be a travel-time curve for a deep source for some velocity profile u(y).

2. Properties of $X_1(p)$ and $T_1(p)$ That Are Known from [1]

1. In considering the parametric Eqs. (1), we determined in [1] the following simple properties of these functions:

a) they are nonnegative for all $p \in [0, \mathscr{P})$;

b) they increase monotonically;

c) they are continuously differentiable everywhere on $[0, \mathscr{P})$;

d) $T_1'(p) - pX_1'(p) = 0$;

e) $X_1(0) = 0$.

2. We examine the functions $X_1(p)$ and $T_1(p)$, when the curve Γ_1 is given. Determination of these functions does not cause difficulties, however. In fact, as was shown in [1], for all $p \in [0, \mathscr{P})$ we have $X_1'(p) > 0$. Therefore, according to property 2.1d,

$$p = \frac{dT_1}{dX_1}.$$

Therefore, from Γ_1 we can always find $X_1(p)$ and $T_1(p)$ if we assign to each point of the curve a value of the parameter p equal to the slope of the tangent to the curve at that point.

Now we shall investigate some analytic relations that we will need to find new properties of $X_1(p)$ and $T_1(p)$.

3. Proof of the Main Equation

In this section we shall prove the equation

$$\int_0^1 \frac{x^k\,dx}{\sqrt{1-\lambda x}} = \frac{k!}{(2k+1)!}\sum_{i=1}^{k+1} \frac{2^i(2k-i+1)!}{(k-i+1)!}\,\frac{1}{(1+\sqrt{1-\lambda})^i},$$

$$0 \leqslant \lambda \leqslant 1, \quad k = 0, 1, 2, \ldots$$

1. First, we obtain two combinatorial relations in a form that will be convenient for future use.

It is simple to verify that

$$\binom{n-1}{m-1} + \binom{n-1}{m} = \binom{n}{m}.$$

Applying this to $\binom{n-1}{m-1}$, we obtain

$$\binom{n-2}{m-2} + \binom{n-2}{m-1} = \binom{n-1}{m-1}$$

$$\cdots\cdots\cdots\cdots\cdots\cdots$$

$$\binom{n-m}{0} + \binom{n-m}{1} = \binom{n-m+1}{1},$$

etc. Since $\binom{n-m}{0} = \binom{n-m-1}{0} = 1$, by successively substituting the lower equalities into the upper, we find that

$$\sum_{i=1}^{m+1} \binom{n-i}{m-i+1} = \binom{n}{m}.$$

Here we let $n = 2k - l + 1$, $m = k - l + 1$, $j = l + i - 2$; we obtain

$$\sum_{j=l-1}^{k} \binom{2k-j-1}{k-j} = \binom{2k-l+1}{k-l+1}.$$

Multiplying the latter equation by $(k-1)!$, we obtain

$$\sum_{j=l-1}^{k} \frac{(2k-j-1)!}{(k-j)!} = \frac{1}{k}\,\frac{(2k-l+1)!}{(k-l+1)!}.$$

This equation can be written as

$$\sum_{j=l-1}^{k} \frac{(2k-j-1)!}{(k-j)!} = 2\,\frac{(2k-1)!}{(2k)!}\,\frac{(2k-l+1)!}{(k-l+1)!}. \tag{1}$$

If we let $l = 2$ here, then

$$\sum_{j=1}^{k} \frac{(2k-j-1)!}{(k-j)!} = \frac{(2k-1)!}{k!}. \tag{2}$$

We shall need Eqs. (1) and (2) in what follows.

2. Let $J_k = \int_0^1 \frac{x^k\,dx}{\sqrt{1-\lambda x}}$, $k = 1, 2, \ldots$. We select a recurrence relation between J_k for $\lambda \neq 0$.

$$J_k = \int_0^1 \frac{x^k\,dx}{\sqrt{1-\lambda x}} = -\frac{2}{\lambda}\int_0^1 x^k\,d\sqrt{1-\lambda x} =$$

$$= -\frac{2}{\lambda}x^k\sqrt{1-\lambda x}\Big|_0^1 + \frac{2k}{\lambda}\int_0^1 x^{k-1}\sqrt{1-\lambda x}\,dx =$$

$$= -\frac{2}{\lambda}\sqrt{1-\lambda} + \frac{2k}{\lambda}\int_0^1 \frac{x^{k-1}(1-\lambda x)}{\sqrt{1-\lambda x}}dx =$$

$$= -\frac{2}{\lambda}\sqrt{1-\lambda} + \frac{2k}{\lambda}J_{k-1} - 2kJ_k.$$

Thus, $(2k+1)J_k = -\frac{2}{\lambda}\sqrt{1-\lambda} + \frac{2k}{\lambda}J_{k-1}$.

3. Now we shall prove by induction the equation which is the goal of this section. We shall assume that $\lambda \neq 0$. Then the equation will also be valid for $\lambda = 0$ by virtue of the continuity of the left and right sides.

When k = 0 the formula is verified directly by calculation:

$$J_0 = \int_0^1 \frac{dx}{\sqrt{1-\lambda x}} = -\frac{2}{\lambda}\sqrt{1-\lambda x}\Big|_0^1 =$$

$$= -\frac{2}{\lambda}(\sqrt{1-\lambda} - 1) = \frac{2}{1+\sqrt{1-\lambda}}.$$

Let it be valid for J_{k-1}: we shall show that it is also valid for J_k. From 3.2,

$$(2k+1)J_k = -\frac{2}{\lambda}\sqrt{1-\lambda} + \frac{2k}{\lambda}J_{k-1}.$$

According to the inductive hypothesis,

$$J_{k-1} = \frac{(k-1)!}{(2k-1)!}\sum_{i=1}^{k}\frac{2^i(2k-i-1)!}{(k-i)!}\frac{1}{(1+\sqrt{1-\lambda})^i}.$$

Let us substitute this expression into the preceding one and add and subtract $2/\lambda$

$$(2k+1)J_k = \frac{2}{\lambda} - \frac{2}{\lambda}\sqrt{1-\lambda} +$$

$$+ \frac{2k}{\lambda}\left(\frac{(k-1)!}{(2k-1)!}\sum_{i=1}^{k}\frac{2^i(2k-i-1)}{(k-i)!}\frac{1}{1+\sqrt{1-\lambda}}\right) - \frac{2}{\lambda}.$$

Let us group the first two and last two terms and use Eq. (2) of 3.1:

$$(2k+1)J_k = \frac{2}{1+\sqrt{1-\lambda}} +$$

$$+ \frac{2}{\lambda}\left[\frac{k!}{(2k-1)!}\sum_{i=1}^{k}\frac{2^i(2k-i-1)!}{(k-i)!}\left(\frac{1}{(1+\sqrt{1-\lambda})^i} - \frac{1}{2^i}\right)\right].$$

The difference in the last parentheses can be rewritten as:

$$\left(\frac{1}{1+\sqrt{1-\lambda}}\right)^i - \left(\frac{1}{2}\right)^i =$$

$$= \left(\frac{1}{1+\sqrt{1-\lambda}} - \frac{1}{2}\right)\sum_{s=0}^{k}\left(\frac{1}{2}\right)^s\left(\frac{1}{1+\sqrt{1-\lambda}}\right)^{i-s-1}.$$

But

$$\frac{1}{1+\sqrt{1-\lambda}} - \frac{1}{2} = \frac{1}{2}\frac{\lambda}{(1+\sqrt{1-\lambda})^2},$$

therefore,

$$\left(\frac{1}{1+\sqrt{1-\lambda}}\right)^i - \left(\frac{1}{2}\right)^i = \lambda\sum_{r=1}^{i}\left(\frac{1}{2}\right)^r\left(\frac{1}{1+\sqrt{1-\lambda}}\right)^{i-r+2}.$$

Substituting into the expression for J_k:

$$(2k+1)J_k = \frac{2}{1+\sqrt{1-\lambda}} +$$

$$+ 2\left[\frac{k!}{(2k-1)!}\sum_{i=1}^{k}\frac{2^i(2k-i-1)!}{(k-i)!}\sum_{r=1}^{i}\left(\frac{1}{2}\right)^r\left(\frac{1}{1+\sqrt{1-\lambda}}\right)^{i-r+2}\right].$$

We let $i - r + 2 = l$ and exchange the order of summation:

$$(2k+1)J_k = \frac{2}{1+\sqrt{1-\lambda}} +$$

$$+ \frac{k!}{(2k-1)!}\sum_{l=2}^{k+1}\frac{2^{l-1}}{(1+\sqrt{1-\lambda})^l}\sum_{i=l-1}^{k}\frac{(2k-i-1)!}{(k-i)!}.$$

Using expression (1) of 3.1 for the latter sum, we obtain

$$(2k+1)J_k = \frac{2}{1+\sqrt{1-\lambda}} +$$

$$+ \frac{k!}{(2k)!}\sum_{l=2}^{k+1}\frac{2^l(2k-l+1)!}{(k-l+1)!}\frac{1}{(1+\sqrt{1-\lambda})^l} =$$

$$= \frac{k!}{(2k)!}\sum_{l=1}^{k+1}\frac{2^l(2k-l+1)!}{(k-l+1)!}\frac{1}{(1+\sqrt{1-\lambda})^l}.$$

We can divide the entire expression by 2k + 1 to obtain the required representation for J_k.

4. Reduction of Integral Equations to Infinite Triangular Systems

We shall show how with the aid of the results of the preceding section, Eqs. (1) of 1.4 can be reduced to an infinite triangular system of linear equations.

1. We calculate the odd moments of $T_1(p)$:

$$\int_0^{\mathscr{P}} p^{2k+1}T_1(p)\,dp = \int_0^{\mathscr{P}} p^{2k+1}\left(\int_0^{\mathscr{P}^{-1}}\frac{dH(r)}{r\sqrt{1-p^2r^2}}\right)dp =$$

$$= \int_0^{\mathscr{P}^{-1}}\frac{dH(r)}{r}\int_0^{\mathscr{P}}\frac{p^{2k+1}dp}{\sqrt{1-p^2r^2}}, \quad k = 0, 1, 2, \ldots$$

For now, we have changed the order of integration formally. If we let $p = v\mathcal{P}$, we obtain

$$\int_0^1 v^{2k+1} T_1(v\mathcal{P})\,dv = \int_0^{\mathcal{P}^{-1}} \frac{dH(r)}{r} \int_0^1 \frac{v^{2k+1}\,dv}{\sqrt{1-\mathcal{P}^2 r^2 v^2}}.$$

If we use the main equation of Section 3, then

$$\int_0^1 v^{2k+1}\, T_1(v\mathcal{P})\,dv =$$

$$= \frac{k!}{2(2k+1)} \sum_{i=1}^{k+1} \frac{(2k-i+1)!}{(k-i+1)!} \int_0^{\mathcal{P}^{-1}} \frac{2^i\,dH(r)}{r\left(1+\sqrt{1-\mathcal{P}^2 r^2}\right)^i},$$

$$k = 0, 1, \ldots$$

We make the substitution using the monotonic function $z = 2\left(1+\sqrt{1-\mathcal{P}^2 r^2}\right)^{-1}$. Then,

$$\int_0^1 v^{2k+1}\, T_1(v\mathcal{P})\,dv =$$

$$= \frac{k!}{2(2k+1)!} \sum_{i=1}^{k+1} \frac{(2k-i+1)!}{(k-i+1)!} \int_1^2 z^i \frac{\mathcal{P}z}{2\sqrt{z-1}}\, dH\left(\frac{2\sqrt{z-1}}{\mathcal{P}z}\right).$$

If we let $\frac{\mathcal{P}z}{4\sqrt{z-1}}\, dH\left(\frac{2\sqrt{z-1}}{\mathcal{P}z}\right) = dH_T(z)$, we obtain

$$\int_0^1 v^{2k+1}\, T_1(v\mathcal{P})\,dv = \frac{k!}{(2k+1)!} \sum_{i=1}^{k+1} \frac{(2k-i+1)!}{(k-i+1)!} \int_1^2 z^i\, dH_T(z).$$

If we introduce the symbols $\beta_i = \int_0^1 v^i T_1(v\mathcal{P})\,dv$, $b_i = \int_0^2 z^i\, dH_T(z)$, $i = 1, 2, \ldots$, we finally have

$$\beta_{2k+1} = \frac{k!}{(2k+1)!} \sum_{i=1}^{k+1} \frac{(2k-i+1)!}{(k-i+1)!}\, b_i, \quad k = 0, 1, 2 \ldots.$$

2. If we calculate the even moments of $X_1(p)$, we obtain exactly the same thing:

$$\alpha_{2k} = \frac{k!}{(2k+1)!} \sum_{i=1}^{k+1} \frac{(2k-i+1)!}{(k-i+1)!}\, a_i, \quad k = 0, 1, 2, \ldots,$$

where

$$\alpha_i = \int_0^1 v^i X_1(v\mathcal{P})\,dv,$$

$$a_i = \int_1^2 z^i\, dH_X(z), \quad i = 0, 1, 2, \ldots,$$

$$dH_X(z) = \frac{\sqrt{z-1}}{z}\, dH\left(\frac{2\sqrt{z-1}}{\mathcal{P}z}\right).$$

3. We must still justify the interchange of the order of integration in 4.1 and the similar change in 4.2. For this, we shall use the Fubini theorem.

The integrals in both paragraphs are Lebesgue-Stieltjes integrals of a nonnegative integrand in positive measure. The following theorem is valid for them.

Let $f(x, y) \geq 0$ and λ be λ-measurable on the rectangle $S\{a \leq x \leq b,\ c \leq y \leq d\}$ and $f(x, y) \geq 0$. Then, from the existence of the iterated integral $\int_a^b \mu(dx) \int_c^d f(x, y)\,\nu(dy)$ follows the λ-summability of $f(x, y)$ in S, i.e., by the Fubini theorem, the existence of another iterated integral and its equality to the first follow.

Here μ and ν are linear positive measures, and λ is a plane Stieltjes measure, which is the product of linear measures.

In our case, the measure on the v axis is a Lebesgue measure, and on the r axis it is given by the nondecreasing function H(r); the product of these measures gives the plane measure λ.

The integrand is positive. It is discontinuous only on zero of λ-measure. In fact, the integrand becomes infinite only at the point ($r = \mathcal{P}^{-1}$, $v = 1$) and (in 4.1) when $r = 0$ and $v \in [0, 1]$. However, H(r) cannot have a discontinuity at zero, since this would be inconsistent with the finiteness of $T_1(p)$ for $p \in [0, \mathcal{P})$. That is, the integrand λ is measurable.

Finally, the existence of one of the iterated integrals follows from the easily verified inequality

$$\int_0^{\mathcal{P}} \frac{p^{2k+1}\,dp}{\sqrt{1-p^2 r^2}} \leqslant \mathcal{P}^{2k}, \quad r \in [0, \mathcal{P}^{-1}]$$

and $\alpha_{2k} \leqslant \mathcal{P}^{2k} \frac{X_1(p)}{p} < \infty$ and $\beta_{2k+1} \leqslant \mathcal{P}^{2k} T_1(p) < \infty$, $p \in [0, \mathcal{P})$, which are obtained with its use.

4. We shall show that the systems of 4.1 and 4.2 are equivalent to the equations

$$T_1(p) = \int_0^{\mathcal{P}^{-1}} \frac{dH(r)}{r\sqrt{1-p^2r^2}}, \qquad X_1(p) = \int_0^{\mathcal{P}^{-1}} \frac{pr\,dH(r)}{\sqrt{1-p^2r^2}},$$

$$p \in [0, \mathcal{P}).$$

Let us consider, for example, 4.2 and the equation for $X_1(p)$. On the one hand, a function H(r) that satisfies the equation also satisfies the system, and the assertion is already proved. On the other hand, let H(r) be a solution of the system of 4.2. If we make transformations in the reverse order, we see that then

$$\int_0^{\mathcal{P}} p^{2k}\left(X_1(p) - \int_0^{\mathcal{P}^{-1}} \frac{pr\,dH(r)}{\sqrt{1-p^2r^2}}\right) dp = 0, \quad k = 0, 1, 2, \ldots$$

Since the expression in parentheses is continuous for $p \in [0, \mathscr{P})$, then

$$X_1(p) = \int_0^{\mathscr{P}^{-1}} \frac{pr\, dH(r)}{\sqrt{1-p^2r^2}}.$$

5. Necessary Conditions Expressible in Terms of Moments

1. From the triangular systems of 4.1 and 4.2:

$$\alpha_{2k} = \frac{k!}{(2k+1)!} \sum_{i=1}^{k+1} \frac{(2k-i+1)!}{(k-i+1)!} a_i,$$

$$\beta_{2k+1} = \frac{k!}{(2k+1)!} \sum_{i=1}^{k+1} \frac{(2k-i+1)!}{(k-i+1)!} b_i, \quad k = 0, 1, 2, \ldots$$

according to the moments of $X_1(p)$ and $T_1(p)$, we can find the values a_i and b_i, $i = 1, 2, \ldots$.

Let us associate with them

$$a_0 = \frac{\mathscr{P}}{2} \lim_{p\to 0} \frac{X_1(p)}{p} = \frac{\mathscr{P}}{2} X_1'(0) = \frac{\mathscr{P}}{2} \int_0^{\mathscr{P}^{-1}} r\, dH(r) = \int_1^2 dH_X(z)$$

and

$$b_0 = \frac{1}{2} \lim_{p\to 0} T_1(p) = \frac{1}{2} T_1(0) = \frac{1}{2} \int_0^{\mathscr{P}^{-1}} \frac{dH(r)}{r} = \int_1^2 dH_T(z).$$

Note that a_i and b_i, $i = 0, 1, 2, \ldots$, are the moments of the nondecreasing functions $H_X(z)$ and $H_T(z)$, $z \in [1, 2]$. Therefore, we can employ the results obtained in the theory of moments to determine the properties of a_i and b_i.

2. Before turning to the properties of a_i and b_i, we should point out that with the aid of the obtained results, we can, by a method different from that in [1], prove the uniqueness of the solution of each of the equations:

$$X_1(p) = \int_0^{\mathscr{P}^{-1}} \frac{pr\, dH(r)}{\sqrt{1-p^2r^2}}, \quad p \in [0, \mathscr{P}),$$

and

$$T_1(p) = \int_0^{\mathscr{P}^{-1}} \frac{dH(r)}{r\sqrt{1-p^2r^2}}, \quad p \in [0, \mathscr{P}).$$

In fact, the moments of $H_X(z)$ and $H_T(z)$ of nonzero order are uniquely determined from the moments of $X_1(p)$ and $T_1(p)$. The zero-order moments are found as in 5.1. From these moments, $H_X(z)$ and $H_T(z)$ are found uniquely, provided that $H_X(1) = H_T(1) = 0$. This follows from the fact that the polynomials form a set that is dense in the space $L_\rho[a, b]$, i.e., in the space of the functions for which the integral $\int_a^b f(x)d\rho(x)$ with the norm $\|f\| = \int_a^b |f(x)|\,d\rho(x)$ exists, where $\rho(x)$ is a monotone nondecreasing function. Finally, the function $H(r)$, where $H(0) = 0$, is uniquely determined by anyone from the combination of the functions $H_X(z)$ or $H_T(z)$.

3. This is examined in the power moment problem.

Given the numbers c_i, $i = 0, 1, 2, \ldots$: What are the necessary and sufficient conditions that these numbers must satisfy in order that there exist a nondecreasing function $\sigma(x)$ such that $c_i = \int_E x^i d\sigma(x)$, where E is some interval, semiaxis, or the entire x axis?

If c_i are given only for $0 \leq i \leq n$, the problem is called "truncated."

For the case of a finite interval $E\{a \leq x \leq b\}$, we obtain the following answer, a proof of which is given in [2].

4. Theorem. In order that an infinite sequence of numbers c_i, $i = 0, 1, 2, \ldots$, be a system of moments for some nondecreasing function $\sigma(x)$ for $x \in [a, b]$, it is necessary and sufficient that for any m the quadratic forms

$$\sum_0^m c_{i+j}x_ix_j, \quad \sum_0^m [(a+b)c_{i+j+1} - abc_{i+j} - c_{i+j+2}]\, x_ix_j,$$

or

$$\sum_0^m (c_{i+j+1} - ac_{i+j})\, x_ix_j, \quad \sum_0^m (bc_{i+j} - c_{i+j+1})\, x_ix_j$$

be nonnegative.

5. In our case the moments are obtained by integrating over the interval [1, 2]. Therefore we obtain the necessary conditions by applying Theorem 5.4 to the numbers a_i and b_i.

Theorem. In order that a curve $\Gamma_1\{X_1(p), T_1(p)\}$, $p \in [0, \mathscr{P})$, be a deep-source travel-time curve, it is necessary that for any m the following quadratic forms be nonnegative:

$$\text{a)} \quad \sum_0^m a_{i+j}x_ix_j \text{ and } \sum_0^m (3a_{i+j+1} - 2a_{i+j} - a_{i+j+2})\, x_ix_j$$

or

$$\sum_0^m (a_{i+j+1} - a_{i+j})\, x_ix_j \text{ and } \sum_0^m (2a_{i+j} - a_{i+j+1})\, x_ix_j;$$

b) $\sum_0^m b_{i+j}x_ix_j$ and $\sum_0^m (3b_{i+j+1} - 2b_{i+j} - b_{i+j+2})\, x_ix_j$

$\sum_0^m (b_{i+j+1} - b_{i+j})\, x_ix_j$ and $\sum_0^m (2b_{i+j+1} - b_{i+j+1})\, x_ix_j$.

6. Insufficiency of Constraints only on Moments of $X_1(p)$

1. In this and the following sections, we shall discover a difference in the properties of $X_1(p)$ and $T_1(p)$. We shall show that the system of conditions 5.5a and 2.1 is not sufficient, while the system of conditions 5.5b and 2.1 is sufficient for Γ_1 to be a travel-time curve. The reason for this is that in spite of the fact that $H_X(z)$, $z \in [1, 2]$, is finite, the integrals $\int_1^{1+\varepsilon} dH_T(z)$ and $\int_0^{\varepsilon} dH(r)$ can diverge for any $\varepsilon > 0$ and, therefore the functions $H_T(z)$ and $H(r)$ cannot exist.

2. Let us give an example. In the equations

$$X_1(p) = \int_0^{\mathscr{P}^{-1}} \frac{pr\, dH(r)}{\sqrt{1 - p^2r^2}}, \quad T_1(p) = \int_0^{\mathscr{P}^{-1}} \frac{dH(r)}{r\sqrt{1 - p^2r^2}},$$

$$p \in [0, \mathscr{P}),$$

we let $\mathscr{P} = 1/2$, $dH(r) = dr/r$.

Since $dH_x(z(r)) = \frac{\mathscr{P}r}{2} dH(r) = \frac{\mathscr{P}}{2} dr = \frac{dr}{4}$, $H_X(z)$ exists and, therefore, conditions 5.5a are satisfied. On the other hand it is obvious that the integrals $\int_1^z dH_T(z)$ and $\int_0^r dH(r)$ diverge.

We shall show that, nevertheless, $T_1(p)$ can be defined such that conditions 2.1 are satisfied. For this, let us use the condition

$$T_1'(p) - pX_1'(p) = 0.$$

We obtain

$$X_1(p) = \int_0^2 \frac{p\, dr}{\sqrt{1 - p^2r^2}} = \arcsin 2p,$$

$$T_1'(p) = pX_1'(p) = \frac{2p}{\sqrt{1 - 4p^2}},$$

hence

$$T_1(p) = \frac{1}{2}(1 - \sqrt{1 - 4p^2}) + C = \frac{2p^2}{1 + \sqrt{1 - 4p^2}} + C.$$

The functions $X_1(p) = \arcsin 2p$, $T_1(p) = \frac{2p^2}{1 + \sqrt{1 - 4p^2}} + C$, $p \in \left[0, \frac{1}{2}\right)$, for any $C > 0$, satisfy conditions 2.1.

7. Sufficient Conditions

Theorem. Let conditions 5.5b and 2.1c, 2.1d, and 2.1e be satisfied. Then the curve $\Gamma_1\{X_1(p), T_1(p)\}$, $p \in [0, \mathscr{P})$, is a travel-time curve.

1. According to theorem 5.4, it follows from condition 5.5b that a nondecreasing function $H_T(z)$ exists such that

$$b_i = \int_1^2 z^i\, dH_T(z), \quad i = 0, 1, 2, \ldots$$

2. If $H_T(z)$ exists, then there exists a finite nondecreasing function $H(r)$ that is defined by the equation $dH_T(z(r)) = dH(r)/2r$, $r \in [0, \mathscr{P}^{-1}\}$ and the condition $H(0) = 0$.

3. The function $H(r)$ satisfies system 4.1; and in accordance with that which was proved, it is equivalent to the equation

$$T_1(p) = \int_0^{\mathscr{P}^{-1}} \frac{dH(r)}{r\sqrt{1 - p^2r^2}}, \quad p \in [0, \mathscr{P}).$$

4. The integral $X_1^*(p) = \int_0^{\mathscr{P}^{-1}} \frac{pr\, dH(r)}{\sqrt{1 - p^2r^2}}$, obviously exists and is differentiable for $p \in [0, \mathscr{P})$. Since $T_1'(p) - pX_1^{*\prime}(p) = 0$ and $X_1^*(0) = 0$, which is easily verified, then

$$X_1^*(p) = \int_0^{\mathscr{P}^{-1}} \frac{pr\, dH(r)}{\sqrt{1 - p^2r^2}} \equiv X_1(p) \text{ for } p \in [0, \mathscr{P})$$

follows from conditions 2.1c, 2.1d, and 2.1e.

8. Conclusions

1. Let

$$\alpha_i = \int_0^1 v^i X_1(v\mathscr{P})\, dv,$$

$$\beta_i = \int_0^1 v^i T_1(v\mathscr{P})\, dv, \quad i = 1, 2, \ldots$$

We determine a_i, b_i, $i = 1, 2, \ldots$, from the equations

$$\alpha_{2k} = \frac{k!}{(2k+1)!} \sum_{i=1}^{k+1} \frac{(2k - i + 1)!}{(k - i + 1)!} a_i,$$

$$\beta_{2k+1} = \frac{k!}{(2k+1)!} \sum_{i=1}^{k+1} \frac{(2k - i + 1)!}{(k - i + 1)!} b_i, \quad k = 0, 1, 2 \ldots$$

We let

$$a_0 = \frac{\mathscr{P}}{2} X_1'(0), \quad b_0 = \frac{1}{2} T_1(0).$$

2. Then in order that a curve

$$\Gamma_1 \{X_1(p), T_1(p)\}, \; p \in [0, \mathscr{P})$$

be a travel-time curve, it is necessary and sufficient that

a) the quadratic forms

$$\sum_0^m b_{i+j} x_i x_j, \quad \sum_0^m (3b_{i+j+1} - 2b_{i+j} - b_{i+j+2}) x_i x_j$$

or

$$\sum_0^m (b_{i+j+1} - b_{i+j}) x_i x_j, \quad \sum_0^m (2b_{i+j+1}) x_i x_j$$

be nonnegative for any $m = 0, 1, 2, \ldots$;

b) the functions $X_1(p)$ and $T_1(p)$ be differentiable for $p \in [0, \mathscr{P})$;

c) $T_1'(p) - pX_1'(p) = 0$;

d) $X_1(0) = 0$.

3. Of the necessary conditions we note:

a) the nonnegativeness of the quadratic forms

$$\sum_0^m a_{i+j} x_i x_j, \quad \sum_0^m (3a_{i+j+1} - 2a_{i+j} - a_{i+j+2}) x_i x_j$$

or

$$\sum_0^m (a_{i+j+1} - a_{i+j}) x_i x_j, \quad \sum_0^m (2a_{i+j} - a_{i+j+1}) x_i x_j$$

for any $m = 0, 1, 2, \ldots$;

b) the monotonic increase and nonnegativeness of $X_1(p)$ and $T_1(p)$, $p \in [0, \mathscr{P})$.

LITERATURE CITED

1. Gerver, M. L., and V. M. Markushevich, "Determining seismic-wave velocity from travel-time curves," this volume, p. 123.
2. Krein, M. G. (1951), "The ideas of P. L. Chebyshev and A. A. Markov in the theorem of limiting values of integrals, and their further development," Usp. Matem. Nauk, 6:3-120.

FINDING A VELOCITY PROFILE FROM A LOVE WAVE DISPERSION CURVE: PROBLEMS OF UNIQUENESS†

M. L. Gerver and D. A. Kazhdan

1. Introduction

In [1] we considered the problem of finding a function $\rho(x)$ from the eigenvalue $s = s(p)$ of an equation $y'' + [p\rho(x) - s]y = 0$. The methods employed in [1] allow us to consider a more general equation $[A(x)y']' + [pB(x) - sC(x)]y = 0$. This in turn makes it possible to study some questions of uniqueness in the inverse problem for Love waves, i.e., determination of the characteristics of the medium from the phase or group velocity.

A. Statement of Problem

The following boundary value problem is considered (see [2]) in the study of Love waves in a spherically symmetric medium:

$$\frac{d}{dr}\left[\mu\left(\frac{dV}{dr}-\frac{V}{r}\right)\right]+\frac{3\mu}{r}\frac{dV}{dr}+\left[\eta^2\rho-\frac{\mu(\xi^2r_2{}^2+{}^9/_4)}{r^2}\right]V=0,$$
$$r_1<r<r_2,$$
$$\mu\left(\frac{dV}{dr}-\frac{V}{r}\right)=0 \quad \text{for } r=r_1 \text{ and } r=r_2. \qquad (1)$$

Here $r_1 < r < r_2$ is an ideally elastic spherical shell whose outside and inside surfaces ($r = r_2$ and $r = r_1$) are free of tangential stresses; the rigidity μ and density ρ are functions only of the radius r, not of the other coordinates, and η and ξ are real parameters.

The transverse wave velocity is simply expressed in terms of $\mu(r)$ and $\rho(r)$: $b(r) = \sqrt{\mu(r)/\rho(r)}$. If the characteristics of the medium $\mu(r)$ and $\rho(r)$, or $\rho(r)$ and $b(r)$, are known, we can determine the eigenvalues $\xi_k^2(\eta)$ of problem (1), $k = 1, 2, \ldots$.

We are interested in the inverse problem: determination of the characteristics of the medium from the eigenvalues.

The parameters η and ξ have the following physical meaning: η is the frequency of oscillation, $\eta/\xi_k(\eta)$ is the phase velocity, and $[d\xi_k(\eta)/d\eta]^{-1}$ is the group velocity (of the k-th harmonic of Love waves). The dependence of the phase and group velocity upon frequency is called the dispersion. The function that expresses this dependence for the k-th harmonic is called the k-th branch of the dispersion curve. Only the first branch is reliably determined experimentally. Therefore it is important to understand which characteristics of the medium can be found when $\xi_1^2(\eta)$ is known.

B. Transformation of Equation

By substituting $V(r) = ry(r)$ and multiplying all terms of the equation by r^3, problem (1) is brought to the form

$$\frac{d}{dr}\left(\mu r^4\frac{dy}{dr}\right)+[\eta^2\rho r^4-(\xi^2r_2^2-{}^9/_4)\mu r^2]\,y=0,$$
$$\mu\frac{dy}{dr}=0 \quad \text{for } r=r_1 \text{ and } r=r_2. \qquad (2)$$

If we introduce the symbols

$$A_0(r)=\mu r^4,\quad B_0(r)=\rho r^4,\quad C_0(r)=\mu r^2,$$
$$p=\eta^2,\quad s=\xi^2r_2{}^2-{}^9/_4,$$

we can rewrite (2) more compactly:

$$\frac{d}{dr}\left(A_0\frac{dy}{dr}\right)+(pB_0-sC_0)\,y=0,\quad \frac{dy}{dr}(r_1)=\frac{dy}{dr}(r_2)=0. \qquad (3)$$

If the first eigenvalue $\xi_1^2(\eta)$ of problem (1), and the radius r_2 are known, then the first eigenvalue $s_1(p)$ of problem (3) is known.

† Translated from Vychislitel'naya Seismologiya, No. 4, pp. 78-94, Moscow, Nauka (1968).

C. Examples of Nonuniqueness

It is clearly not enough to know $s_1(p)$ in order to fully reduce Eq. (3), i.e., to determine the coefficients A_0, B_0, and C_0, or, physically, to find the characteristics of the medium $\mu(r)$ and $\rho(r)$.

For example, $s_1(p)$ is obviously unchanged if we simultaneously multiply A_0, B_0, and C_0 by the same number.

In this example, $\mu(r)$ and $\rho(r)$ are changed, but their ratio $b^2(r)$ is not changed. It would be incorrect to think, however, that $b(r)$ could be uniquely determined from $s_1(p)$; the following examples show this.

Because of the symmetry of the boundary conditions, the function $s_1(p)$ is unchanged when the coefficients A_0, B_0, and C_0 are replaced by functions that are symmetric to them with respect to the middle of the interval $[r_1, r_2]$.

A more general example of nonuniqueness, based on symmetry of the boundary conditions, is as follows. Let the coefficients A_0, B_0, and C_0 of Eq. (3) be periodic functions with period $(r_2 - r_1)/n$ (see Fig. 1a). Along with A_0, B_0, and C_0, let us consider the functions A^0, B^0, and C^0 (see Fig. 1b). On the interval $[r_1 + (m-1)(r_2 - r_1)/n, r_1 + m(r_2 - r_1)/n]$ we take $A^0(r)$, $B^0(r)$, and $C^0(r)$, respectively, to be equal either to the functions $A_0(r)$, $B_0(r)$, and $C_0(r)$ or to the symmetric functions $A_0(r_1 + r_2 - r)$, $B_0(r_1 + r_2 - r)$, and $C_0(r_1 + r_2 - r)$. Thus we can obtain 2^n different equations. When Eq. (3) is replaced by any of these equations, the function $s_1(p)$ is obviously unchanged.

The latter example may give rise to one objection: since $A_0(r) = \mu(r) \cdot r^4$ and $C_0(r) = \mu(r) \cdot r^2$, then $A_0(r)/C_0(r) = r^2$, so that the coefficients A_0 and C_0 cannot both be periodic functions.

This is a valid objection, but the situation is easily rectified: a substitution of the independent variable must be made. Let $x = \ln r$. Then problem (3) takes the form

$$\frac{d}{dx}\left(A(x)\frac{dy}{dx}\right) + [pB(x) - sA(x)]\,y = 0,$$

$$y'(x_1) = y'(x_2) = 0, \qquad (4)$$

where

$$A(x) = \mu(r)\,r^3, \quad B(x) = \rho(r)\,r^5, \quad x_1 = \ln r_1, \quad x_2 = \ln r_2.$$

Functions A and B can already be made periodic on the interval $x_1 \le x \le x_2$.

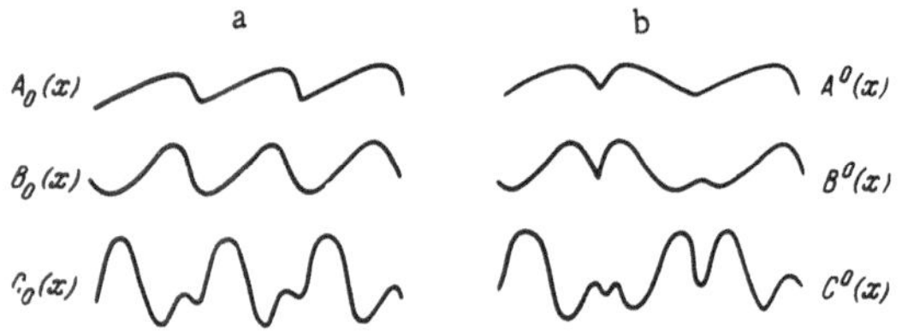

Fig. 1

D. Uniqueness Theorem

As above, we let $x = \ln r$. The ratio $A(x)/B(x)$ we denote by $u^2(x)$. Thus, $u(x) = r^{-1}\sqrt{\mu(r)/\rho(r)} = b(r)/r$. The function $u(x)$ is naturally called the *reduced velocity*. In fact, we can conformally map the ring $r_1 < r < r_2$ onto the sheet $\ln r_1 < x < r_2$. This mapping is performed with the aid of a complex logarithmic function: the point of the ring $re^{i\theta}$ becomes the point of the sheet $\ln r + i\theta = x + i\theta$. The modulus of expansion in this mapping is $1/r$. That is, in the transformed problem $u(x)$ plays the role of the velocity (cf. [3]).

The examples of nonuniqueness considered above show that the reduced velocity $u(x)$ is not uniquely defined by the function $s_1(p)$. The following uniqueness theorem is valid, however.

Let $H(t)$ be the measure of those x on the interval $[x_1, x_2]$ at which $u(x) < t$:

$$H(t) = \text{mes}\,\{x,\ x_1 \leqslant x \leqslant x_2,\ u(x) < t\}.$$

The function $H(t)$ is uniquely defined by $s_1(p)$.

We shall give a more precise formulation.

Let us consider, along with (4), the boundary value problem

$$(\hat{A}(x)y')' + [p\hat{B}(x) - s\hat{A}(x)]\,y = 0,$$

$$y'(x_1) = y'(x_2) = 0. \qquad (\hat{4})$$

Let the coefficients $A(x)$, $B(x)$, $\hat{A}(x)$, and $\hat{B}(x)$ be strictly positive and piecewise twice continuously differentiable functions. (In more detail, the latter condition means that the coefficients have continuous first and second derivatives everywhere on $[x_1, x_2]$, except perhaps at a finite number of points. At a finite number of points, the coefficients and their derivatives can be discontinuous or nonexistent. The first and second derivatives are bounded.)

Let $\hat{u}(x) = \sqrt{\hat{A}(x)/\hat{B}(x)}$, $\hat{H}(t) = \text{mes}\,\{x, x_1 \le x \le x_2, \hat{u}(x) < t\}$. Finally, let $s_k(p)$ and $\hat{s}_k(p)$ be eigenvalues of boundary value problems (4) and $(\hat{4})$, $k = 1, 2, \ldots$. Then the following theorem is valid.

Theorem 1. If the functions $H(t)$ and $\hat{H}(t)$ are not identically equal, then the functions $s_k(p)$ and $\hat{s}_k(p)$ do not agree for any k.

For definiteness, we shall prove that if $s_1(p) \equiv \hat{s}_1(p)$, then $H(t) \equiv \hat{H}(t)$. According to the physical meaning of the problem, $u(x) \not\equiv$ const, and we shall assume that this condition is satisfied (especially since Theorem 1 is quite easily proved when $u(x) \equiv$ const). The proof is done in two steps: in Section 2 it is established that equality of the functions $s_1(p) \equiv \hat{s}_1(p)$ implies coincidence of $s_k(p)$ and $\hat{s}_k(p)$ for an infinite set of k values, and in Section 3 that the function H(t) is recovered from an infinite number of functions $s_k(p)$.

2. Analytic Continuation of $s_1(p)$

A. Asymptotic Properties and Monotonicity of $s_k(p)$

The eigenvalues $s_k = s_k(p)$ of the boundary value problem

$$(Ay')' + (pB - sA)y = 0, \ y'(x_1) = y'(x_2) = 0 \tag{4}$$

possess the following properties.

Lemma 1. Let $\min B/A = C_1$ and $\max B/A = C_2$. Then, for any k, the ratio $s_k(p)/p$ approaches C_1 when $p \to -\infty$ and approaches C_2 when $p \to +\infty$.

Proof. The lemma easily follows from the Sturm oscillation theorem. We use the fact that the k-th eigenfunction has on the interval $[x_1, x_2]$ exactly $(k - 1)$ zeros.

Lemma 2. For any k, the function $s_k(p)$ increases monotonically.

Proof. This lemma also follows from the Sturm theorem, and from the positiveness of functions A and B.

Note. In Section 1D, we assumed that the reduced velocity $u(x) = \sqrt{A(x)/B(x)}$ was not constant on the interval $[x_1, x_2]$. Let $\max u(x) = u_1$ and $\min u(x) = u_2$. Then the constants C_1 and C_2 in Lemma 1 are equal to u_1^{-2} and u_2^{-2}, respectively:

$$0 < C_1 = u_1^{-2} < C_2 = u_2^{-2}.$$

According to Lemma 1, when $p \to +\infty$ the graph l_k of the function $s_k(p)$ has an asymptotic direction. It may or may not have an asymptote. Everything depends upon whether the set $\{x, x_1 \le x \le x_2, B(x)/A(x) = C_2\}$ is of positive or zero measure – or in terms of the function H(t), whether the limit $H(u_2 + 0)$ is positive or zero. If $H(u_2 + 0) > 0$, all graphs l_k have asymptotes; if $H(u_2 + 0) = 0$, then $C_2 p - s_k(p) \to +\infty$ for any k. This fact is proved just as Lemma 1 is. The situation is similar when $p \to -\infty$; here everything depends on the value of $H(u_1 - 0)$.

This is required in the proof of Lemma 3.

B. Structure of the Irreducible Analytic Set Containing the Graph of $s_1(p)$

In Eq. (4) let us replace the real parameters p and s by the complex parameters z and w:

$$(Ay')' + (zB - wA)\, y = 0. \tag{5}$$

Let y (x, z, w) be a solution of Eq. (5) that satisfies the initial conditions $y(x_1, z, w) = 1$ and $y'(x_1, z, w) = 0$. Let $f(z, w) = y'(x_2, z, w)$. Owing to the analytic dependence of the solution upon the parameters, $f(z, w)$ is an entire function.

Let N be the set of its zeros: $(z_0, w_0) \in N$ if and only if $f(z_0, w_0) = 0$. In particular, for any k, N contains a graph l_k of the function $s_k(p)$. Let N_1 be an irreducible analytic subset of N that contains a graph l_1 of the function $s_1(p)$.

Theorem 2. Subset N_1 contains graphs l_k for an infinite set of k values.

Since N_1 is contained in any analytic set $\hat{N}$ that contains l_1, theorem 2 has a corollary.

Corollary. If $s_1(p) \equiv \hat{s}_1(p)$, then $s_k(p) \equiv \hat{s}_k(p)$ for an infinite set of k values.

C. Proof of Theorem 2

Plan of Proof. Having assumed that N_1 contains only a finite number of graphs l_k, we shall show that N_1 is an algebraic curve. On the other hand, it follows from Lemma 3 (see below) that this is impossible. The contradiction proves the theorem.

Lemma 3. No set Γ of graphs l_k is a (real) algebraic curve.

Proof. Let L be an algebraic curve on the plane p, s. We shall complement plane p, s with respect to the projective plane. Then the behavior of L in the vicinity of an infinitely distant line α can be of one of three types (Fig. 2):

A_1 – curve L intersects line α;

A_2 – even-order tangency;

A_3 – odd-order tangency.

At the same time, according to Lemma 1 and the condition $C_1 < C_2$, each of the graphs l_k is placed at infinity by one of two other methods (Fig.

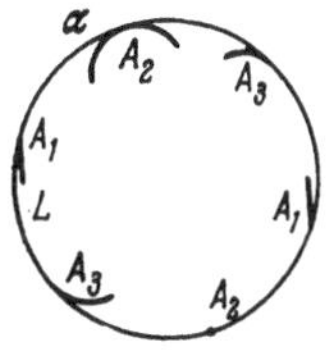

Fig. 2

3; the broken lines show the "deficient parts of l_k," the absence of which proves that Γ is not an algebraic curve).

B_1 – graph l_k approximates line α when $p \to +\infty$ along the asymptote: this will be so if the set $\{x, B(x)/A(x) = C_2\}$ is of positive measure: $H(u_2 + 0) > 0$.

B_2 – when $p \to +\infty$ graph l_k has an asymptotic direction: $s_k(p)/p \to C_2$. But it does not have an asymptote: $C_2p - s_k(p) \to +\infty$ (case of $H(u_2 + 0) = 0$). We emphasize that in this case each of the graphs l_k, when $p \to +\infty$, passes below any line that intersects line α at point B_2. Thus smoothness is lost in moving from one graph l_k to another at point B_2.

When $p \to -\infty$, graph l_k is similarly placed, but it has a different asymptotic direction

$$s_k(p)/p \to C_1 \neq C_2.$$

A number of lemmas are required to verify that set N_1 is an algebraic curve if it contains only a finite number of graphs l_k.

Lemma 4. Let $(z_0, w_0) \in N$. Then the numbers z_0 and w_0 are either both real or both not real.

Proof. We multiply Eq. (5) by $\bar{y}$, the conjugate function of the solution $y = y(x, z_0, w_0)$, and integrate from x_1 to x_2:

$$\int_{x_1}^{x_2} [(Ay')' + (z_0B - w_0A)y]\bar{y}\,dx = 0.$$

The integrals

$$J_1 = \int_{x_1}^{x_2} (Ay')'\bar{y}\,dx = -\int_{x_1}^{x_2} A|y'|^2 dx,$$

$$J_2 = \int_{x_1}^{x_2} B|y|^2 dx, \quad J_3 = \int_{x_1}^{x_2} A|y|^2 dx$$

are all real. Therefore the equality $J_1 = w_0J_3 - z_0J_2$ is possible only if the numbers z_0 and w_0 are either both real or both not real.

Lemma 5. The function $f(z, w)$ has for any z a half order of increase in w.

Proof. The substitution $y_0 = y\sqrt{A}$ brings Eq. (5) to the form

$$y_0'' + [q(x, z) - w]\,y_0 = 0, \qquad (5')$$

where

$$q(x, z) = \left(\frac{A'}{2A}\right)^2 - \frac{A''}{2A} + z\frac{B}{A}.$$

The requirements imposed on functions A and B (strict positiveness and piecewise double continuous differentiability) ensure the boundedness and piecewise continuity of $q(x, z)$ for any z.

Let y_1 and y_1' be the initial conditions that are satisfied by the solution $y_0(x, z, w)$:

$$y_0(x_1, z, w) = y_1, \; y_0'(x_1, z, w) = y_1'.$$

It is easy to verify (by double differentiation) that $y_0(x, z, w)$ satisfies the relation

$$y_0(x, z, w) = y_1 \operatorname{ch}\{\sqrt{w}(x - x_1)\} + y_1' \frac{\operatorname{sh}\{\sqrt{w}(x - x_1)\}}{\sqrt{w}} -$$

$$-\frac{1}{\sqrt{w}} \int_{x_1}^{x} \operatorname{sh}\{\sqrt{w}(x - t)\}\, q(t, z)\, y_0(t, z, w)\,dt.$$

Thus the function $f(z, w) = y'(x_2, z, w) = y_0'(x_2, z, w) A^{-1/2}(x_2) - \frac{1}{2}y_0(x_2, z, w) A^{-3/2}(x_2) A'(x_2)$ has for any z a half order of increase in w. A more detailed proof of this can be found in [4].

The following lemma is similarly proved.

Lemma 5. The function $f(z, w)$ has for any w a half order of increase in z.

To prove this we first make the substitution

$$x^* = \int_{x_1}^{x} \sqrt{\frac{B(t)}{A(t)}}\,dt, \quad y^*(x, z, w) = y(x, z, w)\sqrt{A(x)B(x)}.$$

This substitution is described in greater detail in [5].

The functions g(z, w) and g*(z, w). If we use Lemma 5 and 5* and take into account the relationship between the increase of the entire function of one variable and the distribution of its zeros (see [6]), it is easy to construct two entire functions $g(z, w)$ and $g^*(z, w)$ that vanish in the set N_1 (and only in it) and have an order of increase of not more than one-half (the former in w for any z and the latter in z for any w).

Let us describe, for example, the construction of $g(z, w)$.

Let Z be the set of those z for which $(z, 0) \in N_1$ and let $g^0(z)$ be any entire function for which Z is a set of zeros.

Let us arbitrarily fix $\tilde{z} \bar{\in} z$. The function $f(\tilde{z}, w)$ has a half order of increase in w. Therefore, its zeros are "rather widely" distributed on the plane w (see [6]). The points w at which $(\tilde{z}, w) \in N_1$ are spaced no less widely. Therefore there exists an entire function $\tilde{g}(w)$ that is equal to zero at these and only at these points and has an order of increase not greater than one-half. Such a function (see [6]) is uniquely defined except for a constant factor. Let us take this factor such that $\tilde{g}(0) = g^0(\tilde{z})$.

We let $g(\tilde{z}, w) = \tilde{g}(w)$ for $\tilde{z} \bar{\in} Z$; at points (z, w), $z \in Z$, we extend the definition of $g(z, w)$ by continuity.

Since $g(z, w)$ and $g^*(z, w)$ have common zeros,

$$g(z, w) = g^*(z, w)\,\Phi(z, w),$$

where $\Phi(z, w)$ is an entire function that is nowhere equal to zero.

Note. So far we have not made use of the assumption that N_1 contains only a finite number of graphs l_k, but the following discussion will be based on this assumption.

Lemma 6. The function $g(z, w)$ is a polynomial in w:

$$g(z, w) = \sum_{k=0}^{n} a_k(z)\, w^k,$$

the coefficients $a_k(z)$ are entire functions of z.

Proof. We represent $g(z, w)$ as a series in w with coefficients that are entire functions of z:

$$g(z, w) = \sum_{k=0}^{\infty} a_k(z)\, w^k.$$

We arbitrarily fix the real number z_0. The function $g_0(w) = g(z_0, w)$ has, according to Lemma 4, only real zeros. If we assume that N_1 contains only n graphs l_k, we find that $g_0(w)$ has n zeros. By definition, the order of increase of $g_0(w)$ is not greater than one-half. Therefore, $g_0(w)$ is a polynomial of degree n, i.e., $a_k(z_0) = 0$ when $k > n$. Since $a_k(z)$ is an entire function and z_0 is any real number, then $a_k(z) \equiv 0$ when $k > n$. The lemma is proved.

Note 1. Since $s_k(p)$ are monotonic functions (Lemma 2), it is quite similarly proved that

$$g^*(z, w) = \sum_{k=0}^{n} a_k^*(w)\, z^k,$$

where a_k^* are entire functions of w.

Corollary. Let W^* be the set of those w at which $a_k^*(w) \neq 0$. Then for any $w \in W^*$ there exist not more than n values of z such that $(z, w) \in N_1$.

Note 2. Since $g(z, 0) = a_0(z)$, then $a_0(z)$ has n zeros. That is, $a_0(z)$ is the product of a polynomial and the inverse function $\varphi(z)$. We can assume that $\varphi(z) \equiv 1$. If this is not so, then instead of $g(z, w)$ we must consider the function $g(z, w) \cdot \varphi^{-1}(z)$. In other words we shall assume that $a_0(z)$ is a polynomial.

Notation. Let

$$a_k(z)/a_n(z) = b_k(z),\quad 0 \leqslant k \leqslant n;$$

$$e(z, w) = g(z, w)/a_n(z) = w^n + \sum_{k=0}^{n-1} b_k(z)\, w^k.$$

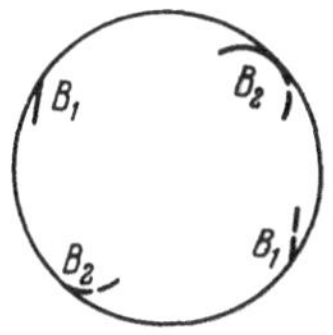

Fig. 3

We consider the implicit function $w(z)$, which is defined by the equation $e(z, w) = 0$. The set of values that it takes when $|z| > m$ we shall denote by W_m.

Lemma 7. If at least one of the meromorphic functions $b_k(z)$ is not rational, then W_m is everywhere dense for any m.

Proof.

Case 1. W_m contains points outside of any circle $|w| < R$.

We shall assume the converse: for $|z| > m$ let all the roots of the polynomial $w^n + \sum_{k=0}^{n-1} b_k(z) w^k$ be less than R in absolute value. Then the functions $b_k(z)$, which are symmetric functions of the roots, are bounded when $|z| > m$. This is impossible, however, if at least one of them is not rational.

Case 2. W_m contains points in any vicinity of a zero.

If $W_m \ni 0$, this is obvious; if $W_m \not\ni 0$, we must apply the result of the previous paragraph to the function $\omega(z) = w^{-1}(z)$, which is a root of the polynomial $\omega^n + \sum_{k=0}^{n-1} \frac{b_{n-k}(z)}{b_0(z)} \omega^k$.

Case 3. W_m contains points in any vicinity of an arbitrary point w_0. For the proof we must apply the result of paragraph 2 to the function $\omega_0(z) = w(z) - w_0$, which is a root of the polynomial

$$\omega_0^n + \sum_{k=0}^{n-1} \frac{1}{k!} \frac{\partial^k e(z, w_0)}{\partial w^k} \omega_0^k.$$

Lemma 8. If W_m is everywhere dense for any m, then a value $w^0 \in W^*$ can be found which the function $w(z)$, defined by the equation $e(z, w) = 0$, takes at an infinite number of points z.

Proof. Let (z_0, w_0) be any zero of the function $e(z, w)$ such that $w_0 \in W^*$ (we recall that the complement of W^* is a discrete set of zeros of the function a_n^*). We describe about points z_0 and w_0 circles U_0 and V_0 (Fig. 4). Circle V_0 is made so small that $V_0 \subset W^*$ and that for any $w' \in V_0$ there exists a $z' \in U_0$ such that $e(z', w') = 0$. This can obviously be done.

We make the number m_1 so great that circle U_0 lies inside the circle $|z| < m_1$. In circle V_0 we find the point $w_1 \in W_{m_1}$. We take z_1 such that:

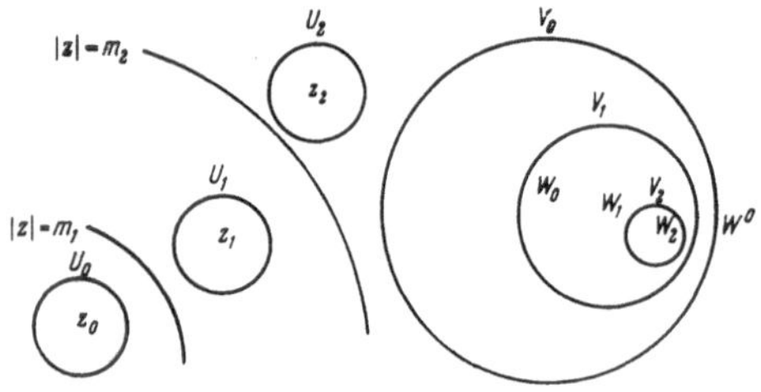

Fig. 4

1) $e(z_1, w_1) = 0$, and 2) $|z_1| > m_1$. We describe circles U_1 and V_1 about points z_1 and w_1. We take circle U_1 outside of $|z| = m_1$ and make circle V_1 so small that it lies inside V_0 and that for any $w' \in V_1$ there exists a $z' \in U_1$ such that $e(z', w') = 0$.

Repeating this process, in the k-th step we find the number m_k, points w_k and z_k, and circles U_k and V_k which satisfy the following conditions: the circle $|z| < m_k$ contains U_{k-1}; $w_k \in V_{k-1} \cap W_{m_k}$; $e(z_k, w_k) = 0$, $|z_k| > m_k$; $U_k \cap \{|z| = m_k\} = 0$; $V_k \subset V_{k-1} \cap w(U_k)$.

Let w^0 be any point of the intersection $\bigcap\limits_{k=0}^{\infty} V_k$. By definition a sequence z_k^0, $z_k^0 \in U_k$, can be found such that $w(z_k^0) = w^0$. Since circles U_k do not overlap, all the z_k^0 are different. The lemma is proved.

Corollary 1. All functions $b_k(z)$ are rational.

Proof. Having assumed the converse, we arrive, using Lemmas 7 and 8, at a contradiction of the corollary of Lemma 6.

Corollary 2. All functions $a_k(z)$ are polynomials.

Proof. According to Note 2 to Lemma 6, the function $a_0(z)$ is a polynomial. That is, $a_n(z) = a_0(z)/b_0(z)$ is rational and (being entire) is a polynomial. Therefore, $a_k(z) = a_n(z) \cdot b_k(z)$ is also a polynomial, $0 < k < n$.

Corollary 3. The function $g(z, w)$ is a polynomial with respect to the set of variables, and thus N_1 is an algebraic curve.

Since the result contradicts Lemma 3, it must be concluded that N_1 contains an infinite number of graphs l_k; hence Theorem 2 is proved.

3. Recovery of the Function $H(t)_k$ from an Infinite Number of Graphs l_k

Summary. Let us consider a solution of Eq. (4), $y(x, p, s)$ that satisfies the initial conditions $y(x_1, p, s) = 1$, $y'(x_1, p, s) = 0$. Let $N(p, s)$ be the number of zeros of $y(x, p, s)$ on the interval $[x_1, x_2]$.

We consider the behavior of the function $N(p, s)$ when $s = cp$ (c is any real number) and when $p \to +\infty$. It will be shown that

$$\lim_{p\to+\infty} \frac{N(p, cp)}{\sqrt{p}} = \frac{1}{\pi}\int_{x_1}^{x_2} \mathrm{Re}\sqrt{\frac{B(x)}{A(x)} - c}\,dx. \quad (6)$$

In Section 2 we showed that an infinite number of functions $s_k(p)$ are uniquely defined by the function $s_1(p)$. If we know the graphs l_k for an infinite number of functions $s_k(p)$, we shall be able to find the limit of the left side of formula (6) for all $c \in (C_1, C_2)$. This would be quite simple to do if we know not only the graphs l_k but also their numbers. Since the numbers of the graphs that are defined by the function $s_1(p)$ are unknown, to find this limit we must use still another asymptotic relation:

$$\frac{N(0, s)}{\sqrt{-s}} \underset{s\to\infty}{\sim} \frac{x_2 - x_1}{\pi}.$$

After we find the left side of (6), we also know the function

$$J(c) = \int_{x_1}^{x_2} \mathrm{Re}\sqrt{\frac{B(x)}{A(x)} - c}\,dx, \quad c \in (C_1, C_2).$$

The function $H(t)$ is expressed in terms of it by the formula

$$H(t) = -\frac{2}{\pi}\int_{t^{-2}}^{C_2} \frac{J'(c)\,dc}{\sqrt{c - t^{-2}}}.$$

This, briefly, is the contents of Section 3. Let us move on to a more detailed discussion.

Lemma 9. Let $q(x)$ be a piecewise continuous bounded function on the interval $a \le x \le b$. We consider the equation

$$y'' + pq(x)y = 0, \quad (7)$$

where p is a positive real number. Let $y_p(x)$ be a solution of this equation that is not identically zero. Let $N(p)$ be the number of zeros of $y_p(x)$ on the interval $[a, b]$. Then

$$\lim_{p\to+\infty} \frac{N(p)}{\sqrt{p}} = \frac{1}{\pi}\int_a^b \mathrm{Re}\sqrt{q(x)}\,dx. \quad (8)$$

Proof.

1. The lemma is obviously valid if $q(x) = q = \text{const}$: when $q < 0$, both sides of (8) are equal to zero; when $q > 0$, the distance between adjacent zeros of $y_p(x)$ is $\pi/\sqrt{pq}$. Thus when $p \to +\infty$, $N(p) \sim \sqrt{pq}\,(b - a)/\pi$, which is equivalent to formula (8).

2. Now let $q(x)$ be a step function: the interval $[a, b]$ is divided into a finite number of intervals

$\Delta_1, \ldots, \Delta_n$, and $q(x) = q_i$ when $x \in \Delta_i$. Let $N_i(p)$ be the number of zeros of $y_p(x)$ on the interval Δ_i. Then, obviously,

$$\frac{N(p)}{\sqrt{p}} \sim \sum_{i=1}^{n} \frac{N_i(p)}{\sqrt{p}} \sim \sum_{i=1}^{n} \frac{\operatorname{Re}\sqrt{q_i}}{\pi} |\Delta_i| = \frac{1}{\pi}\int_a^b \operatorname{Re}\sqrt{q(x)}\,dx.$$

Thus the lemma is also valid for step functions.

3. Finally, let $q(x)$ be any piecewise continuous bounded function. We divide $[a, b]$ into n equal intervals $\Delta_1, \ldots, \Delta_n$ and let $q^+(x) = \sup q(x)$ and $q^-(x) = \inf q(x)$ when $x \in \Delta_i$. Let $y_p^+(x)$ be a solution of Eq. (7) in which $q^+(x)$ takes the place of $q(x)$. We denote by $N^+(p)$ the number of zeros of $y_p^+(x)$ on $[a, b]$. We similarly define $N^-(p)$. According to the Sturm oscillation theorem,

$$N^+(p) \geqslant N(p) \geqslant N^-(p).$$

According to the preceding paragraph of the proof,

$$\frac{N^+(p)}{\sqrt{p}} \sim \frac{1}{\pi}\int_a^b \operatorname{Re}\sqrt{q^+(x)}\,dx, \quad \frac{N^-(p)}{\sqrt{p}} \sim \frac{1}{\pi}\int_a^b \operatorname{Re}\sqrt{q^-(x)}\,dx.$$

As n approaches infinity, we obtain the desired equation: the lemma is proved.

Corollary. The asymptotic formulas

$$\frac{N(p, cp)}{\sqrt{p}} \underset{p\to+\infty}{\sim} \frac{1}{\pi}\int_{x_1}^{x_2} \operatorname{Re}\sqrt{\frac{B(x)}{A(x)} - c}\,dx \tag{9}$$

(for any real number c) and

$$\frac{N(0, s)}{\sqrt{-s}} \underset{s\to-\infty}{\sim} \frac{x_2 - x_1}{\pi} \tag{10}$$

are valid for the function $N(p, s)$ introduced at the beginning of Section 3.

Proof. Substitution of the independent variable

$$\bar{x} = \int_{x_1}^{x} \frac{dt}{A(t)}$$

brings Eq. (4) to the form

$$\frac{d^2y}{d\bar{x}^2} + [p\bar{B}(\bar{x}) - s\bar{A}(\bar{x})]\,y = 0, \tag{$\bar{4}$}$$

where

$$\bar{B}(\bar{x}) = A(x)B(x), \quad \bar{A}(\bar{x}) = A^2(x), \quad 0 \leqslant \bar{x} \leqslant X = \bar{x}(x_2).$$

We arbitrarily fix the real number c. Let $\bar{B}(\bar{x}) - c\bar{A}(\bar{x}) = q(\bar{x})$. Then when $s = cp$, Eq. ($\bar{4}$) takes the form

$$\frac{d^2y}{d\bar{x}^2} + pq(\bar{x})\,y = 0.$$

According to Lemma 9,

$$\frac{N(p, cp)}{\sqrt{p}} \underset{p\to+\infty}{\sim} \frac{1}{\pi}\int_0^X \operatorname{Re}\sqrt{\bar{B}(\bar{x}) - c\bar{A}(\bar{x})}\,d\bar{x} = $$

$$= \frac{1}{\pi}\int_{x_1}^{x_2} \operatorname{Re}\sqrt{\frac{B(x)}{A(x)} - c}\,dx.$$

Formula (9) is proved.

Now we let $p = 0$ in Eqs. (4) and ($\bar{4}$). According to Lemma 9,

$$\frac{N(0, s)}{\sqrt{-s}} \underset{s\to-\infty}{\sim} \frac{1}{\pi}\int_0^X \operatorname{Re}\sqrt{\bar{A}(\bar{x})}\,d\bar{x} = \frac{1}{\pi}\int_{x_1}^{x_2} dx = \frac{x_2 - x_1}{\pi}.$$

Formula (10) is also proved.

Lemma 10. The function

$$J(c) = \int_{x_1}^{x_2} \operatorname{Re}\sqrt{\frac{B(x)}{A(x)} - c}\,dx, \quad C_1 < c < C_2$$

is uniquely defined by an infinite number of graphs l_k.

Proof. We arbitrarily fix $c \in (C_1, C_2)$. According to Part A of Section 2, the line $s = cp$ intersects each graph l_k; we denote by p_k the abscissa of the point of intersection. At points (p, s) that belong to a graph l_k, the function $N(p, s)$ obviously takes one and the same value: $N(p, s_k(p)) = k - 1$. Thus,

$$N(p_k, cp_k) = N(0, s_k(0)).$$

That is, if we know an infinite number of graphs l_k

$$l_{k_1}, l_{k_2}, \ldots, l_{k_m},$$

then even if we do not know their numbers, we can determine

$$\lim_{k\to\infty} \frac{N(p_k, cp_k)}{\sqrt{p_k}}.$$

According to formula (10), this limit is

$$\frac{x_2 - x_1}{\pi} \lim_{m\to\infty} \sqrt{-s_{k_m}(0)/p_{k_m}}$$

Therefore, according to formula (9),

$$J(c) = (x_2 - x_1) \lim_{m\to\infty} \sqrt{-s_{k_m}(0)/p_{k_m}}$$

(on the right side of the latter equation, the value of p_{k_m} is a function of c). The lemma is proved.

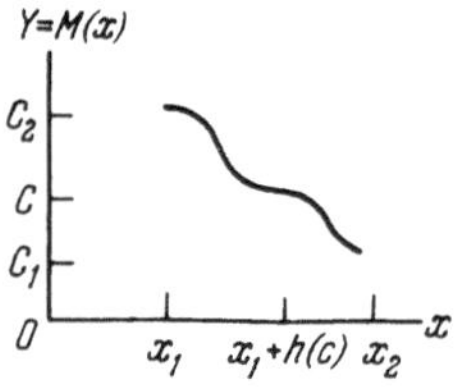

Fig. 5

L e m m a 11. The function H(t) is expressed in terms of J(c) by the formula

$$H(t) = -\frac{2}{\pi}\int_{t^{-2}}^{C_2} \frac{J'(c)\,dc}{\sqrt{c - t^{-2}}}.$$

P r o o f. Let

$$h(c) = \operatorname{mes}\left\{x,\ x_1 < x < x_2,\ \frac{B(x)}{A(x)} > c\right\}.$$

Thus, $h(c) = H(1/\sqrt{c})$.

Let M(x) be a nonincreasing function that is equimeasurable with B(x)/A(x):

$$\operatorname{mes}\{x,\ x_1 < x < x_2,\ M(x) > c\} = h(c).$$

The function J(c) is expressed in terms of M(x) and h(c) by the formula

$$J(c) = \int_{x_1}^{x_2} \operatorname{Re}\sqrt{M(x) - c}\,dx = \int_{x_1}^{x_1+h(c)} \sqrt{M(x) - c}\,dx \qquad (11)$$

(Fig. 5). Let us use the relation

$$\int_a^b \frac{dY}{\sqrt{(b-Y)(Y-a)}} = \pi. \qquad (12)$$

It follows from (12) that

$$h(c) = \frac{1}{\pi}\int_{x_1}^{x_1+h(c)} \pi dx = \frac{1}{\pi}\int_{x_1}^{x_1+h(c)}\int_c^{M(x)} \frac{dY\,dx}{\sqrt{[M(x) - Y](Y - c)}}.$$

Changing the order of integration (Fig. 5) and taking (11) into account, we obtain

$$h(c) = \frac{1}{\pi}\int_c^{C_2} \frac{dY}{\sqrt{Y - c}} \int_{x_1}^{x_1+h(Y)} \frac{dx}{\sqrt{M(x) - Y}} = -\frac{2}{\pi}\int_c^{C_2} \frac{J'(Y)\,dY}{\sqrt{Y - c}},$$

or, finally,

$$H(t) = h(t^{-2}) = -\frac{2}{\pi}\int_{t^{-2}}^{C_2} \frac{J'(Y)\,dY}{\sqrt{Y - t^{-2}}},$$

Q.E.D.

LITERATURE CITED

1. Gerver, M. L., and D. A. Kazhdan (1967), "On finding the function $\rho(x)$ from the eigenvalue $s = s(p)$ of the equation $y'' + [p\rho(x) - s]y = 0$," Matem. Sb., 73:115.
2. Vil'kovich, E. V., A. L. Levshin, and M. G. Neigauz (1966), "Love waves in a vertically inhomogeneous medium (allowance for sphericity, parameter variations, and absorption)," Computational Seismology, No. 2, Moscow, Nauka, pp. 130-149 [English translation in: Seismic Love Waves, New York, Consultants Bureau (1967)].
3. Gerver, M. L., and V. M. Markushevich (1965), "Study of ambiguity in determining seismic-wave propagation velocity from travel-time curves," Dokl. Akad. Nauk SSSR, Vol. 163, No. 6.
4. Titchmarsh, E. C. (1962), "Eigenfunction Expansions Associated with Second-Order Differential Equations, Vol. 1, Oxford, Clarendon Press.
5. Petrovskii, I. G. (1961), "Lectures on Partial Differential Equations," Moscow, Fizmatgiz, p. 182.
6. Markushevich, V. M. (1950), Theory of Analytic Functions, Moscow and Leningrad, Gostekhizdat.

PART IV

STATISTICAL APPLICATIONS

EFFECT OF STATION AND SOURCE FACTORS ON THE ACCURACY OF SEISMIC PARAMETER DETERMINATION†

V. F. Pisarenko and T. G. Rautian

1. Introduction

In the solution of many experimental seismological problems, the ultimate aim is to obtain an average value of some parameter y or an averaged function y(x) of some independent variable x. The desired value y or function y(x) is isolated in a random scatter of individual measurements. Therefore only results that have been averaged for a sufficiently large volume of material are reliable and accurate.

The concepts of variance and the confidence interval are used in mathematical statistics to estimate the reliability of results [1]. Assume, for example, that we must estimate the mean value y_0 of a parameter y from N observations $y_1, \ldots, y_N$ that have a random scatter about y_0. For this, we shall use the arithmetic mean

$$\bar{y} = \frac{1}{N} \sum_{i=1}^{N} y_i. \quad (1)$$

We shall let σ^2 be the variance of the random scatter of an individual observation y_i and assume that all observations are independent. Then the variance of y is

$$\sigma_{\bar{y}}^2 = \sigma^2 / N.$$

Usually σ^2 is unknown, and instead we take its estimate

$$s^2 = \frac{1}{N-1} \sum_{i=1}^{N} (y_i - \bar{y})^2. \quad (2)$$

Knowing s^2 and N, we construct a confidence interval for the unknown mean y_0 (see [1], p. 212).

In practice, observations are often such that the result of an individual measurement y_i is a function of one or more systematic but unknown factors (for example, the local characteristics of the seismic recording station). Then formula (2) is unsuitable for estimating the accuracy of an arithmetic mean. The principal aim of this paper is to develop a method for estimating the accuracy of a result of averaging when the scatter is related to systematic factors.

2. Statement of Problem

Let us assume that y, the parameter under study, is a function of two factors: in our example these will be effects due to the stations at which the earthquakes were recorded, and effects due to the individual earthquakes. The result can easily be extended to include a greater number of factors.

Let y_{ik} be the result of measurement of parameter y for the k-th earthquake at the i-th station. We shall assume that there cannot be more than one observation for each i–k pair, i.e., our experiment is performed without repetition. We shall represent the set of individual measurements as a table in which the row number corresponds to the earthquake number and the column number corresponds to the station number. The number of rows M in the table is the number of earthquakes considered and the number of columns L is the number of seismic stations. When the table is completely filled, the total number of observations is N = ML. It often happens, however, that, for some reason, the table has gaps – not every earthquake was recorded at all L stations and not every station recorded all M earthquakes. We shall let

† Translated from Vychislitel'naya Seismologiya, No. 1, pp. 160-186, Moscow, Nauka (1966).

v_k be the number of stations that recorded the k-th earthquake and let u_i be the number of earthquakes recorded at the i-th station. Obviously the total number of measurements N is

$$N = \sum_{k=1}^{M} v_k = \sum_{i=1}^{L} u_i.$$

The purpose of this investigation is to obtain the mean value of y from the N individual observations and to estimate the accuracy of this mean.

We introduce the following assumptions.

a) The effect of each factor is independent of that of all the others, i.e., there is no interaction between factors. This means that an individual measurement y_{ik} has the form

$$y_{ik} = y_0 + \xi_i + \eta_k + \varepsilon_{ik}, \tag{3}$$

where ξ_i is the effect of the i-th station on parameter y_{ik}, η_k is the effect of the k-th earthquake, ε_{ik} is random Gaussian noise, and y_0 is the mean value of the parameter y.

b) The ξ_i, η_k, and the Gaussian noise, have zero mean values:

$$\frac{1}{L}\sum_{i=1}^{L} \xi_i \to 0 \quad \text{and} \quad \frac{1}{M}\sum_{k=1}^{M} \eta_k \to 0$$

when

$$L \to \infty, \; M \to \infty.$$

c) The random noise ε_{ik} is independent and has the same unknown variance σ_ε^2 for different stations and for different earthquakes.

The accuracy problem is made up of several parts. The first problem is to determine whether or not one or another factor (in our example, the station number and earthquake number) affects the measured value.

The second problem is to estimate the accuracy of the mean, with allowance for this effect, if it exists. The accuracy of the mean is naturally a function of the accuracy of the individual measurements. But it is also a function of the influence of the factors, of the total number of measurements N, and of the numbers L and M.

Finally, the third problem is as follows. Let us assume that our aim is to obtain a final result with fully defined accuracy and with the minimum possible number of individual measurements. For this, we shall study a preliminary sampling of data and answer three questions: does a factor influence exist? how great is it? and what is the variance of an individual measurement? Then we must determine which and how many observations are needed to achieve the required accuracy. Further measurements are planned on the basis of these conclusions from the preliminary data sampling.

Methods for solving these problems are given in Sections 3-5. Specific examples are obtained in a study of seismic-wave periods. In Section 6, we shall discuss the case of an empirical function (the dependence of the period of P waves upon epicentral distance is studied as an illustration).

3. Detection of the Effect of Factors on the Value under Study

The methods of the analysis of variance are used to establish whether or not a factor affects a value under study. These methods are described in a number of handbooks [2-4], so we shall dwell on them only briefly.

First let us consider the case when the table of data is entirely filled, so that $u_i = M$, $v_k = L$, and $N = ML$.

We shall calculate the total average $\bar{y}$ and the marginal averages $y_{\cdot k}$ for each row and $y_{i\cdot}$ for each column of the table. In our example, these are the average values of y for each focus and each station.

$$\bar{y} = \frac{1}{N}\sum_{i=1}^{L}\sum_{k=1}^{M} y_{ik}; \tag{4}$$

$$y_{i\cdot} = \frac{1}{M}\sum_{k=1}^{M} y_{ik}, \quad i = 1, \ldots, L; \tag{5}$$

$$y_{\cdot k} = \frac{1}{L}\sum_{i=1}^{L} y_{ik}, \quad k = 1, \ldots, M. \tag{6}$$

The values $y_{i\cdot}$ ($i = 1, \ldots, L$) are the estimates for $y_0 + \xi_i$ and $y_{\cdot k}$ are the estimates for $y_0 + \eta_k$. In fact,

$$y_{i\cdot} = y_0 + \xi_i + \frac{1}{M}\sum_{k=1}^{M} \eta_k + \frac{1}{M}\sum_{k=1}^{M} \varepsilon_{ik},$$

$$y_{\cdot k} = y_0 + \eta_k + \frac{1}{L}\sum_{i=1}^{L} \xi_i + \frac{1}{L}\sum_{i=1}^{L} \varepsilon_{ik}. \tag{7}$$

follows from formulas (3), (5), and (6).

The third and fourth terms of (7) tend toward a zero average value. If, however, the second term is small in comparison with, for example, the fourth, when there are insufficient earthquakes M (or stations L) the sums in the third or fourth term can be rather far from zero and comparable with the second term or even greater than it. Under these conditions the effect of factors ξ and η (station and focus) is negligible in comparison with random deviations, and they cannot be ascertained from the intermediate averages $y_{i\cdot}$ and $y_{\cdot k}$.

Fisher's test is used to establish the degree of the effect. We form the sums

$$\sum_{i=1}^{L}\sum_{k=1}^{M}(y_{ik}-\bar{y})^2;\quad \sum_{i=1}^{L}M(y_{i\cdot}-\bar{y})^2;\quad \sum_{k=1}^{M}L(y_{\cdot k}-\bar{y})^2 \tag{8}$$

and the difference between these sums

$$\sum_{i=1}^{L}\sum_{k=1}^{M}(y_{ik}-y_{i\cdot}-y_{\cdot k}+\bar{y})^2=\sum_{i=1}^{L}\sum_{k=1}^{M}(y_{ik}-\bar{y})^2-$$

$$-\sum_{i=1}^{L}(y_{i\cdot}-\bar{y})^2M-\sum_{k=1}^{M}L(y_{\cdot k}-\bar{y})^2. \tag{9}$$

Then we form the ratio

$$F_\xi=\frac{\sum_{i=1}^{L}(y_{i\cdot}-\bar{y})^2M}{L-1}\cdot\frac{N-L-M+1}{\sum_{i=1}^{L}\sum_{k=1}^{M}(y_{ik}-y_{i\cdot}-y_{\cdot k}+\bar{y})^2},$$

which has Fisher's F-distribution with $L - 1$ and $N - L - M + 1$ degrees of freedom. If we compare the value of this ratio with the tabulated value F^* for the same degrees of freedom and the given significance level, we can determine whether or not the effect of the station factor is significant in comparison with Gaussian noise.

The significance of the effect of an earthquake's characteristics is similarly checked. For this, we use the ratio

$$F_\eta=\frac{\sum_{k=1}^{M}L(y_{\cdot k}-\bar{y})^2}{M-1}\cdot\frac{N-L-M+1}{\sum_{i=1}^{L}\sum_{k=1}^{M}(y_{ik}-y_{i\cdot}-y_{\cdot k}+\bar{y})^2}.$$

If the table is not completely filled, expressions (5) and (6) for the intermediate averages are modified:

$$y_{i\cdot}=\frac{1}{u_i}\sum_{k=1}^{M}y_{ik}; \tag{10}$$

$$y_{\cdot k}=\frac{1}{v_k}\sum_{i=1}^{L}y_{ik}. \tag{11}$$

Summation here and below is only over the pairs of i, k for which there are corresponding observations.

To use Fisher's test, we form the sums

$$\sum_{i=1}^{L}\sum_{k=1}^{M}(y_{ik}-\bar{y})^2;\quad \sum_{i=1}^{L}u_i(y_{i\cdot}-\bar{y})^2;\quad \sum_{k=1}^{M}v_k(y_{\cdot k}-\bar{y})^2. \tag{12}$$

Instead of Eq. (9), we have

$$\sum_{i=1}^{L}\sum_{k=1}^{M}(y_{ik}-y_{i\cdot}-y_{\cdot k}+\bar{y}^2)=$$

$$=\sum_{i=1}^{L}\sum_{k=1}^{M}(y_{ik}-\bar{y})^2-\sum_{i=1}^{L}u_i(y_{i\cdot}-\bar{y})^2-\sum_{k=1}^{M}v_k(y_{\cdot k}-\bar{y})^2+$$

$$+2\sum_{i=1}^{L}\sum_{k=1}^{M}(y_{i\cdot}-\bar{y})(y_{\cdot k}-\bar{y}). \tag{13}$$

The fourth term, which arises because the table is not completely filled, makes analysis of variance too complicated. We can, however, use an approximate method and ignore this fourth term. If we compare the mathematical expectations of all terms of (13), it is easy to see that if

$$\sum_{i=1}^{L}\sum_{k=1}^{M}\frac{1}{u_iv_k}-1<L-1;$$
$$\sum_{i=1}^{L}\sum_{k=1}^{M}\frac{1}{u_iv_k}-1<M-1;$$
$$\sum_{i=1}^{L}\sum_{k=1}^{M}\frac{1}{u_iv_k}-1<N-L-M+1, \tag{14}$$

then ignoring the last term on the right side of (13) is justified. Then the Fisher test for the degree of influence of the factors is applied by comparing the ratios

$$F_\xi=\frac{\sum_{i=1}^{L}u_i(y_{i\cdot}-\bar{y})^2}{L-1}\cdot\frac{N-L-M+1}{\sum_{i=1}^{L}\sum_{k=1}^{M}(y_{ik}-x_{i\cdot}-y_{\cdot k}+\bar{y})^2};$$

$$F_\eta=\frac{\sum_{k=1}^{M}v_k(y_{\cdot k}-\bar{y})^2}{M-1}\cdot\frac{N-L-M+1}{\sum_{i=1}^{L}\sum_{k=1}^{M}(y_{ik}-y_{i\cdot}-y_{\cdot k}+\bar{y})^2}$$

with the tabulated values F^* for the corresponding degrees of freedom: $L - 1$ (or $M - 1$) and $N - L - M + 1$.

Let us consider an example: the oscillation period T_p in a group of maximum displacements of longitudinal waves from earthquakes. The dependence of the period upon the different independent variables (for example, earthquake magnitude or epicentral distance) has been eliminated. The way in which the elimination was accomplished is not important to us at this point; we shall discuss it later (in Section 6). For convenience let us take as y not the period itself but its logarithm multiplied by 100. It is obvious that this will in no way affect our findings about the influence of the station and focus characteristics on y.

TABLE 1. Earthquake Observation Data $y = 100 \log T_p$

Earthquake No.	Station No. 1	2	3	4	5	6	v_k	$y_{\cdot k}$
1	–24	+38	—	–2	—	+28	4	+10.0
2	–10	+5	—	+5	—	+14	4	+3.5
3	–16	–20	—	–33	–25	+7	5	–17.4
4	–34	+9	—	+7	–24	+5	5	–7.4
5	+11	+28	—	+11	–18	+16	5	+9.6
6	+8	+21	—	0	—	+12	4	+10.2
7	–27	+6	—	–30	—	0	4	–12.7
8	–14	0	–36	+19	+2	–8	6	–6.2
9	–28	–8	–28	–20	—	+1	5	+16.6
10	–32	–9	—	–13	—	+20	4	–8.5
11	+26	–26	—	—	+6	–1	4	+1.2
12	–3	—	—	+7	–17	+12	4	–0.2
13	–23	–20	—	–58	–65	+12	5	–22.8
14	–15	+14	—	–16	—	+35	4	+4.5
15	–28	–8	—	–44	—	–19	4	–24.7
16	–3	+6	—	–23	—	+14	4	–1.5
17	–3	+30	—	–6	—	+20	4	+10.2
18	+22	—	–11	–56	—	+18	4	–6.7
19	–50	—	–28	–22	–79	–9	5	–37.6
20	+23	—	—	+12	–8	+15	4	+10.5
u_i	20	16	4	19	9	20	$N = 88$	
$y_{i\cdot}$	–11	+6.6	–25.7	–13.8	–25.3	+9.6	$\bar{y} = -5.85$	

Assume that we have N = 88 individual observations of y, which are shown in Table 1. These measurements were made for M = 20 earthquakes recorded at L = 6 seismic stations. The table of observations has gaps (not all stations recorded all 20 earthquakes): u_i varied from 20 to 4. Accordingly, not all earthquakes were recorded by all six stations: v_k varied from 6 to 4. The u_i and v_k values, and the means $\bar{y}$, $y_{i\cdot}$, and $y_{\cdot k}$ are given in Table 1.

First of all let us check whether we have correctly used the approximate method of estimating the F ratios, i.e., whether or not condition (14) is satisfied. For this, we calculate the value $\sum_{i=1}^{L}\sum_{k=1}^{M}\frac{1}{u_i v_k} - 1$. It turns out to equal 1.66, and this is clearly less than L – 1 = 5, M – 1 = 19, and N – L – M + 1 = 63.

We calculate the sums

$$\sum_{i=1}^{L} u_i (y_{i\cdot} - \bar{y})^2 = 13{,}990; \; \sum_{k=1}^{M} v_k (y_{\cdot k} - \bar{y})^2 = 15{,}960;$$

$$\sum_{i=1}^{L}\sum_{k=1}^{M} (y_{ik} - y_{i\cdot} - y_{\cdot k} + \bar{y})^2 = 20{,}015.$$

Now we calculate F_ξ and F_η:

$$F_\xi = 8.81; \; F_\eta = 2.64.$$

Turning to the table of the F distribution for degrees of freedom $f_1 = 5$ and $f_2 = 63$, we find that $F^*_{0.005}(5.60) = 3.76 < F_\xi$, i.e., we can with probability $1 - \alpha > 0.995$ assert that the effect of the station characteristics is considerable in comparison with random Gaussian noise.

In exactly the same way we find for degrees of freedom $f_1 = 19$ and $f_2 = 63$ that the tabulated value $F^*_{0.005}(19.60) = 2.39 < F_\eta$. Therefore, with 0.995 probability we can assert that the effect of the focus is considerable in comparison with the random error.

Thus we have established that the effects of station characteristics and of the characteristics of an individual earthquake must be taken into account in estimating the accuracy of $\bar{y}$.

4. Estimation of Accuracy of Mean in Presence of Influencing Factors

We shall represent the result of an individual measurement y_{ik} in the form of sum (3). Then for an estimate of the mean y_0 of y we have

$$\bar{y} = y_0 + \sum_{i=1}^{L} \frac{u_i}{N}\xi_i + \sum_{k=1}^{M} \frac{v_k}{N}\eta_k + \sum_{i=1}^{L}\sum_{k=1}^{M} \frac{\varepsilon_{ik}}{N}. \tag{15}$$

Now we calculate the variance of this mean. Since the variances of ε_{ik}, ξ_i, and η_k are σ_ε^2, σ_ξ^2, and σ_η^2, respectively, the variance of the term $(y_i/N)\xi_i$ is $(\sigma_\xi^2/N^2)u_i^2$. We obtain similar expressions for the variances of the terms of the second and third sums. They are expressed by the values σ_η^2/N^2, v_k^2, and σ_ε^2/N^2. Since all of the terms of expression (15) are unknown, the variance of the sum is equal to the sum of the variances, and for the variance of the mean $\bar{y}$ we obtain the expression

$$\sigma_{\bar{y}}^2 = \frac{1}{N^2}\left(\sigma_\xi^2 \sum_{i=1}^{L} u_i^2 + \sigma_\eta^2 \sum_{k=1}^{M} v_k^2\right) + \frac{\sigma_\varepsilon^2}{N}. \tag{16}$$

The variances σ_ε^2, σ_ξ^2, and σ_η^2 are unknown, so we must obtain estimates for them. We shall do this by the method of moments (see [5], p. 540). From the available data y_{ik} we form the sums

$$\sum_{i=1}^{L}(y_{i\cdot}-\bar{y})^2;\quad \sum_{k=1}^{M}(y_{\cdot k}-\bar{y})^2;$$

$$\sum_{i=1}^{K}\sum_{k=1}^{M}(y_{ik}-y_{i\cdot}-y_{\cdot k}+\bar{y})^2.$$

If we equate these sums to their mathematic expectations, we obtain a system of three equations in three unknowns σ_ξ^2, σ_η^2, and σ_ε^2:

$$\left.\begin{aligned}
&\sigma_\xi^2\left(L+\frac{L}{N^2}\sum_{i=1}^{L}u_i^2-2\right)+\sigma_\eta^2\left(\frac{L}{N^2}\sum_{k=1}^{M}v_k^2-\sum_{i=1}^{L}\frac{1}{u_i}\right)+\\
&\qquad+\sigma_\varepsilon^2\left(\sum_{i=1}^{L}\frac{1}{u_i}-\frac{L}{N}\right)=\sum_{i=1}^{L}(y_{i\cdot}-\bar{y})^2;\\
&\sigma_\xi^2\left(\frac{M}{N^2}\sum_{i=1}^{L}u_i^2-\sum_{k=1}^{M}\frac{1}{v_k}\right)+\sigma_\eta^2\left(M+\frac{M}{N^2}\sum_{k=1}^{M}v_k^2-2\right)+\\
&\qquad+\sigma_\varepsilon^2\left(\sum_{k=1}^{M}\frac{1}{v_k}-\frac{M}{N}\right)=\sum_{k=1}^{M}(y_{\cdot k}-\bar{y})^2;\\
&\sigma_\xi^2\left(\frac{1}{N}\sum_{i=1}^{L}u_i^2-M\right)+\sigma_\eta^2\left(\frac{1}{N}\sum_{k=1}^{M}v_k^2-L\right)+\\
&\qquad+\sigma_\varepsilon^2\left(N-L-M+2\sum_{i=1}^{L}\sum_{k=1}^{M}\frac{1}{u_iv_k}\right)=\\
&\qquad=\sum_{i=1}^{L}\sum_{k=1}^{M}(y_{ik}-y_{i\cdot}-y_{\cdot k}+\bar{y})^2.
\end{aligned}\right\} \tag{17}$$

When the table is entirely filled, these equations are greatly simplified:

$$\left.\begin{aligned}
&\sigma_\xi^2(L-1)+\sigma_\eta^2\frac{L-1}{M}=\sum_{i=1}^{L}(y_{i\cdot}-\bar{y})^2;\\
&\sigma_\eta^2\left(\frac{M-1}{L}\right)+\sigma_\eta^2(M-1)=\sum_{k=1}^{M}(y_{\cdot k}-\bar{y})^2;\\
&\sigma_\varepsilon^2(N-L-M+1)=\sum_{i=1}^{L}\sum_{k=1}^{M}(y_{ik}-y_{i\cdot}-y_{\cdot k}+\bar{y})^2.
\end{aligned}\right\} \tag{18}$$

The solutions of systems (17) and (18) are the estimates of the variances σ_ξ^2, σ_η^2, and σ_ε^2. If we call then s_ξ^2, s_η^2, and s_ε^2 and substitute them into (16), we obtain an expression for the variance of the mean:

$$s_{\bar{y}}^2 = \frac{s_\xi^2 \sum_{i=1}^{L} u_i^2 + s_\eta^2 \sum_{k=1}^{M} v_k^2 + s_\varepsilon^2 N}{N^2}. \tag{19}$$

It is more convenient to analyze this expression for $s_{\bar{y}}^2$ if we put it into a form that is similar to the form for the case of independent observations. For this, we introduce a number n_{ef}. It shows what number of independent observations is equivalent to a given series of N observations in which, besides the random error ε, the influences of factors ξ and η manifest themselves. Equivalence means that the variance of the mean $s_{\bar{y}}^2$ is smaller by a factor of n_{ef} than the variance of an individual measurement $s^2 = s_\xi^2 + s_\eta^2 + s_\varepsilon^2$.

$$n_{ef} = \frac{s^2}{s_{\bar{y}}^2}.$$

Hence, we obtain an equation for determining n_{ef}:

$$\frac{s_\xi^2+s_\eta^2+s_\varepsilon^2}{n_{ef}} = \frac{s_\xi^2 \sum_{i=1}^{L} u_i^2 + s_\eta^2 \sum_{k=1}^{M} v_k^2 + s_\varepsilon^2 N}{N^2}. \tag{20}$$

If we introduce the coefficients $\gamma_\xi = \sigma_\xi^2/\sigma^2$, $\gamma_\eta = \sigma_\eta^2/\sigma^2$, and $\gamma_\varepsilon = \sigma_\varepsilon^2/\sigma^2$, so that $\gamma_\xi + \gamma_\eta + \gamma_\varepsilon = 1$, which are the specific weights of each of the variances in the total variance of an individual measurement, then expression (20) takes the form

$$n_{ef} = \frac{N^2}{\gamma_\xi \sum_{i=1}^{L} u_i^2 + \gamma_\eta \sum_{k=1}^{M} v_k^2 + \gamma_\varepsilon N}. \tag{21}$$

When the table of observations is completely filled, this expression is simplified:

$$\frac{1}{n_{ef}} = \frac{\gamma_\xi}{L} + \frac{\gamma_\eta}{M} + \frac{\gamma_\varepsilon}{LM}. \tag{22}$$

Now, if we know n_{ef} and use the variances of an individual measurement, we can give a final estimate of the accuracy of the mean $\bar{y}$, i.e., the confidence interval Δy

$$\bar{y}+\Delta y \geqslant y_0 \geqslant \bar{y}-\Delta y,$$

within which the true mean y_0 is found with a given probability $1-\alpha$.

If n_{ef} is great ($n_{ef} \geq 30$), we can assume that the value

$$\frac{\Delta y}{s}\sqrt{n_{ef}-1}, \tag{23}$$

where $s = \sqrt{s_\xi^2 + s_\eta^2 + s_\varepsilon^2}$, has a normal distribution. Hence, we obtain a confidence interval of level α for the unknown mean y_0:

$$\bar{y} = \frac{s\lambda_\alpha}{\sqrt{n_{ef}-1}} \leqslant y_0 \leqslant \bar{y} + \frac{s\lambda_\alpha}{\sqrt{n_{ef}-1}}, \tag{24}$$

where λ_α is the quantile (obtained from tables) of a normal distribution of level α. The unknown parameter lies within interval (24) with probability $1 - \alpha$. In particular, for $\lambda_\alpha = 1$ we obtain the confidence level $1 - \alpha = 0.68$, for $\lambda_\alpha = 2$, $1 - \alpha = 0.95$, and for $\lambda_\alpha = 3$, $1 - \alpha = 0.997$.

If n_{ef} is not too great ($n_{ef} < 30$), it is better to use the Student t-distribution to obtain the confidence interval. We can assume that (23) has Student's t-distribution with $n_{ef} - 1$ degrees of freedom. For y_0 we obtain the confidence interval of level α

$$\bar{y} - \frac{st_\alpha^{(n_{ef}-1)}}{\sqrt{n_{ef}-1}} \leqslant y_0 \leqslant \bar{y} + \frac{st_\alpha^{(n_{ef}-1)}}{\sqrt{n_{ef}-1}}, \tag{25}$$

where $t_\alpha^{n_{ef}-1}$ is the quantile (obtained from tables) of level α for Student's t-distribution with $n_{ef} - 1$ degrees of freedom.

Example. Let us consider estimation of the variance in the same example that we used above (see Table 1). In order to set up a system of equations, we must first calculate all of the coefficients in it. On the basis of the data in Table 1, we obtain

$$\sum_{i=1}^{L} u_i^2 = 1540; \quad \sum_{i=1}^{L} \frac{1}{u_i} = 0.575; \quad \sum_{i=1}^{L}\sum_{k=1}^{M} \frac{1}{u_i v_k} = 2.66;$$

$$\sum_{k=1}^{M} v_k^2 = 394; \quad \sum_{k=1}^{M} \frac{1}{v_k} = 4.62.$$

We also calculate the sums on the right sides of Eqs. (17):

$$\sum_{i=1}^{L} (y_{i\cdot} - \bar{y})^2 = 1268; \quad \sum_{k=1}^{M} (y_{\cdot k} - \bar{y})^2 = 3534;$$

$$\sum_{i=1}^{L}\sum_{k=1}^{M} (y_{ik} - y_{i\cdot} - y_{\cdot k} + \bar{y})^2 = 20\,015.$$

Substituting them into (17), we obtain equations with numerical coefficients:

$$5.18s_\xi^2 - 0.27s_\eta^2 + 0.51s_\varepsilon^2 = 1268;$$
$$-0.71s_\xi^2 + 19.0s_\eta^2 + 4.39s_\varepsilon^2 = 3534;$$
$$-2.8s_\xi^2 - 1.53s_\eta^2 + 66.3s_\varepsilon^2 = 20015,$$

which allow us to find the variances:

$$s_\xi^2 = 220; \quad s_\eta^2 = 122; \quad s_\varepsilon^2 = 314;$$
$$s^2 = 656; \quad s = 25.6.$$

Hence, the coefficients γ are:

$$\gamma_\xi = 0.32; \quad \gamma_\eta = 0.18; \quad \gamma_\varepsilon = 0.50.$$

Now we can calculate n_{ef} and estimate the accuracy of the mean:

$$n_{ef} = \frac{88^2}{0.32 \cdot 1540 + 0.18 \cdot 394 + 0.50 \cdot 88} = 12.2;$$

$$s_{\bar{y}}^2 = \frac{656}{12.2} = 53.8; \quad s_{\bar{y}} = 7.3.$$

Thus our series of 88 measurements is equivalent to as few as 12 independent observations. Since $n_{ef} < 30$, Student's t-distribution is used to estimate the confidence interval. For the confidence level $1 - \alpha = 0.7$, the quantile $t_{0.3}^{(11)} = 1.09$, i.e., the width of the confidence interval $\Delta y = 1.09\,(25.6/\sqrt{11.2}) = 8.4$. For a confidence level of 0.95, we obtain $\Delta y = 2.2\,(25.6/\sqrt{11.2}) = 16.8$.

Thus we see that in our case, when the effects of the station characteristics and focus are rather strong in comparison with random errors, the accuracy of the mean is appreciably lower than it would have been in the case of observations that were really independent. In fact, if all 88 measurements were independent, the variance of the mean would be

$$s_{\bar{y}}^2 = \frac{656}{88} = 8.2; \quad s_{\bar{y}} = 2.86,$$

while due to the effect of factors ξ and η, the standard deviation would be greater by a factor of 2.5 and the variance would be greater by a factor of 6.5. Thus, estimation of the accuracy with allowance for the effect of the factors in our example allowed us to avoid overly optimistic conclusions about the accuracy of the result.

5. Optimal Design of Observations

The results of Section 4 allow us to select the material for measurements so that maximum accuracy is obtained for a minimum number of measurements. Various situations can occur in practice:

a) The number of measurements is in no way limited. What is the minimum number of observations and specifically what observations are required to obtain a preassigned accuracy?

b) The total number of observations N is limited, but L and M can vary in any way. What is the maximum accuracy (minimum variance) attainable under these conditions and for what L and M can it be achieved?

c) The number of possible values of one of the factors (for example, L) is limited by the experimental conditions, but the number of M values can be arbitrarily great. What are the most suitable methods for increasing accuracy and for what M and N can this be done?

d) Both factors (M and L) are limited. What is the maximum possible accuracy and how can it be obtained with the minimum number of measurements?

e) Assume that a particular accuracy has been assigned, and that both factors are limited in the number of their values ($L < n_{ef}$ and $M < n_{ef}$). Can we obtain the desired accuracy, and if so, how?

All of these questions can be answered by analyzing formula (21). For simplicity we assume that we have a partially filled table of observations in which each earthquake was recorded by the same number of stations $v_k = q = \text{const}$, $L \geq q \geq 1$, and that each station, therefore, recorded the same number $u_i = ML/q$ earthquakes, where

$$M \geqslant \frac{M}{L} q \geqslant 1.$$

Here the total number of observations N equals Mq, and the formula for n_{ef} has the form

$$\frac{1}{n_{ef}} = \frac{\gamma_\xi}{L} + \frac{\gamma_\eta}{M} + \frac{\gamma_\varepsilon}{Mq}. \qquad (26)$$

When q = 1 the table is minimally filled and when q = L it is entirely filled.

Thus in proceeding to a series of observations, we assigned a specific accuracy, i.e., the width of the interval Δy within which the true mean y_0 is contained with probability $1 - \alpha$.

In order to answer the questions raised above, we must know the variance s^2 of an individual measurement. For this, in turn, we must know the variances s_ξ^2, s_η^2, and s_ε^2. An estimate of s^2 can be obtained from a preliminary data sampling. We substitute the estimate of the variance $\tilde{s}^2 = \tilde{s}_\xi^2 + \tilde{s}_\eta^2 + \tilde{s}_\varepsilon^2$ into expression (23) and obtain n_{ef}, which provides us with the required accuracy:

$$n_{ef} = \frac{\tilde{s}^2 \lambda_\alpha^2}{(\Delta y)^2} + 1.$$

Now let us see how this n_{ef} is obtained with a minimum total number of measurements. It follows from (26) that the N is equal to n_{ef} if M = L = N. Then

$$\frac{1}{n_{ef}} = \frac{\gamma_\xi}{N} + \frac{\gamma_\eta}{N} + \frac{\gamma_\varepsilon}{N} = \frac{1}{N}.$$

In other words, to achieve the specified accuracy with a minimum volume of measurements, we must take the number of stations and the number of earthquakes equal to the desired n_{ef}, measure each earthquake only at one station, and use for each station only one earthquake, so that a diagonal table of observations is obtained. In this case, L = M = N and $N = n_{ef}$, which is natural since all observations of a diagonal table are independent.

On the other hand if the total number of measurements N is limited (for example, by the time available for performing the experiment), the highest accuracy can be obtained by taking all observations as independent, i.e., using a diagonal table of observations.

Thus, the case of $n_{ef} = N$ is the most advantageous for a minimum of measurements and a maximum of accuracy.

In practice the amount of experimental material is usually limited. Let us assume the following: the number of stations L is limited, but there can be any number of earthquakes M, and an increase in N due to q, i.e., closer filling of the table, is possible. Under these conditions, we strive to increase as much as possible the accuracy of the result without specifying any particular n_{ef}; then we increase M and q. It is more advantageous to increase M, since this reduces the second and third terms of formula (26). For a fixed M an increase in q reduces only the third term. It follows from (26) that even when M is increased without limit, n_{ef} increases only to some limit L/γ_ξ; after $\frac{1}{M}\left(\gamma_\eta + \frac{\gamma_\varepsilon}{q}\right)$ becomes less than L/γ_ξ, a further increase in M has very little effect on n_{ef}, although the number of measurements increases. In striving for the maximum possible accuracy, therefore, it makes sense to increase M not without bound but only until M becomes greater by a factor of 2-3 than $\gamma_\xi/[L(\gamma_\eta + \gamma_\varepsilon)]$.

Let the experimental conditions limit both L and M. Then we can attempt to increase accuracy by filling the table, i.e., by increasing q. But this is meaningful only when γ_ε/M is greater than or comparable with $\gamma_\xi/L + \gamma_\eta/M$. In practice it is advisable to increase q only until q is greater by a factor of 2-3 than

$$\frac{\gamma_\xi}{\gamma_\xi + \gamma_\eta \frac{L}{M}}.$$

Let us assume the following: the accuracy of the results is preassigned, L or M is limited, and either one or both of the inequalities $L < n_{ef}$ or $M < n_{ef}$ is satisfied. First, it should be determined whether or not the required n_{ef} can be obtained at all. If L/γ_ξ or M/γ_η is less than n_{ef} the specified accuracy is unattainable. It cannot be increased by increasing q or the second unlimited parameter. To increase accuracy, the experiment must be set up so that M (or L) is increased to the level

$$n_{ef} > M > \gamma_\eta n_{ef} \quad \text{or} \quad n_{ef} > L > \gamma_\xi n_{ef}. \tag{27}$$

Let us assume that condition (27) is satisfied for one of the factors, for example L, and that M is in no way limited. Then from (26) we obtain the required M when q = 1.

$$M = \frac{\gamma_\eta + \gamma_\varepsilon}{\frac{1}{n_{ef}} - \frac{\gamma_\xi}{L}}. \tag{28}$$

If condition (27) is satisfied for both factors, or M is less than that required by (28), the required n_{ef} can be achieved only by increasing q. Let us calculate the required q: from Eq. (26) we have

$$q = \frac{\gamma_\varepsilon}{M\left(\frac{1}{n_{ef}} - \frac{\gamma_\xi}{L}\right) - \gamma_\eta}. \tag{29}$$

Thus by analyzing (21) we can make a preliminary estimate of the variance, and limit the sampling of experimental material so that with a minimum number of measurements we can estimate the mean with the necessary or maximum possible accuracy. Note that a preliminary estimate of the variance of an individual measurement $\tilde{s}^2$ is used in formula (20) for calculating n_{ef}. Assume that L, M, and N have been selected on the basis of this n_{ef} and that the corresponding measurements have been made. Using these measurements we can, with formulas (17), refine the estimates of s_ξ^2, s_η^2, and s_ε^2. Then the refined value $s^2 = s_\xi^2 + s_\eta^2 + s_\varepsilon^2$ can be substituted into formula (20) or (22) and n_{ef} can be refined. For the refined values of s^2 and n_{ef}, the confidence interval calculated by (24) or (25) may differ somewhat from the specified interval Δy. If it is greater than Δy and this does not suit us, we must add a certain number of measurements. This refines $\bar{y}$ and reduces the confidence interval to the required value Δy.

Let us consider an example, again the study of the period T_p in earthquakes. In Section 4 we saw that if we have 88 measurements for 20 earthquakes, we obtain a comparatively small $n_{ef} \simeq 12$, because the number of stations used – six in all – was very small, and the effect of the station characteristics was great: $\gamma_\xi = 0.32$. In an optimum version of 12 stations and 12 earthquakes, 12 measurements, i.e., seven times less data, would be sufficient to obtain the same accuracy.

Let us see what amount of data is necessary to obtain higher accuracy, for example, for a confidence interval $\Delta y = 2.5$ at $1 - \alpha = 0.70$. At this confidence level, $\Delta y = s_{\bar{y}}$. In our example the variance of an individual measurement was $s^2 = 656$. If we specify the variance $s_{\bar{y}}^2 = 6.25$, we find that the necessary n_{ef} is:

$$n_{ef} = \frac{656}{6.25} + 1 = 106 \simeq 100.$$

The minimum number of observations that provide this accuracy is $N \sim 100$, if M = 100, L = 100, and q = 1. If less than 100 stations are available to us, we can still obtain the desired accuracy provided that $L \leq \gamma_\xi$ and $n_{ef} = 32$. Assume that we have 50 stations: then when q = 1 it is necessary to have data for 190 earthquakes:

$$N = M = \frac{0.50 + 0.18}{0.01 - \frac{0.32}{50}} \simeq 190.$$

Thus the number of measurements required for the given accuracy increases almost by a factor of two. Now let us see what happens when we do not have the required number of earthquake records. We calculate the necessary q:

$$q = \frac{0.50}{90\,(0.01 - 0.32 \cdot 0.02) - 0.18} \approx 3.5.$$

Thus each earthquake must now be recorded by three or four stations. In this case, (90/50) 3.5 = 6.3, i.e., six or seven records from different earthquakes must be used at each station. The total number of measurements is 3.5 • 90 = 315, which is more than triple n_{ef}. It no longer makes sense to increase q further, since this will increase accuracy very slowly. In fact, if we measured all 90 earthquakes at all 50 stations (q = L), the total number of measurements would increase to 4500, i.e., by a factor of 15. But the effective number of independent observations would be only 115, i.e., 15% more than for q = 3.5. It is clear such an increase in the volume of measurements is entirely unjustified.

6. Estimation of Accuracy of Empirical Functions

A frequent practical research problem is that of obtaining an experimental dependence of y upon some independent variable x: for example, epicentral distance, earthquake energy or magnitude, etc. The problem of estimating the accuracy of the curve so obtained arises in such cases. The above methods of estimating the accuracy $\bar{y}$ can be extended to the case of y(x). This can be done in different ways, depending upon how the experimental data were averaged.

Let us examine two of the more common cases.

a) The experimental points are approximated by some function. The form of the function is usually selected from theoretical considerations. This can be a linear function $y = a + bx$, a polynomial of some degree n, or, finally, a linear combination of known functions with unknown coefficients or parameters. Then the relation is found by the method of least squares [6]. The accuracy of the estimate of each of the coefficients, which is expressed by its variance, is usually inversely proportional to the number of individual measurements used in their determination. If all measurements are independent, then this number is also the total number of observations (that is, points on the experimental curve). But if one or another factor exerts an influence, N must be replaced by n_{ef}.

b) The form of the averaging function is unknown beforehand and no assumptions are made about it. In this case the experimental curve is usually sought as follows. The entire range of variation of the independent variable is divided into intervals $\Delta^j x$, and then the mean values $\bar{x}^j$ and $\bar{y}^j$ in these intervals are calculated by averaging all experimental data in these intervals. The averaging curve is drawn through these mean points.

The accuracy estimate for the latter case has the following meaning. The accuracy of the mean $\bar{y}^j$ in each interval $\Delta^j x$ is determined by the confidence interval Δy^j. To estimate it we must calculate the variance of an individual observation s^2 and its components s_ξ^2, s_η^2, and s_ε^2, and we must know n_{ef}^j for each interval $\Delta^j x$. If there is a basis for assuming that the variance is a function of x, to determine these factors we must perform all of the procedures described in Sections 3 and 4 for each of the intervals $\Delta^j x$ separately. If there is no such basis, it is simpler to assume that s_ε^2, s_ξ^2, s_η^2, and s^2 are independent of x. Then the variance values that are common to the entire curve are estimated. Naturally, dependence upon the independent variable should be ruled out here. This can be done by taking all y_{ik} that are used to estimate the variances from one interval $\Delta^j x$, so that the values of the independent variable can be considered fixed. If this method is for some reason inconvenient, we can use experimental values y_{ik} for any values of the independent variable x but with the difference $y_{ik}^s - \bar{y}_{ik}^j$ in the interval to which the given value y_{ik}^j pertains taken as y_{ik}. Then the confidence interval Δy^j for each $\bar{y}^j$ is calculated by formula (20) or (22) on the basis of the variance s^2 of an individual measurement (which is independent of x) and n_{ef}^j, which is determined separately for each interval according to the amount of data L^j, M^j, and N^j in that interval.

The set of confidence intervals makes up the "confidence corridor" within which the real curve lies. To be sure, it must be borne in mind that the probability, i.e., that the entire curve as a whole is not beyond the limits of the "confidence corridor," is naturally lower than the corresponding probability for an individual interval.

Several questions arise after estimation of the confidence intervals.

The first question is whether or not the intervals $\Delta^j x$ were properly selected from the point of view of the relationship between accuracy (Δy^j) and detail ($\Delta^j x$). In fact, overly fine division into intervals reduces accuracy by reducing the number of data in each interval. Conversely, important features of the curve shape can be lost when the intervals $\Delta^j x$ are too wide. It sometimes makes sense to increase the intervals $\Delta^j x$ by joining adjacent ones in order to decrease the confidence interval Δy^j. It should be borne in mind, however, that the joining of $\Delta^j x$ does not always increase accuracy. This happens when the effects of the factors ξ and η are great, and only N is increased when intervals are joined (L and M do not increase), because the same stations and the same earthquakes were used in each of the joined intervals. Then n_{ef}^j, which is mainly a function of L^j and M^j, is not increased. Although detail has been lost we have practically no gain in accuracy.

The second question concerns a special check of the reliability of one or another feature of the experimental curve, for example, a sharp peak, a depression, etc. In this case, Student's t-test is used, which makes it possible to estimate the probability that a difference in y between adjacent points $\bar{y}^j$ and $\bar{y}^{j+1}$ actually exists. For this, we form the ratio

$$t = \frac{|\bar{y}^j - \bar{y}^{j+1}|}{s\sqrt{\frac{1}{n_{ef}^j} + \frac{1}{n_{ef}^{j+1}}}}. \tag{30}$$

TABLE 2. Calculation of "Confidence Corridor" of Curve y(x)

j	$\Delta^j x$	$\bar{y}^j$	N^j	L^j	M^j	$\frac{\gamma_\varepsilon}{N^j}$	$\frac{\gamma_\xi}{L^j}$	$\frac{\gamma_\eta}{M^j}$	$\frac{1}{n^j_{ef}}$	n^j_{ef}	$\frac{s}{\sqrt{n^j_{ef}-1}}$	$t^j_{0.3}$	$\Delta\bar{y}^j$
1	400—600	21.5	20	6	15	0.0250	0.053	0.0120	0.090	11.1	8.0	1.093	8,8
2	600—900	29.0	20	6	16	0.0310	0.053	0.0110	0.095	10.5	8.3	1.097	9.1
3	900—1500	39.5	37	7	28	0.0135	0.046	0.0064	0.066	15.6	6.7	1.075	7.2
4	1500—2100	48.0	112	7	85	0.0045	0.046	0.0021	0.053	18.9	6.0	1.067	6.4
5	2100—2600	51.5	98	8	73	0.0051	0.040	0.0025	0.048	20.8	5.7	1.063	6,1
6	2600—3100	53.0	72	8	65	0.0069	0.040	0.0028	0.050	20.0	5.9	1.066	6.3
7	3100—3800	49.0	38	8	29	0.0130	0.040	0.0032	0.059	17.0	6.4	1.071	6.9
8	3800—4500	64.0	35	8	28	0.0140	0.040	0.0064	0.060	16.7	6.5	1.072	7.0
9	4500—5500	67.0	77	7	60	0.0035	0.046	0.0030	0.056	18.5	6.1	1.068	6.5
10	5500—7000	68.5	72	3	52	0.0039	0.107	0.0035	0.117	8.5	9.4	1.113	10.5
11	7000—10000	68.0	60	3	46	0.0083	0.107	0,0039	0.119	8.3	9.5	1.116	10.6

We find the solution by comparing (30) with the tabulated value t^*_α that corresponds to degrees of freedom $n^j_{ef} + n^{j+1}_{ef} - 2$ for some significance level α. If $t < t^*_\alpha$, the means $\bar{y}^j$ and $\bar{y}^{j+1}$ are identical and the feature of the curve is insignificant. If $t > t^*_\alpha$, they differ with probability $1 - \alpha$.

Finally, if it turns out that the difference has a low probability for the available n^j_{ef} and n^{j+1}_{ef} and we must establish the fact of the difference or identity of $\bar{y}^j$ and $\bar{y}^{j+1}$ with greater reliability, we can use formula (30) to estimate the n_{ef} that would provide us with a reliable solution, i.e., would give us by Student's t-test a probability of difference $1 - \alpha$ that was fairly close to 1. Assuming for simplicity that $n^j_{ef} = n^{j+1}_{ef} = n_{ef}$, we obtain from (30)

$$n_{ef} = \frac{2s^2 t^2_\alpha}{(\bar{y}^j - \bar{y}^{j+1})^2}\,. \qquad (31)$$

Then if we know n_{ef} and s^2_ξ, s^2_ε, s^2_η, and s^2, we can plan the selection of additional material for measurements in accordance with the recommendations of Section 5.

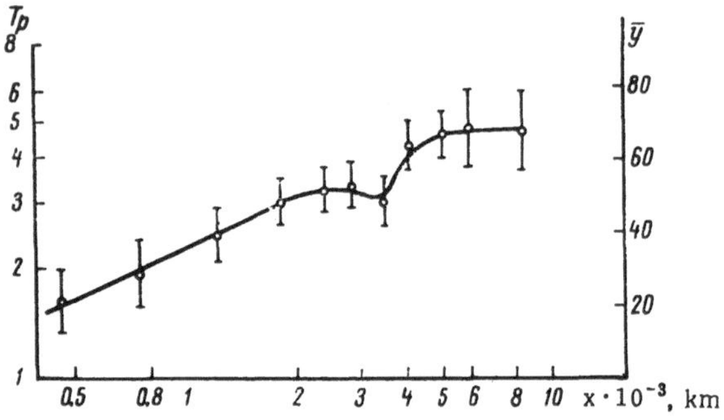

Fig. 1. Experimental curve of period T_p versus epicentral distance. The points are the means $\bar{y}^j = 100 \log T_p$ in the intervals of the independent variable $\Delta^j x$ and the vertical lines are confidence intervals with the level $1 - \alpha = 0.70$.

Let us apply our results to the study of periods T_p. In the example considered above, we already used data on the wave period T_p in earthquakes, and for convenience we let $y = 100 \log T$. The period is a function of the distance x and the magnitude m. In estimating the variance, we assumed that it was not a function of x or m, and we used records of earthquakes with epicentral distances of from 1000 to 4000 km and magnitudes of from $4\,^1\!/_4$ to $5\,^3\!/_4$.

In estimating the variance, the effect of x on y was ruled out, since the difference $y^j_{ik} - \bar{y}^j$ was used as y_{ik}.† The effect of the magnitude was eliminated by selecting different $\bar{y}^j$, depending upon which magnitude interval ($4\,^1\!/_2$-5 or $5\,^1\!/_4$-6) the given earthquake fell into.

Using the variance estimates and γ so obtained ($s^2 = 656$, $\gamma_\xi = 0.32$, $\gamma_\eta = 0.18$, and $\gamma_\varepsilon = 0.50$), we can find the confidence intervals for all $\bar{y}^s$. The experimental curve y(x) is plotted from the points $\bar{y}^s$ in Fig. 1. The confidence interval $\Delta\bar{y}^j$ at the confidence level $1 - \alpha = 0.7$ is shown for each point $\bar{y}^j$ by a vertical line. The values used to find the "confidence corridor" are shown in Table 2. Let us see, for example, whether there is any point in joining intervals 5 and 6 or 8 and 9. Let us calculate Δy^j for such combined intervals (Table 3). If we compare the obtained Δy^j for the joined Δx with the old forms, we see that the gain in accuracy (increase in the confidence interval) is only 5-10%, while the detail of the curve is considerably im-

† Here, naturally, we must obtain $\bar{y} = 0$. In our example in Section 4, $\bar{y} \neq 0$, since we determined $\bar{y}^j$ from all available data while the variance estimate was made with a comparatively small data sampling ($N = 88$, $n_{ef} = 12$), which was limited to the 20 earthquakes that were recorded by not less than four stations.

TABLE 3. Parameters for Finding "Confidence Corridor"

$j;\ j+1$	N	L	M	γ_{ε}/N	γ_{ξ}/L	γ_{η}/M	$1/n_{ef}$	n_{ef}	$\frac{s}{\sqrt{n_{ef}-1}}$	$t_{0.3}$	Δy^j
5; 6	170	8	105	0,0029	0.04	0.0017	0,045	22.2	5.55	1.063	5.9
8; 9	112	8	45	0,0045	0.04	0.0040	0.048	20.8	5.75	1.064	6.1

paired. The depression in the curve at x = 3500 km becomes less distinct and it would be more difficult to relate it to specific layers of the earth.

Now let us see if the obtained accuracy justifies the number of measurements that were made in our example. We see that 35-38 measurements were made in Δx intervals 7 and 8 while about 100 were made in 4 and 5. Meanwhile, n_{ef} is 17 in the first case and 19-21 in the second, i.e., 20% greater, and accordingly the confidence interval is approximately 10% smaller. It is clear that for such an insignificant increase in accuracy there is no point in making such a great increase in the number of measurements. It would be sufficient to make 35-40 measurements in each interval. Much greater accuracy would be gained by increasing the number of stations used. For example, if 15 stations could be used instead of six or eight, for N = 35 and M = 25-30 we would obtain

$$n_{ef}=\frac{1}{\frac{0.5}{35}+\frac{0.32}{15}+\frac{0.18}{28}}\approx 24.$$

If we compare this result with $n_{ef} = 16.7$ for the eighth interval for the same M and N, we see that n_{ef} would be increased by 40% and the confidence interval would be reduced by 20% for the same number of measurements.

Let us make another estimate. In Fig. 1, a decrease in period is observed at distances of 2500-3500 km and a sharp increase is observed at distances of 3500-4500 km. Let us see whether, within our accuracy, the existence on the true curve y(x) of this descent and ascent can be considered reliable. We shall use Student's t-test and assume that the existence of these details on the experimental curve is positively certain if the probability of their existence is not less than 80%. Let us estimate t for intervals 6 and 7 and for intervals 7 and 8 (see Table 2):

$$t^{6;\,7}=\frac{4}{25.6\sqrt{0.050+0.059}}\approx 0.47;$$

$$t^{7;\,8}=\frac{15}{25.6\sqrt{0.059+0.060}}\approx 1.70.$$

The probability that for these t values there is a difference between y^6 and y^7 and also between y^7 and y^8 is estimated with the aid of Student-distribution tables for degrees of freedom of 35 and 31.7, respectively. These probabilities are $(1-\alpha)_{6;7} = 0.36$ and $(1-\alpha)_{7;8} = 0.90$.

Thus the sharp increase in period at epicentral distances of 3500-4500 km can be considered real. The existence of a decrease in period at distances of 2500-3500 km remains doubtful. To answer this question with greater certainty, we must appreciably increase the observation accuracy and, accordingly, n_{ef}. Let us see what n_{ef} would provide us with the required accuracy. For a reliability $1-\alpha = 0.8$.

$$n_{ef}=2\frac{s^2t^2_{0.2}}{(y^6-y^7)^2}1\approx 130.$$

If we are satisfied with the somewhat lower reliability $1-\alpha = 0.7$, then $n_{ef} = 90$. It is obvious that we cannot achieve such accuracy if we have only eight stations, since the limiting n_{ef} in this case is 8/0.32 = 25. The minimum number of stations with which we can obtain the required accuracy, depending upon the specified reliability, is $L = \gamma\xi_{ef}$, i.e., $L_{min} = 0.32 \cdot 130 \approx 42$ or $L_{min} = 0.32 \cdot 90 \approx 28$. Using the results of Section 5, let us calculate the number of observations N (we assume M = N) that are required to obtain, with different numbers of stations $L > L_{min}$, $1-\alpha = 0.8$ or 0.7, and q = 1 (Table 4).

TABLE 4. Number of Observations Versus Reliability

$1-\alpha$	Number of observations for number of stations					
	30	45	60	90	110	130
0.8	—	2700	340	170	1153	130
0.7	1700	170	117	105	95	79

After these estimates, we should if possible make long-term measurements and, having increased L or M and supplemented the most interesting parts of the curve with these new data, refine the corresponding $\bar{y}^j$. We can also improve accuracy or detail by combining or subdividing the intervals $\Delta^j x$.

After doing this, we again draw the experimental curve, not through each point $\bar{y}^j$ but smoothing it somewhat within the "confidence corridor," preserving only the details that prove to be significant according to Student's t-test.

7. Conclusion

1. In any experimental study we encounter situations in which the value under study is subject to the influence of a number of factors. Some of them are systematic and have physical meaning. In research into seismic wave dynamics, these can be earthquake magnitude, epicentral distance, focal depth, etc. The effects of such factors on the values under study can be defined explicitly as functions or one or more independent variables. Other factors are of a random nature and manifest themselves as an increase in the scatter of individual measurements. These random deviations may indeed be manifestations of some physical causes, but since these causes are still unknown to us we relate the corresponding deviations to some external random factors that do not have physical meaning – for example, the seismic station number, the earthquake number, the observer, the instrument, etc.

In a formal investigation, the effect of systematic factors is studied by isolating experimental functions of one or another independent variable. The effect of random factors is detected by analysis of variance, using Fisher's F-test. Some residual deviations – Gaussian noise and "pure noise" – cannot be related to any factor.

2. The specific weight γ_i of the effect of one or another factor (focus characteristics, station characteristics, or Gaussian noise) on the scatter of the value under study is determined by the formula

$$\gamma_i = \frac{s_i^2}{s^2},$$

where s_i^2 is the variance related to the given spread and s^2 is the total variance of an individual measurement, which is the sum of the partial variances

$$s^2 = s_\varepsilon^2 + s_\xi^2 + s_\eta^2.$$

3. Assume that we wish to obtain a result of greater generality, i.e., one which would be valid for any stations and any earthquakes, within the limits of accuracy. In this sense, accuracy is determined simply by the total number of individual data N and their variance s^2. It is a function, too, of γ_i and of the number of stations L and of the number of earthquakes M. The accuracy of the mean is characterized by the variance of the mean, which is calculated by the formula

$$s_{\bar{y}}^2 = \frac{s^2}{n_{ef}}, \tag{32}$$

where n_{ef} is the effective number of independent observations, which is equivalent to a given series of N measurements ($1/n_{ef} = \gamma_\xi/N + \gamma_\xi/L + \gamma_\eta/M$).

For example, if the number of stations L is small, there exists an accuracy limit that cannot be surpassed even if the number of earthquakes is increased without bound. The corresponding limit $n_{ef} = L/\gamma_\xi$.

4. Analysis of formula (32) allows us to choose the most advantageous ways of achieving the maximum or desired accuracy for a minimum number of measurements, according to the specific situation. The most advantageous situation is that in which the number of stations and the number of earthquakes are the same. Each earthquake is observed at only one station and each station records only one earthquake. In this case we have a diagonal table of observations and all measurements statistically independent, and so the number of observations N is equal to n_{ef}. But if because of the experimental conditions one of the factors is limited, for example, $L < n_{ef}$ but $L > n_{ef}\gamma_\xi$, the required accuracy can be obtained by increasing the number of values of the second parameter $M > n_{ef}$. Here the total number of measurements is unavoidably increased: $N > n_{ef}$. It is disadvantageous to increase the filling of the table of observations, i.e., to measure each earthquake at almost all stations and to use all earthquakes recorded at each station, especially if the effects of the station and earthquake are great, i.e., $\gamma_\xi > 0.2$ and $\gamma_\eta > 0.2$. This leads to an excessively large volume of measurements but does not, for all practical purposes, increase the accuracy of the mean.

Thus having estimated from some preliminary data sampling the variance of an individual measurement s^2 and the degree of the influence of one or another factor γ_ξ, γ_η, or γ_ε, we can optimally organize the selection of material for future measurements.

5. Similar considerations apply when the dependence of some value y upon an independent variable x is studied. In particular, if the dependence is sought in the form of a specific function whose parameters can be obtained by the method of least squares, then the accuracy of the coefficients found will be determined not by the total number of experimental points N but by n_{ef}. If the form of the experimental relation is unknown beforehand and is sought by obtaining the mean values $\bar{y}^j$ in a number of intervals of the independent variable $\Delta^j x$, the accuracy of the experimental curve is characterized by the variance of each of these values $s^2_{\bar{y}^j}$. The latter is not a function of the number of individual points N^j in a given interval of the independent variable but of n_{ef}. The formulas given here make it possible to estimate how many additional measurements are required to obtain a specified accuracy.

LITERATURE CITED

1. Smirnov, N. V., and I. V. Dunin-Barkovskii (1959), Brief Course in Mathematical Statistics with Engineering Applications, Moscow, Fizmatgiz.
2. Brownlee, K. A. (1965), Statistical Theory and Methodology in Science and Engineering, New York, John Wiley and Sons.
3. Sheffé, H. (1959), Analysis of Variance, New York, John Wiley and Sons.
4. Romanovskii (1947), Experimental Applications of Mathematical Statistics, Moscow and Leningrad, Gostekhizdat.
5. Cramér, H. (1951), Mathematical Methods of Statistics, Princeton, Princeton University Press.
6. Linnik, Yu. V. (1962), The Method of Least Squares and Fundamentals of the Theory of Observation Processing, Moscow, Fizmatgiz.

A STATISTICAL METHOD FOR DETERMINING EARTHQUAKE FOCAL DEPTH FROM THE RECORD OF ONE SEISMIC STATION†

V. F. Pisarenko and A. A. Poplavskii

1. Introduction

In many cases in practice, the focal depth of an earthquake can be found from the record of one seismic station. The additional requirement of promptness often arises here. In particular, this is necessary in the seismological urgent-report service and is of particular value in improving the quality of work of the tsunami warning service. The present paper was also written in connection with the problem of tsunami forecasting, since knowledge of the focal depth of an earthquake can be useful in determining its tsunami danger [2].

Focal depth can be determined most reliably and sufficiently rapidly from a single station, using the delay of the secondary waves pP and sP relative to the first arrival, provided that these arrivals can be identified on the record. Since this is not always the case, it is desirable to find methods that will allow focal depth to be determined from a single record, regardless of whether or not the arrivals of waves reflected near the epicenter are isolated on it. From the statistical point of view, this is a problem in pattern recognition [3].

We shall describe an algorithm for recognizing two patterns that are given by two finite samples of multidimensional vectors. The statistical quantity on which the recognition is based is the likelihood ratio, in which the multivariate probability densities are replaced by their estimates, which are obtained from the finite samples. The algorithm has been used for the practical separation of shallow (focal depth less than 25 km) and deep Kurile Islands earthquakes. The experimental materials were seismograms recorded at the Yuzhno-Sakhalinsk seismic station.

2. Selection of Criteria for Recognition

Three parts of an earthquake record are considered: the longitudinal waves, the interval between longitudinal and transverse waves, and the transverse waves. The earliest train of oscillations composed of five extrema with apparent periods of not over 12-13 sec and with the largest amplitudes was taken as the longitudinal waves. The oscillation train that followed the moment of arrival of a transverse wave composed of four or five extrema with an average period of not over 18-20 sec was taken as the transverse waves. Often, transverse wave arrivals were not isolated; in these cases the arrival time was taken from the travel-time curve. All oscillations that followed longitudinal waves but were earlier than transverse waves were classed in the "interval between P and S" (below, this is indicated as P → S).

Each of the above-mentioned parts is characterized by the following values:

T_{min} – the minimum period of apparent oscillations,

T_{max} – the maximum period of apparent oscillations,

$T(A_{max})$ – oscillation period with maximum amplitude,

$T(\frac{A}{T}_{max})$ – oscillation period with maximum energy (the ratio A/T is proportional to the oscillation energy),

n – the number of periods measured on a seismogram.

† Translated from Vychislitel'naya Seismologiya, No. 5, Moscow, Nauka (1971), pp. 28-54.

Records of Far Eastern earthquakes are distinguished by high complexity and consideration of four periods is justified, although sometimes some of them coincide. For the amplitudes of oscillations that have periods T_{min}, T_{max}, etc., we have a number of other criteria:

$$\log\frac{A}{T}(T_{\min}),\quad \log\frac{A}{T}(T_{\max}),\quad \log\frac{A}{T}(A_{\max}),\quad \log\left(\frac{A}{T}\right)_{\max},$$

that characterize the energy of the corresponding oscillations. Values that characterize relationships between oscillation amplitudes and periods T_{min}, T_{max}, $T(A_{max})$, and $T((A/T)_{max})$ were also selected as criteria:

$$\log\frac{A(T_{\min})}{A_{\max}},\quad \log\frac{A(T_{\max})}{A_{\max}},\quad \log\frac{A\left(\frac{A}{T}\max\right)}{A_{\max}}.$$

All the above-mentioned eleven values can be assigned some physical meaning. For example, T_{min} and T_{max} approximately determine the limits of the spectrum, $T(A_{max})$ determines the position of the principal maximum in the spectrum, and n gives the total number of the most intense maxima and describes roughly how many peaks the spectrum contains.

Since each of the above values must be measured on three different parts of a record, we have 36 criteria that characterize the initial part of an earthquake seismogram. Several values that describe the change of individual elements of a record in moving from one part of the record to another were also selected as criteria:

$$\frac{T_{\min P}}{T_{\min S}},\quad \frac{T_{\max P}}{T_{\max S}},\quad \frac{T(A_{\max})_P}{T(A_{\max})_S},\quad \frac{T\left(\frac{A}{T}\max\right)_P}{T\left(\frac{A}{T}\max\right)_S},$$

$$\log\frac{A_{\max P\to S}}{A_{\max P}},\quad \log\frac{A_{\max S}}{A_{\max P}}.$$

It makes sense to take into account the criteria that characterize the envelope of a record. For this, the seismogram is divided into two parts: the first contains the longitudinal waves and the interval between P and S, and the second contains the initial part of S. Both the rise and fall of envelope amplitude were taken into account in the first part, but only the rise was taken into account in the second.

In the first part of the seismogram we measured the apparent value of the first displacement a_1 (in mm), the maximum displacement a_2, the three greatest displacements a_3, a_4, and a_5 following a_2 and the time intervals between them $\tau_{i,i+1}$. Only the first displacement b_1 and one of the subsequent displacements b_2 within the transverse wave train, and also the time interval between them τ_s were measured in the second region. These values make up criteria that characterize the relative value of the first amplitude of the envelope $\log a_1/a_2$, the steepness of rise of the envelope $(a_2 - a_1)/\tau_3$, the attenuation rate $(a_2 - a_5)/(\tau_{23} + \tau_{34} + \tau_{45})$ in the first region, and, finally, the rate of amplitude rise $(b_2 - b_1)/\tau_s$ in the second region. A total of 45 criteria were obtained for the initial part of an earthquake record.

The statistical processing method used determines the correlation between each pair of criteria. In order to study their dependence on the principal earthquake parameters, we introduced into the group of criteria the magnitude M, epicentral distance Δ, and focal depth h. The latter three criteria were not used in recognition, however.

Records of the Yuzhno-Sakhalinsk station were chosen as the material for study. The epicentral distances of the earthquakes of the Kurile–Kamchatka zone whose tsunami hazard must be determined do not exceed 2000 km. This was taken as the upper limit of distance; the lower limit was not established. The seismograms were randomly selected. It was important only that a sufficient number (on the order of 30 for each interval) of good-quality records be selected. A complete tabulation of the material can be found in Table 4. The earthquake data were taken mainly from the "Bulletin of the Far East Seismic-Station Network." The Δ values in the catalog refer to the Yuzhno-Sakhalinsk station.

3. Algorithm for Recognition of Two Patterns

Information on the two patterns that we wish to learn to recognize is given by two finite samplings of multidimensional vectors. In our problem, the first pattern is the set of earthquakes with focal depth $h < 30$ km and the second pattern is the set of earthquakes with $30 \text{ km} \leq h < 80$ km. A multidimensional vector characterizing an individual earthquake consists of the parameters described in Section 2. We shall assume that the set of vectors of each pattern obeys an unknown distribution function. We shall assume that other than the two finite samples of each pattern, we have no other *a priori* information on these distributions. The samples will be called the instruction material. We shall denote the dimensionality of the vectors in question by r, the vectors of the first pattern by $\bar{x}^{(1)}, \ldots, \bar{x}^{(N)}$, and the vectors of the second pattern by $\bar{y}^{(1)}, \ldots, \bar{y}^{(M)}$. The recognition algorithm is a function ψ whose argument is an r-dimensional vector of the parameters $\bar{z} = (z_1, \ldots, z_r)$ of the unknown pattern. This function must show to which pattern the vector $\bar{z}$ should be related. For definiteness, we shall assume that if $\psi(\bar{z}) = 1$, the algorithm relates vector $\bar{z}$ to the first pattern; if $\psi(\bar{z}) = 0$, to the second.

Thus, the function ψ divides the entire r-dimensional space into two parts. If the vector being checked $\bar{z}$ falls into the first part, it is decided that $\bar{z}$ pertains to the first pattern; otherwise, it pertains to the second. It is sometimes more advantageous to consider algorithms that can "refuse to make decisions" for certain $\bar{z}$.

This possibility could have been included in our algorithm, but it would have complicated analysis of the problem. We would have had to introduce a new parameter – the risk ratio, or the risk of refusing a decision as opposed to the risk of an incorrect decision.

On the other hand, the function ψ depends upon the instruction material, i.e., it divides the entire r-dimensional space into two parts, according to vectors x and y that have been given as instruction material. The following are examples of recognition algorithms.

a) Linear combination of the parameters

$$\psi(\mathbf{z}) = \begin{cases} 1, & \text{if} \quad a_1 z_1 + \ldots + a_r z_r > a_0, \\ 0, & \text{if} \quad a_1 z_1 + \ldots + a_r z_r \leqslant a_0, \end{cases}$$

where $a_0, a_1, \ldots, a_r$ are constants, which are selected by some rule according to the values of the instruction vectors x and y;

b) The "average distance" algorithm

$$\psi(\mathbf{z}) = \begin{cases} 1, & \text{if} \quad \frac{1}{N} \sum_{i=1}^{N} R(\mathbf{z}, \mathbf{x}^{(i)}) > \frac{1}{M} \sum_{i=1}^{M} R(\mathbf{z}, \mathbf{y}^{(i)}), \\ 0 - \text{otherwise}, \end{cases}$$

where $R(\bar{z}, \bar{x})$ is the distance between vectors $\bar{z}$ and $\bar{x}$.

We introduce the value γ, which characterizes the recognition quality of some algorithm ψ for given instruction material:

$$\gamma = \frac{\alpha}{2N} \sum_{i=1}^{N} [1 - \psi_i(\mathbf{x}^{(i)})] + \frac{1}{2M} \sum_{i=1}^{M} \tilde{\psi}_i(\mathbf{y}^{(i)}). \qquad (1)$$

Here, ψ_i denotes a recognition algorithm that uses all of the given instruction material except the vector $\bar{x}^{(i)}$, while $\tilde{\psi}_i$ uses all of the material except the vector $\bar{y}^{(i)}$. Thus the algorithm ψ is checked for each vector of the instruction material; each time, the vector being checked is discarded from the instruction material.† The first sum in (1) gives the number of incorrect decisions made in checking the vectors of the first pattern; the second sum gives this number for the second pattern. These sums are divided by twice the number of vectors in the corresponding pattern, and in addition the first sum is multiplied by the constant α, which expresses numerically the significance of erroneous decision, a vector of the first pattern to the second pattern as opposed to the opposite error of relating a vector of the second pattern to the first pattern. In our problem, we assumed that both these errors were equally significant and therefore let $\alpha = 1$. Thus in our problem γ is an estimate of the probability of an incorrect decision, and this estimate was obtained from the same material on which instruction was based. Note that this is an unbiased estimate, i.e., its mathematical expectation coincides with the true probability of an incorrect decision.

Thus, using γ we can compare two recognition algorithms for given instruction material and give preference to the one for which γ is smaller.

If we knew the probability densities for the first and second patterns, the optimum recognition algorithm (which is well known from mathematical statistics: see, for example, [5]), would give on the average a minimum γ and would be specified by the ratio of these densities (the likelihood ratio). This assertion is proved in the well-known Neyman–Pearson theorem. In our algorithm we use the ratio of the estimates of the probability densities of the first and second patterns. The estimates are obtained from the instruction material. We shall use consistent estimates of the multivariate probability densities, i.e., estimates that converge to the true values as the quantity of instruction material increases. Thus our algorithm will converge to the optimum algorithm as the quantity of instruction material is increased. Of course, it does not follow that our algorithm will be optimum for a small amount of instruction material.

The method used for estimation of the probability density function is an extension to the multivariate case of a method described by Parzen [6]. This method, which is sometimes called the "potential method" (see [7, 8]), consists of the following: We construct some function $\varphi(x_1, \ldots, x_r)$, called the "potential," which has the following properties:

$$\varphi(x_1, \ldots, x_r) \geqslant 0, \rho\varphi(x_1, \ldots, x_r) \to 0 \quad \text{for} \quad \rho \to \infty,$$

† The idea of this "cyclic rejection" of vectors being checked was proposed in [4].

where

$$\rho^2 = x_1^2 + \ldots + x_r^2,$$

$$\int_{-\infty}^{\infty} \ldots \int \varphi(x_1, \ldots, x_r)\, dx_1, \ldots, dx_r = 1. \quad (2)$$

The following probability density functions are examples of potentials:

$$(2\pi)^{-r/2} \exp\left[\frac{-x_1^2}{2} - \ldots - \frac{x_r^2}{2}\right],$$

$$\pi^{-r}(1 + x_1^2)^{-1} \ldots (1 + x_r^2)^{-1}. \quad (3)$$

The following function can be taken as an estimate of the multivariate probability density of the first pattern:

$$f(x_1, \ldots, x_r) = N^{-1}\sigma_1^{-1} \ldots \sigma_r^{-1} \times$$

$$\times \sum_{k=1}^{N} \varphi\left(\frac{x_1^{(k)} - x_1}{\sigma_1}, \ldots, \frac{x_r^{(k)} - x_r}{\sigma_r}\right), \quad (4)$$

where $x_1^{(k)}, \ldots, (x_r^{(k)})$ are the coordinates of the k-th instruction vector of the first pattern $\bar{x}^{(k)}$, and $\sigma_1, \ldots, \sigma_r$ are positive numbers called smoothing constants. Parzen proved [6] for the univariate case that estimate (4) converged to the true probability density when $N \to \infty$ if the smoothing constant approached zero as $N^{-1/2}$. Parzen's results are easily extended to the multivariate case. We shall make such an extension, for simplicity not taking the most general case.

Theorem. Given a sampling of N independent vectors $\bar{x}^{(1)}, \ldots, \bar{x}^{(N)}$ that obey an unknown probability density, let $\varphi(x_1, \ldots, x_r)$ be a function that satisfies conditions (2) and for any $i = 1, \ldots, r$

$$\sigma_i(N) \to 0; \quad N\sigma_1(N) \ldots \sigma_r(N) \to \infty; \quad (5)$$
$$[\max_k \sigma_k(N) / \min_k \sigma_k(N)] < C \quad \text{for } N \to \infty.$$

Then estimate (4) converges in mean-square sense to the true probability density at any point where this density is continuous.

The proof is similar to Parzen's proof [6], so we shall omit it. Thus an estimate of the form of (4) is, under conditions that as a rule occur in practical applications, a consistent estimate of the multivariate probability density. When the dimensionality r is high, however, the convergence of estimate (4) to the true value is greatly slowed. It follows from conditions (5) that $\sigma_i(N)$ must be of the order of $N^{-1/2}$ and, at the same time, must be very small. Thus when the dimensionality r increases, the number of observations must increase approximately as N^r to keep the accuracy of the estimate at about the same level. This makes it impossible to use estimate (1) for $r > 3$ if the number of observations does not exceed 100. We shall use estimate (1) only for $r \leq 3$.

The question arises of how to use pattern recognition for all or some coordinates in our problem when their number is greater than 3.

We proceeded as follows. We consider all possible combinations of three coordinates, and for each combination calculate its corresponding γ value. Then we sort out several nonoverlapping triplets that have minimum γ values. The likelihood ratios for the sorted-out triplets are multiplied and the product is considered as the likelihood substitute of the set of triplets. Note that multiplication of the likelihood ratios would give exactly the likelihood ratio of the entire set if all the triplets were independent, for both the first and second patterns (here, the likelihood ratio is not the ratio of the true probability densities of the first and second patterns, which are unknown, but the ratio of their estimates, which are obtained from given instruction material, i.e., the estimate of the likelihood ratio). Note, too, that multiplication of the estimates of the likelihood ratios improves recognition, as a rule, even if these estimates are dependent (although not when they are too greatly dependent, to be sure). This circumstance was used in [11], where likelihood estimates corresponding to individual parameters were multiplied.

Our recognition algorithm is constructed as follows.

All possible triplets are sorted. The corresponding γ is calculated for each triplet and several nonoverlapping triplets with the smallest γ values are found. The estimates of the likelihood ratios for the chosen triplets are multiplied. The product is considered as a substitute for the likelihood ratio for the set of triplets. If it is greater than some threshold constant, we assume that the vector being checked pertains to the first pattern; otherwise, to the second.

Now let us consider the problem of obtaining an estimate of the probability of an incorrect decision for a given recognition algorithm. It is simplest and most reliable to check the algorithm on additional material for which it is known which vectors belong to which pattern. The results of this check (correct and incorrect responses given by the algorithm) form a series of Bernoulli tests with an unknown probability of incorrect response. Methods of estimating this probability and construction of confidence intervals for it are well known in statistics (see [9], for example). But this requires additional material (or use of part of the available material for control instead of instruction), which is always undesirable. The question arises as to whether the probability of incorrect response can be estimated using the same material with which instruction is performed. We have been unable to find a completely satisfactory answer to this question.

The difficulty here is this. In sorting (individual parameters, pairs, or triplets), we select the combination of parameters with the smallest γ. For each combination of parameters, γ is a random number embodying a random deviation from its mean value, the quantity which is of interest to us. If we select from among the many possible combinations, those combinations of parameters with the smallest γ, we isolate with high probability those combinations for which γ is less than its mean value. Here, the situation is similar to that encountered in the theory of extreme values (see [10]): if we select the smallest of a series of identically distributed independent random values, then with high probability our selection will have a value less than the mean. In our case, the situation is complicated by two factors: first, the random values γ corresponding to different combinations of the parameters are not distributed in the same way; second, values of γ corresponding to overlapping combinations (combinations that have one or more identical parameters) are surely statistically dependent. Moreover, γ values that correspond to nonoverlapping combinations can also be dependent, due to dependence of the parameters. All of this makes it impossible to use the theory of extreme values to estimate the downward bias of γ relative to its mean value. In order to have some idea of the magnitude of this bias, we conducted an experiment with two artificially created patterns. Each pattern consisted of 46-dimensional independent normalized Gaussian random numbers from random number tables. Just as in our problem, the first pattern had 26 vectors and the second had 43. Various constants A were added to the vectors of the second pattern and the combinations of parameters were sorted. The true mean γ_t for single, paired, and triple combinations is easily calculated. By comparing it with the calculated value γ, we can obtain an idea of the possible deviation of γ downward from its mean value due to sorting. Of course, in our problem the parameter distribution is not Gaussian, so the deviations may not be quite the same as those obtained with the artificial example, but there is no reason to expect that the deviations in our problem will be much greater than in the artificial example. Note that in the artificial example the parameters are independent and have the same distribution. These two circumstances (especially the latter) can only increase the deviation in the artificial example as compared to the deviation in our problem. Therefore, the first can give a somewhat exaggerated estimate for the second. The numerical results for recognition of the artificial patterns are given in Table 1.

It is apparent from this table that in our case the reduction of the minimum γ as compared with the true value is approximately 0.06. Considering the above, this value can be taken as an overestimate of the discrepancy between the true probability of an incorrect decision and the calculated value γ.

4. Recognition of Shallow and Deep Earthquakes

As we have already indicated, the instruction material consisted of 26 46-dimensional vectors for the first pattern and 43 vectors for the second pattern. The parameters that make up the vectors are described in Section 2. First we calculated the 49-dimensional correlation matrices (unnormalized and normalized) for each of the patterns. Remember that we added the magnitude, depth, and epicentral distance to the 46 parameters that take part in the recognition. Note, by the way, that the focal depth h was weakly correlated with all parameters. The highest sample correlation coefficient (in absolute value) was obtained for the pair (h, $\log(a_1/a_{\max})$): -0.307; for which the 95% confidence interval was $(-0.65, +0.10)$.

TABLE 1

True error probability	Singles		Pairs		Triplets	
	min γ	A	min γ	A	min γ	A
0.2	0.12	1.70	0.14	1.49	0.14	0.97
0.3	0.24	1.05	0.23	0.74	0.24	0.60

The correlation matrices are required, in the first place, to obtain the normalizing constants σ_4, $\ldots, \sigma_{49}$ (formula (4)), and secondly so that in selecting the nonoverlapping triplets with the smallest γ, those triplets could be selected whose parameters were not very strongly correlated. Thirdly, correlation matrices are among the most important statistical characteristics of multidimensional samplings and are useful in interpreting results. What conclusions can be drawn from an analysis of the correlation matrices we have obtained? First of all, the assertion made in Section 1 about the impossibility of establishing close linear relationships between the individual elements of an earthquake record and the focal depth h is confirmed. The 95% confidence intervals for the correlation coefficients of h with all parameters of the record contain a zero, which indicates that not one of the parameters has a significant correlation with h. For the earthquake magnitude, the sample correlation coefficients for the second pattern were considerably greater than for the first. The mean absolute value of the sample correlation coefficient of M was 0.2 for the first pattern and 0.44 for the second. In the first pattern, even such parameters as log (A/T) had a much greater dispersion and were less closely related to M than in the second pattern. Hence the rather unexpected conclusion that earthquakes in the crust and immediately below the crust should be distinguished when plotting calibration curves (log [A(Δ)/T] for fixed magnitude) from body waves. Note also that in the second pattern the parameters themselves were more strongly correlated between each other than in the first pattern: for example, eight pairs of parameters in the second pattern have sample correlation coefficients greater than 0.8; six pairs, coefficients greater than 0.9. In the first pattern, only three pairs have correlation coefficients greater than 0.8, and not one pair has a coefficient greater than 0.9. This, of course, is due to the fact that the magnitude is more strongly related to the form of the record for the second pattern. Correlation matrices contain a great deal of information that is useful in analysis of the dynamic characteristics of earthquake records, but their detailed analysis is beyond the scope of this paper.

Next we analyzed the single parameters and all possible pairs of the parameters. The value γ was calculated for each of the 46 parameters and for each pair. This analysis allowed us to discard, before the final sorting of the triple combinations of parameters, those parameters that had the greatest γ values and that were apparently not related to focal depth. In addition, for the good parameters, it is useful to know how they "recognize" the patterns singly and in paired combinations. The results of analysis of the single parameters are given in Table 2, which indicates the parameters that have the smallest γ. The mean values and standard deviations of these parameters for the first and second patterns are also given there.

Table 3 shows the best nonoverlapping pairs. As a result of the analysis for singles and pairs, we discarded 24 parameters all of which recognized the patterns very poorly in singles as well as in pairs. Their γ value was not less than 0.4; the remaining parameters had, as a rule, values from 0.25 to 0.3. The indices of the discarded parameters were 4, 8, 14, 18, 19, 24-32, 34, 37-41, 43, 46, 47, and 49. The best triplets among the remaining 22 parameters were selected. In all, $C_{22}^3 = 1540$ triplets were tried. The best triplet (11, 36, 42) had $\gamma = 0.16$. Parameters 36, 42, 17, 6, and 11 were encountered most often in the 150 triplets with the smallest γ values. The four nonoverlapping triplets with the smallest γ values were chosen. In addition to (11, 36, 42), these were: (15, 17, 45), $\gamma = 0.21$: (9, 22, 48), $\gamma = 0.23$; and (6, 10, 12), $\gamma = 0.24$.

Thus only the twelve parameters in these triplets remained for the final recognition. The set of these four triplets gave $\gamma = 0.15$. If neces-

TABLE 2

Parameter No.	γ	Mean		Standard deviation	
		first pattern	second pattern	first pattern	second pattern
42	0.27	−0.64	−1.26	0.49	0.49
17	0.29	4.75	11.60	5.97	9.59
11	0.32	−0.45	−0.09	0.42	0.44
16	0.34	13.5	17.5	9.3	6.8
12	0.34	−0.16	0.14	0.43	0.40
13	0.34	2.9	3.8	1.2	1.1
22	0.34	0.11	0.44	0.41	0.45
28	0.35	11.5	15.4	6.4	6.1
29	0.36	10.6	14.5	7.1	6.9
15	0.36	16.3	18.9	7.6	5.8

TABLE 3

Parameter	1	2	3	4	5	6	7	8	9	10	11	12	13	14	15	16	17	18	19	20	21	22	23	24	25
1	4.8	1720	20	1	7	2	2	0	−0.48	−0.22	−0.80	−0.22	4	1	10	2.5	2.5	0	−0.33	−0.11	−1.40	−0.11	8	−0.87	0.12
2	5.5	1550	25	1	6	4	4	0	−0.55	−0.09	−0.37	−0.09	5	2	9	5	2	−0.29	−0.11	0.01	−0.51	0.11	8	−1.0	0.24
3	5.5	1075	20	3	3	3	3	0	0.45	0.45	0.45	0.45	1	1	25	25	3	−0.30	0.02	0.05	0.31	0.92	15	−0.95	0.64
4	5.8	1530	25	1.5	8	8	6.5	−0.03	−0.28	−0.32	−0.32	−0.25	4	1	13	7	5	−0.01	−1.16	−0.19	−1.16	−0.06	16	−0.52	0.25
5	5.2	1845	5	1.5	4.5	4.5	4.5	0	−0.57	−0.50	−0.50	−0.50	2	4	8	7	7	0	−0.52	−0.42	−0.66	−0.42	4	−0.20	0.09
6	6	730	0	1	6	6	6	0	−0.09	0.35	0.35	0.35	3	1	26	26	26	0	−0.03	0.41	0.41	0.41	24	−0.42	4.55
7	4.0	535	25	1	8	2	1	−0.11	0.56	0.36	−0.54	0.56	4	1	10	2	1	−0.08	0.66	0.44	−1.16	0.66	12	−0.50	1.0
8	4.6	630	0	1.5	8	8	1.5	−0.50	−0.68	−0.92	−0.92	−0.68	2	1	10	2	2	0	−0.28	−0.11	−1.40	−0.11	11	−0.30	0.06
9	4.5	625	20	1	6	6	1	−0.48	−0.48	−0.80	−0.80	−0.48	3	2	7	6.5	6.5	0	−0.35	−0.16	−0.19	−0.16	9	−0.12	0.02
10	4.9	720	20	1	5	1	1	0	0.05	0.05	−0.92	0.05	3	1	23	23	23	0	0.05	0.09	0.09	0.09	8	−0.87	0.10
11	2	460	5	1	5	1.5	1.5	0	−0.32	−0.26	−0.80	−0.26	4	1	5.5	3	3	0	−0.47	−0.39	−0.77	−0.39	9	−0.11	0.01
12	4.7	825	20	2	5	5	5	0	−1.16	−0.02	−0.02	−0.02	2	1	25	25	3	−0.46	−0.15	−0.49	−0.49	−0.03	16	−0.70	0.40
13	4.6	620	20	1	5	5	1	−0.42	−0.66	−0.92	−0.92	−0.66	2	1	24	20	3	−0.70	−0.92	−0.57	−0.74	−0.43	14	−0.63	0.06
14	4.8	140	20	1	3	3	1	−0.40	0.42	0.34	0.34	0.42	3	1	28	28	1	−1.22	0.75	0.50	0.50	0.75	12	−0.26	1.05
15	4	270	20	2	2	2	2	0	−0.32	−0.32	−0.32	−0.32	1	1	8	8	2	−0.24	0.04	−0.24	−0.24	0.12	6	−0.60	0.10
16	4.5	635	20	4	8	7	4	−0.07	−0.77	−0.96	−1.06	−0.77	3	1	15	7	7	0	−0.17	−0.11	−0.92	−0.11	13	−0.75	0.04
17	4.5	700	15	1.5	1.5	1.5	1.5	0	−0.17	−0.17	−0.17	−0.17	1	1	5	2	2	0	−0.22	−0.13	−0.85	−0.13	6	−0.40	0.30
18	4.8	390	10	1	4	4	4	0	−1.16	−0.80	−0.80	−0.80	2	1	13	11	1	−0.32	−0.15	−0.85	−1.22	−0.15	6	−0.35	0.25
19	4.8	680	20	1	4.5	4.5	4.5	0	−1.16	−0.85	−0.85	−0.85	3	1	21	21	4	−0.70	−0.28	−0.28	−0.28	−0.26	10	−0.28	0.10
20	5.8	840	20	1	10	6	6	0	−0.22	0.45	−0.60	0.45	5	1	22	22	5	−0.38	−0.06	0.17	0.17	0.44	20	−0.53	1.62
21	4.2	575	20	1	5	5	1	−0.19	−0.10	−0.60	−0.60	−0.10	2	1	22	21	1	−0.80	0.18	−0.49	−0.42	0.18	11	0.56	0.09
22	5.5	620	20	1	5.5	5	5	0	−0.85	−0.44	−0.55	−0.44	3	1	26	23	2	−0.77	−0.15	−0.32	−0.52	−0.04	14	−0.60	0.14
23	2	780	0	1	9	9	1	−0.62	0.47	0.13	0.13	0.47	3	1	17	8	2	−0.27	0.72	0.51	0.09	0.85	14	−0.80	0.30
24	5.5	670	20	6	7	6	6	0	−0.33	−0.33	−0.51	−0.33	2	1	27	27	4	−0.82	−0.35	−0.21	−0.21	−0.19	10	−0.25	0.22
25	4.9	865	20	1	5.5	5.5	5.5	0	−0.28	−0.23	−0.23	−0.23	5	1	8	4	4	0	−0.48	0.05	−0.89	0.05	9	−0.50	0.50
26	5	355	20	1	7	6	1	−0.28	0.35	−0.15	−0.48	0.35	4	1	16	16	1.5	−0.68	0.50	−0.02	−0.02	0.98	9	−0.22	−0.20

TABLE 4

Parameter	1	2	3	4	5	6	7	8	9	10	11	12	13	14	15	16	17	18	19	20	21	22	23	24	25
1	5	2110	30	2	8	8	8	0	−0.55	−0.50	−0.50	−0.50	4	2	10	6	2	−0.34	−0.54	−0.60	−1.16	−0.46	4	−0.49	0.09
2	5.4	1460	40	1.5	8	7.5	1.5	−0.19	0.24	−0.28	−0.80	+0.24	4	1	14	6	4	−0.16	−0.14	0.22	−1.0	0.24	20	−0.57	0.22
3	4.5	700	30	1	2	1.5	1.5	0	−0.35	−0.31	−1.00	−0.31	3	1	8	3	1	−0.11	−0.17	−0.41	−1.06	−0.76	8	−0.10	0.20
4	4.6	835	60	1	7	7	1.5	−0.16	−0.40	−0.48	−0.48	0.04	3	1	22	22	14	−0.82	0.004	−0.39	−0.39	0.20	13	−0.58	0.38
5	4.8	600	40	1	6	6	6	0	−0.44	0.15	0.15	0.15	4	1	20	20	4.5	0	−0.12	0.03	0.03	0.21	8	−0.52	0.67
6	5	520	50	1	7	2	2	0	−0.44	0.02	−0.70	0.02	3	1.5	6	2.5	2.5	0	−0.60	0.22	−0.51	0.22	6	−0.67	0.06
7	5.5	715	40	1	5	5	5	0	−0.12	0.14	0.14	0.14	3	1	20	20	20	−0.08	−0.12	0.69	0.69	0.69	8	−0.68	0.39
8	5	650	50	3	7	7	7	0	−0.92	−0.28	−0.28	−0.28	4	1	22	20	20	0	−0.82	−0.19	−0.19	−0.11	12	−0.40	0.10
9	5.6	840	40	1	11	9	9	0	0.12	0.36	−0.19	0.36	5	1	23	23	23	0	−0.18	0.63	0.63	0.63	16	−0.74	8.15
10	5.2	615	40	1	6	6	1	−0.61	0.35	0.19	0.19	0.35	3	1	18	18	2	−0.49	0.65	0.25	0.25	0.70	12	−0.54	0.24
11	4.6	640	40	1	5	5	1.5	−0.08	−0.66	−0.85	−0.85	−0.40	4	1	20	19	1	−1.06	−0.06	−0.26	−0.38	−0.05	14	−0.30	0.04
12	4.8	670	30	1	6	6	6	0	−1.16	−0.44	−0.44	−0.44	3	1	7	6	1	−0.34	0.16	−0.28	−0.44	0.16	8	−0.18	0.04
13	4.8	615	30	1	6	6	6	0	−0.35	−0.23	−0.23	−0.23	5	1	19	19	2	−0.61	−0.82	−0.12	−0.12	0.17	11	−0.22	−0.05
14	5	720	30	3	4	4	4	0	−0.70	−0.26	−0.26	−0.26	2	1	21	21	3	−0.77	−0.42	+0.04	0.04	0.12	11	−0.74	0.12
15	5	640	30	1	12	12	6	0	−1.10	−0.14	−0.64	−0.14	6	1	20	18	1	−0.85	0.42	0.02	−0.04	0.42	14	−0.52	0.04
16	4.8	595	70	1	6	6	6	0	−0.58	−0.05	−0.05	−0.05	4	1	20	19	4	−0.39	−0.34	−0.20	−0.33	0.08	16	−0.40	0.12
17	5	690	40	1	6	6	1	−0.68	−0.52	−0.62	−0.62	−0.52	3	1	20	20	1	−1.22	−0.05	−0.08	−0.08	−0.05	10	−0.78	0.07
18	5.2	930	30	1	6	5	5	0	−0.96	−0.28	−0.28	−0.03	3	1	22	22	4	−0.39	−0.28	−0.37	−0.37	−0.07	12	−0.48	0.33
19	6.2	855	40	1	10	8	8	0	0.22	0.95	0.28	0.95	5	2	24	22	22	0	0.60	1.39	1.14	1.39	19	−1.10	5.88
20	5	650	50	3	7	7	7	0	−0.92	−0.28	−0.28	−0.28	4	1	20	20	20	0	−0.82	0	0	0	9	−0.40	0.10
21	5.5	715	40	1	5	5	5	0	−0.12	0.14	0.14	0.14	3	1	20	20	20	0	−0.12	0.69	0.69	0.69	6	−0.68	0.39
22	5.8	650	30	1	12	12	5	−0.35	−0.12	0.62	0.62	0.65	5	1	18	18	18	0	−0.12	0.88	0.88	0.88	6	−0.62	0.70
23	4.5	520	65	1	3	3	3	0	−0.06	0.09	0.09	0.09	3	1	4	3	1	−0.20	0.27	−0.01	−0.26	0.27	4	−0.50	0.03
24	6.2	610	30	1	12	12	12	0	−0.22	0.94	0.94	0.94	5	1	19	19	19	0	0.88	1.42	1.42	1.42	10	−1.15	6.86
25	5.2	850	40	1	12	6	6	0	−0.82	0	−0.17	0	5	1	24	23	3	−0.60	−0.82	−0.11	−0.21	0.17	14	−0.62	0.85
26	5.0	865	40	1	5	5	5	0	−0.44	0.21	0.21	0.21	2	1	10.5	6	1	−0.49	0	−0.28	−0.85	0	8	−0.62	1.07
27	5.6	630	30	1	3	3	1	0	−0.06	−0.08	−0.08	−0.04	3	1	24	22	18	−0.04	0.38	0.34	0.34	0.38	15	−0.51	0.18
28	6.0	620	40	1	11	11	11	0	−0.82	0.57	0.57	0.57	6	1	21	19	18	−0.61	0.42	0.57	0.53	0.63	12	−0.60	0.49
29	5.8	630	35	1	6	5	5	0	−0.45	0	−0.33	0	4	1	26	26	26	0	0.18	0.72	0.90	0.90	16	−0.69	0.45
30	6.5	620	30	1	9	9	9	0	−0.12	0.52	0.52	0.52	3	1	24	24	24	−0.20	0.48	0.89	0.97	0.97	17	−0.81	0.89
31	5	525	50	1	7	6	1	−0.77	−0.06	−0.07	−0.25	−0.06	4	1	19	28	1	−1.10	0.42	0.26	0.23	0.42	8	−0.70	0.34
32	4.5	730	40	1	6.5	6	6	0	−0.82	−0.44	−0.50	−0.44	3	1	6	5	2	−0.29	−0.22	−0.24	−0.33	−0.13	5	−1.08	0.02
33	5.2	645	40	1	6	6	0.5	−0.27	−0.42	−0.15	−0.15	0.66	3	1	20	19	19	−1.39	0.35	0.43	0.28	0.43	7	−1.15	0.34
34	5.8	695	60	1	6.5	5.5	5.5	0	0.05	0.77	0.51	0.77	4	1	20	20	20	0	0.54	1.15	1.15	1.15	14	−0.20	1.75
35	5.9	720	40	2	6	6	6	0	−0.43	−0.22	−0.02	−0.02	3	1	24	24	24	0	−0.22	0.28	0.28	0.28	15	−0.38	0.17
36	5.5	790	40	1	5	5	4.5	−0.02	−0.42	0.16	0.16	0.23	4	1	25	25	25	0	−0.18	0.46	0.46	0.48	14	−0.54	1.32
37	6	610	60	1	20	18	18	0	0.35	0.89	0.42	0.89	5	1	22	20	22	0	−0.12	1.22	1.22	1.22	10	−0.75	0.07
38	5.4	655	30	1	7.5	7.5	5	−0.005	−0.82	−0.18	−0.18	−0.01	5	1	24	21	2	−0.02	−0.72	0.07	−0.17	0.17	12	−0.73	0.20
39	6.1	595	30	1	7	7	7	0	−0.44	0.37	0.37	0.37	3	1	22	19	19	0	0.16	0.91	0.52	0.91	21	−0.91	1.26
40	5.8	840	60	1	8	7.5	6	−0.01	−0.42	0.41	−0.46	0.47	6	1	24	24	24	−0.35	−0.24	−0.44	0.44	0.44	16	−0.74	1.0
41	6	840	30	1	11	6.5	6.5	0	−0.42	0.77	0.64	0.77	5	1	23	23	21	−0.03	0.48	0.79	0.79	0.80	18	−0.95	3.31
42	6.2	560	40	1	5	5	1	−0.55	0.48	0.30	0.30	0.48	2	1	23	22	21	0	0.64	0.99	0.72	0.99	7	−0.70	1.55
43	5.8	1120	60	1.5	6	5	2	−0.38	0.07	0.21	−0.44	0.24	4	1	20	6	2	−0.07	0.15	0.59	−0.13	1.01	12	−1.16	0.91

Parameter	26	27	28	29	30	31	32	33	34	35	36	37	38	39	40	41	42	43	44	45	46	47	48	49
1	0.18	1.5	8	8	2	−0.33	−1.06	−0.77	−0.77	−0.49	3	0.777	0.67	0.88	0.25	1	−0.56	−0.04	−0.68	−0.80	−1.11	−0.16	0.24	0.07
2	2.00	4	9	9	5	−0.13	−0.28	0	0	0.12	4	0.777	0.25	0.67	0.44	0.80	−1.20	−0.11	−0.85	−0.25	−0.65	0	0.19	0.44
3	1.14	3.5	25	25	25	0	0.33	0.74	0.74	0.74	7	−1.58	0.86	0.12	0.12	0.12	−0.09	−0.09	−1.71	0	−1.26	0	0.78	1.20
4	0.24	2.0	9	9	9	0	−1.00	−0.15	−0.15	−0.15	2	2.13	0.75	0.89	0.89	0.72	−1.28	−1.80	−1.65	−0.70	−1.50	0	−0.07	0.22
5	0.19	1	10	10	10	0	−0.58	−0.58	−0.58	−0.58	2	0.777	1.5	0.45	0.45	0.45	−0.54	0	−0.37	−0.18	−1.11	0	0.27	0.27
6	0.78	2	22	22	22	0	−0.25	0.06	0.06	0.42	5	−0.22	0.50	0.27	0.27	0.27	−1.22	−0.35	−1.87	−0.52	−2.29	0	0.70	0.54
7	0.16	1	8	7	1	−0.02	0.23	−0.60	−0.74	0.23	3	0.06	1.0	1.0	0.30	1	−0.10	−0.87	−0.08	−0.88	−0.02	−0.10	0.09	−0.41
8	0.14	1	8	3	2	−0.04	−0.46	−0.24	−0.82	−0.11	4	0.15	1.5	1.0	2.67	0.75	−0.48	0	−0.48	−0.57	−0.75	−0.15	0.22	0.28
9	0.10	3	3	3	3	0	−0.18	−0.18	−0.18	−0.18	1	0.20	0.33	2.0	2.0	0.33	−0.48	0	−0.70	−0.24	0	0	0.70	0.32
10	0.05	1	20	20	3	−0.80	−0.06	−0.04	−0.04	0	4	0.13	1.0	0.25	0.05	0.33	−0.44	−0.26	−1.50	0	−1.66	0	1.41	1.22
11	0.10	1	12	2	1.5	0	−0.16	−0.16	−1.10	−0.15	6	0.80	1.0	0.42	0.75	0.75	−0.43	−0.06	−1.08	−0.12	−0.51	−0.11	0.18	0.21
12	0.46	1	17	17	17	0	−1.16	−0.31	−0.31	−0.31	4	0.42	2.0	0.29	0.29	0.29	−1.67	0	−1.61	0	−1.92	0	0.24	0.24
13	0.09	3	4	3	3	0	−0.38	−0.38	−0.80	−0.38	2	0.33	0.33	1.25	1.67	0.33	−0.40	0	−1.73	−0.11	0	−0.30	1.03	0.38
14	0.92	1	5	5	5	0	0.30	0.61	0.61	0.61	2	19.6	1.0	0.6	0.60	0.20	−0.51	0	−1.20	0	−1.03	0	1.13	0.50
15	0.02	1	1.5	1.5	1.5	0	0.02	0.33	0.33	0.33	2	2.67	2.0	1.33	1.33	1.33	0	0	−0.65	0	0.50	0	0.70	0.55
16	0.071	1	6.5	6.5	1	−0.52	0.03	−0.25	−0.25	0.03	4	0.777	4.0	1.23	1.08	4.0	−0.07	0	−0.95	−0.49	−0.56	0	0.88	0.72
17	0.09	1	3	3	3	0	−0.82	−0.47	−0.47	−0.47	3	−0.33	1.5	0.50	0.50	0.50	−0.16	−0.16	−0.84	−0.37	−1.00	0	0.15	0
18	0.45	1	17	17	4	−0.19	−0.28	−0.52	−0.52	−0.09	10	0.25	1	0.24	0.24	1.0	−0.78	0	−0.30	−0.30	−1.0	0	0.37	0.92
19	0.06	1	20	20	20	0	−0.42	−0.13	−0.13	−0.13	7	0.25	1	0.22	0.22	0.22	−0.84	0	−1.73	0	−1.69	0	1.18	1.32
20	0.22	1	12	12	2	−0.24	−0.52	−0.37	−0.37	0.17	4	1.95	1	0.83	0.50	3.0	−1.35	−0.75	−2.22	0	−1.24	0	0.38	−0.42
21	0.02	1	19	19	1.5	−0.14	0.13	−0.57	−0.57	0.39	6	2.0	1	0.26	0.26	0.67	−0.18	0	−0.79	−0.12	−0.58	0	0.88	0.62
22	0	1	9	3	3	0	−1.16	−0.01	−0.92	−0.01	4	2.2	1	0.61	1.67	1.67	−1.26	−0.08	−1.20	−0.16	−1.46	−0.42	0.79	0.21
23	1.35	2	12	12	2	−0.11	0.99	0.32	0.32	0.99	2	9.16	0.50	0.75	0.75	0.50	−0.62	0	−1.27	−0.10	−0.50	0	0.33	0.32
24	0.09	1	6.5	6	6	0	−0.06	0.15	−0.31	0.15	3	2.5	6.0	1.08	1.0	1.0	0	−0.12	−2.22	0	−0.96	−0.43	0.78	0.47
25	0.10	2.5	20	20	20	0	−0.52	−0.50	−0.50	−0.50	4	0.777	0.4	0.28	0.28	0.28	−1.50	0	−1.18	−0.65	−0.95	0	0.15	0.29
26	0.34	1	14	14	1.5	−0.37	0.42	0.56	0.56	1.16	6	−1.0	1	0.50	0.43	0.67	−0.58	−0.26	−0.90	0	−1.29	0	0.56	1.08

Parameter	26	27	28	29	30	31	32	33	34	35	36	37	38	39	40	41	42	43	44	45	46	47	48	49
1	0.08	4	10	6	6	0	−0.85	−0.12	−0.72	−0.12	5.0	0.37	0.50	0.80	1.33	1.33	−0.92	0.00	−0.48	−0.40	−0.95	−0.40	−0.22	0.26
2	0.22	1	10	7	7	0	−1.16	−0.21	−0.40	−0.21	4.0	1.03	1.50	0.80	1.07	0.21	−0.75	−0.51	−1.30	−1.04	−1.63	−0.04	0.40	0.04
3	0.05	1	4	3.5	3.5	0	−0.66	−0.51	−0.96	−0.51	4.0	−0.13	1.0	0.5	0.43	0.43	−0.24	−0.54	−0.74	−0.20	−0.74	−0.44	0.20	0.20
4	0.19	1	6	6	1	−0.31	−0.09	−0.30	−0.30	−0.09	4.0	0.53	1.0	1.17	1.16	1.5	−0.76	0.00	−0.05	0.00	−0.63	0.00	0.59	0.12
5	0.36	1	20	20	3	−0.28	−0.12	0.09	0.09	0.64	4.0	−1.5	1.0	0.3	0.3	2.0	−1.45	0.00	−1.49	0.00	−1.49	0.00	0.40	0.40
6	0.30	1	10	10	2	−0.11	−0.28	−0.33	−0.33	0.22	4.0	−0.9	1.0	0.7	0.2	1.0	−0.82	−0.16	−1.15	−0.37	−0.97	0.00	0.32	0.37
7	0.27	1	26	26	26	0	−0.12	1.03	1.03	1.03	4.0	0.73	1.0	0.19	0.19	0.19	−0.99	−0.75	−2.15	0.00	−2.60	0.00	1.16	1.61
8	0.09	3	19	19	19	0	−0.27	0.15	0.15	0.15	5.0	0.777	1.0	0.37	0.37	0.37	−1.08	0.00	−2.19	−0.03	−1.22	0.00	0.63	0.87
9	−0.04	1.5	20	20	20	0	−0.66	1.16	1.16	1.16	4.0	1.36	0.67	0.55	0.45	0.45	−1.84	−0.47	−0.229	0.00	−2.98	0.00	−1.11	1.15
10	0.18	1	9	9	1	−0.27	0.66	−0.02	−0.02	0.66	4.0	−0.06	1.0	0.67	0.67	1.0	−1.42	0.00	−1.07	0.00	−0.54	0.00	1.04	0.97
11	0.11	1	20	20	3	−0.28	−0.28	−0.47	−0.47	0.08	5.0	0	1.0	0.25	0.25	0.50	−0.52	0.00	−0.73	−0.39	−0.84	−0.35	0.43	0.69
12	0.05	1	20	20	2	−0.68	−0.42	−0.51	−0.51	−0.20	6.0	0.777	1.0	0.30	0.30	3.0	−1.48	−0.18	−0.35	−0.18	−0.72	−0.11	0.18	−0.62
13	−0.05	1	9	6.5	1	−0.47	0.42	0.08	−0.32	0.42	4.0	2.0	1.0	0.67	0.92	6.0	−0.92	0.00	−0.132	−0.240	−1.20	−0.28	−0.08	0.34
14	0.25	1	20	20	20	0	−0.42	−0.04	−0.04	−0.04	5.0	0.25	3.0	0.20	0.20	0.20	−0.57	−0.14	−1.21	−0.41	−1.08	0.00	0.33	0.30
15	−0.06	1	7	5	4	−0.16	−0.28	0.23	−0.28	0.26	5.0	1.20	1.0	1.71	2.40	1.50	−1.49	−0.19	−0.34	−0.39	−1.22	−0.33	0.03	0.28
16	0.08	1	10	7	3	0	−0.89	−0.05	−0.41	0.26	5.0	0	1.0	0.6	0.86	2.0	−1.63	−0.45	−1.57	−0.57	−1.70	−0.22	−0.05	0.07
17	0.05	1	18	18	3	−0.57	−0.35	−0.12	−0.12	0.09	4.0	0.13	1.0	0.34	0.33	0.33	−0.70	0.00	−0.63	−0.40	−1.48	0.00	0.18	0.48
18	0.35	2	6	6	2	−0.43	−0.58	−0.62	−0.62	−0.58	3.0	0.777	0.5	1.0	0.83	2.5	−1.81	−0.15	−1.02	−0.36	−0.46	0.00	−0.09	−0.51
19	1.06	4	14	14	14	0	0.59	0.85	0.85	0.85	3.0	3.40	0.25	1.25	0.57	0.57	−1.02	−0.59	−1.09	−0.50	−0.58	0.00	0.10	−0.09
20	0.09	3	19	19	19	0	−0.27	0.15	−0.27	0.15	5.0	0.777	1.0	0.37	0.37	0.37	−0.97	0.00	−1.51	−0.53	−0.66	0.00	0.14	0.33
21	0.27	1	26	26	26	0	−0.12	1.03	1.03	1.03	4.0	0.69	1.0	0.19	0.19	0.19	−0.98	−0.75	−1.50	0.00	−1.62	0.00	0.52	0.64
22	0.48	1	22	22	22	0	−0.12	0.97	0.97	0.97	4.0	0.64	1.0	0.54	0.54	0.23	−1.49	−0.07	−1.78	0.00	−1.85	0.00	0.29	0.21
23	0.16	1	10	25	1	−0.27	0.27	0.14	−0.54	0.27	4.0	0.70	1.0	0.30	1.2	3.00	−0.80	0.00	−0.82	−0.43	−0.27	−0.07	−0.10	−0.03
24	2.94	1	8	8	8	0	0.72	0.80	0.80	0.80	2.0	0.777	1.0	1.50	1.50	1.50	−1.90	0.00	−2.13	−0.22	−0.01	0.00	0.38	0.11
25	0.14	2	19	19	19	0	−0.43	0.05	0.05	0.05	6.0	0.67	0.5	0.63	0.32	0.32	−1.63	−0.63	−1.48	−0.33	−0.90	0.00	−0.15	−0.03
26	1.01	2	18	18	18	0	−0.48	0.07	0.07	0.07	7.0	0.75	0.5	0.28	0.28	0.28	−1.40	−0.29	−0.78	−0.38	−1.0	0.00	−0.42	−0.10
27	0.24	2	20	20	20	0	0.11	0.95	0.95	0.95	5.0	2.00	0.5	0.15	0.15	0.05	−0.63	−0.23	−1.03	−0.41	−1.22	0.00	0.62	1.23
28	1.56	2	14	14	2	−0.77	−0.27	0.18	0.18	0.27	5.0	0.22	0.5	0.78	0.78	5.50	−2.11	0.00	−1.58	−0.25	−0.57	0.00	1.00	−0.14
29	0.69	1	22	22	22	0	0.27	0.66	0.66	0.66	6.0	1.33	1.0	0.27	0.23	0.23	−1.54	−0.24	−1.30	−0.12	−1.39	0.00	0.46	0.55
30	2.80	1	14	3	3	0	−0.82	0.73	−0.06	0.73	5.0	2.87	1.0	0.64	3.00	3.00	−1.75	0.00	−2.27	−0.12	−2.04	−0.30	0.29	−0.11
31	0.06	1	5	4	4	0	0.18	0.52	0.29	0.52	4.0	2.00	1.0	1.4	1.50	0.25	−1.56	0.00	−0.86	−0.42	−0.95	−0.02	0.30	0.40
32	0	1	8	8	1	−0.74	−0.42	−0.58	−0.58	−0.42	5.0	0.05	1.0	0.81	0.27	6.0	−1.15	−0.02	−0.84	−0.75	−0.84	0.00	0.12	−0.01
33	0.21	1	20	20	17.5	−0.22	−0.12	0.49	0.51	0.51	4.0	0.777	1.0	0.30	0.30	0.03	−0.28	−0.13	−1.86	−0.31	−2.01	0.00	1.08	1.17
34	0.18	1	17	17	17	0	0.57	0.79	0.79	0.79	4.0	2.50	1.0	0.38	0.32	0.32	−1.47	−0.19	−1.83	−0.30	−2.24	0.00	0.94	0.51
35	0.14	2	18	18	16	−0.01	−0.52	0	0.00	0.05	6.0	0.31	1.0	0.33	0.33	0.38	−0.90	−0.14	−1.88	−0.22	−1.78	0.00	0.91	0.51
36	0.15	2	24	20	20	−0.22	−0.13	0.54	0.54	0.65	4.0	0.31	0.5	0.21	0.21	0.23	−1.42	0.00	−1.68	0.00	−1.78	0.00	0.95	1.02
37	0.51	1	20	20	20	0	0.18	1.03	1.03	1.03	6.0	0.67	1.0	1.0	0.90	1.00	−2.10	−0.42	−2.41	0.00	−2.16	0.00	0.42	0.19
38	0.09	1	15	15	2	−0.15	−0.22	−0.26	0.26	0.47	5.0	0.46	1.0	0.50	0.50	0.25	−1.69	0.00	−2.39	−0.19	−1.61	0.00	0.700	0.22
39	2.58	3	20	20	20	0	0.79	1.24	1.24	1.24	4.0	12.25	0.33	0.35	0.35	0.35	−1.73	−0.07	−2.04	−0.32	−1.28	0.00	0.98	1.33
40	0.31	2	23	18	18	0	−0.13	0.16	0.16	0.44	6.0	2.50	0.5	0.35	0.42	0.33	−1.78	−0.82	−2.12	0.00	−1.55	−0.18	0.56	0.44
41	0.50	2	18	18	18	0	−0.43	1.40	1.40	1.40	3.0	5.50	0.5	0.61	0.36	0.36	−2.10	−0.35	−1.89	0.00	−2.81	0.00	0.47	1.07
42	2.26	1	8	8	8	0	0.57	0.76	0.76	0.76	2.0	0.777	1.0	0.63	0.62	0.12	−1.00	0.00	−1.69	−0.25	−1.09	0.00	1.32	0.66
43	0.31	1	18	18	2	−0.69	−0.16	−0.08	−0.08	0.20	4.0	2.73	1.5	0.33	0.28	1.0	−1.31	−0.59	−1.42	−0.20	−1.33	0.00	0.46	0.27

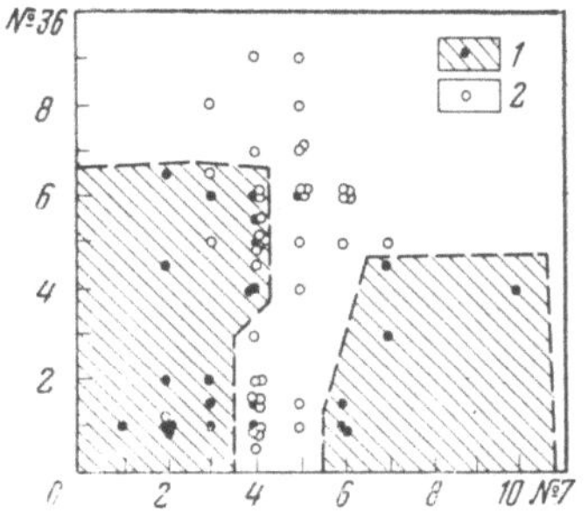

Fig. 1

sary, a smaller number of parameters can be used. For example, we can use only the triplet (11, 36, 42) or one of the better pairs: (9, 42), (7, 36), (21, 45). These combinations of parameters for the instruction material are shown in Figs. 1-3.

The following conclusions can be drawn from an analysis of the recognition results.

1. The parameters associated with the P wave were found to be the most useful for recognition. Then followed the parameters associated with the interval P → S. This was somewhat unexpected, since it seemed *a priori* that the latter must be considerably better than the former. The P parameters appeared seven times, the P → S parameters five times, and the S parameters once in the four triplets.

2. The main difference in the wave pictures given by the earthquakes of the two patterns was that the contribution of the long-period part of the spectrum was increased for the deep earthquakes. In other words, on the earthquake records of the second pattern, over the entire region of P and P → S, the amplitudes of the oscillations that had short periods were weaker than on the records of the first pattern. The oscillations with the higher T_{max} values became noticeable more often.

3. A second important difference is that expansion of the spectrum toward the long-period side in the second pattern is accompanied by an increase in the number of peaks (local maxima) in the spectrum. This is reflected in an increase in parameters 13 and 36.

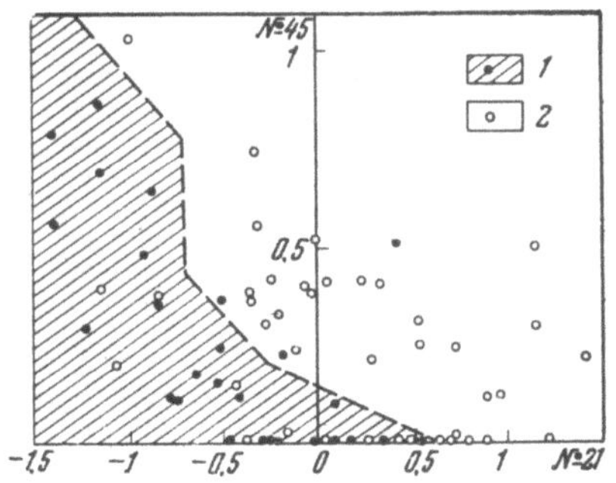

Fig. 2

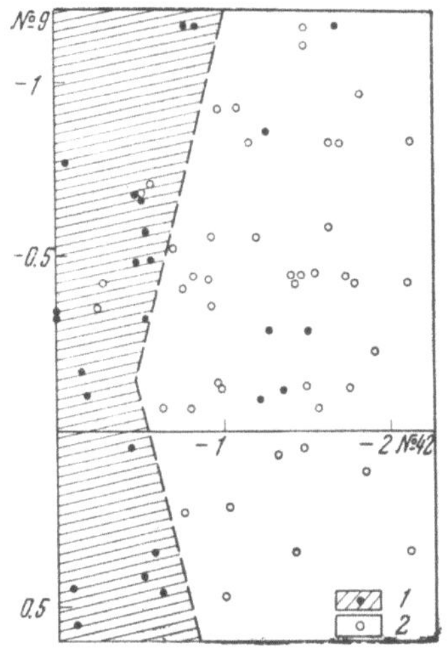

Fig. 3

4. The differences associated with criteria 12 and 22 are most likely due to the fact that in the second interval the earthquakes have greater magnitudes. This can exert an indirect influence on criteria 17, 11, and 16, since it is known that an increase in magnitude increases the absolute values of all these criteria, other conditions being equal.

These differences in wave patterns, which are primarily related to focal depth, can be qualitatively explained as follows.

As is well known, the strongest secondary waves are the waves reflected near the epicenter. Evidently along with the first-arrival longitudinal waves, they play a decisive role in the formation of the wave pattern on the part of the record we considered. When an earthquake occurs in the crust, the time delays τ_j of such waves are rather small (for h ≈ 15 km, $\tau(_pP) \approx 3.1$ sec and $\tau(_sP) \approx 5.4$ sec). In addition there are a number of other secondary waves that exist only when the focus is in a layer; these have even smaller time delays. In this case, in the region of P and P → S, the short-period part of the spectrum must be more noticeable (T < 5-6 sec).

The following changes occur upon entering the second depth interval: the time delays of the waves reflected near the epicenter increase sharply, and the waves that have short delays relative to P disappear.

Thus, secondary arrivals are removed from the P region of the record, which results in a relative increase in the long-period part of the spectrum in P and P → S. Similar reasoning about the secondary waves that follow S explains the behavior of the corresponding criteria.

The absolute value of each term under the sum in (6)

$$\Phi(\omega) = f(\omega)\exp\{i\varphi(\omega)\}\sum_{j=0}^{n} a_j \exp[-i\omega\tau_j]. \quad (6)$$

reaches its maximum when $\omega\tau_j = 2\pi n$ ($n = 0, 1, 2, \ldots$). If the arrivals of secondary waves are sufficiently rare (which happens in the second depth interval), the positions of the principal maxima of the absolute value of the sum

$$\sum a_j l^{-i\omega\tau_j}$$

are approximately determined by the same conditions ($\omega\tau = 2\pi n$). Hence the cutoff of the spectrum in the parts of the record in question.

In our case, the joint use of the better triplets (i.e., using the product of the ratios of the corresponding individual triplets as the likelihood ratio) caused almost no change in the γ values. For example, the pair of triplets (42, 6, 11) and (45, 17, 15) has $\gamma = 0.15$. The set of three triplets (42, 36, 11), (45, 17, 15), and (12, 10, 6) has $\gamma = 0.15$. This slight decrease in γ has two causes: first, we have already indicated that the product of the likelihood ratios corresponding to individual triplets would give the likelihood ratio of the entire set if the parameters were statistically independent. Of the four triplets given above, the parameters of (42, 36, 11) and (48, 22, 9) and (42, 36, 11) and (12, 10, 6) are most strongly related. In the first pair, the maximum sample correlation coefficient for parameter pairs from different triplets is 0.697; in the second, it is 0.826. Secondly, our instruction material may contain earthquakes with incorrectly determined depths or earthquakes that are unlike the typical earthquake of the given depth interval. It is obviously either impossible or very difficult to improve the recognition of such earthquakes. Earthquake No. 6 and possibly Nos. 20 and 26 of the first pattern and earthquake No. 3 and possibly Nos. 1, 2, and 10 of the second pattern may be such untypical earthquakes. For Nos. 6 and 3, all four of the better triplets give an incorrect answer; for the other earthquakes, three of the four triplets give an incorrect answer. Note that Nos. 6 and 20 differ from the other earthquakes of the first pattern, since they have the highest magnitudes in the first pattern. Earthquake No. 26 has a rather small epicentral distance, and No. 3 has the lowest magnitude in the second pattern.

Thus, in our case, the set of four triplets gave approximatelv the same γ as each of the triplets individually. It may be hoped, however, that with a great deal of additional material the set of triplets would give an appreciably lower γ than would each of the triplets individually.

Now let us discuss the selection of the potential function and smoothing constants σ_i in formula (4). We tested both functions (5) as potentials and found that they both gave almost the same γ for all of the cases we considered. We took the second function in (5) as the potential, since it is easier to calculate. Our tests showed that the constants σ_i in formula (4) must be equal to the sample standard deviations of the corresponding parameters, which can be obtained from the correlation matrix (if these standard deviations differ considerably for different patterns, the smaller of them should be selected). The smoothing constants were varied for some individual parameters, pairs, and triplets. It was found that as a rule γ varied within ± 0.05, and an increase or decrease in the deviation of the smoothing constants from the standard deviation either did not considerably change γ or else increased it. Of the 38 combinations of parameters checked, there was a slight decrease in γ, compared with the standard deviation, in only five cases when the smoothing constants were increased and in three cases when they decreased. Thus we can assume that γ is negligibly dependent on the smoothing constants, and the sample standard deviations of the parameters can serve as these constants.

We used the discriminant function – a well-known algorithm in statistics – for comparison with our recognition algorithm (see [12]). This method makes the rather strong assumptions that the parameters are normally distributed, the correlation matrices of both patterns are the same, and that the distinction consists in the difference between the mean values. Under these assumptions, we seek a linear combination of parameters – called the discriminant function – that gives the minimum probability of an incorrect decision. If we let R be the common correlation matrix for both patterns and let $\bar{m}_1$ and $\bar{m}_2$ be the vector columns of the mean values of the parameters of the first and second patterns, respectively, the discriminant function for a vector $\bar{z}$ being checked is written as follows:

$$D(\mathbf{z}) = \mathbf{z}^* R^{-1}(\mathbf{m}_1 - \mathbf{m}_2) - \frac{1}{2}(\mathbf{m}_1 + \mathbf{m}_2)^* R^{-1}(\mathbf{m}_1 - \mathbf{m}_2).$$

Here, R^{-1} is the inverse of matrix R and $\bar{z}^*$ is the transpose of vector $\bar{z}$ (i.e., the row vector). When the above assumptions are valid, the value $D(\bar{z})$ has a normal distribution with variance

$$\sigma^2 = (\mathbf{m}_1 - \mathbf{m}_2)^* R^{-1} (\mathbf{m}_1 - \mathbf{m}_2).$$

The mean value of $D(\bar{z})$ for a vector $\bar{z}$ of the first pattern is $\sigma^2/2$; for the second pattern, $-\sigma^2/2$. To make a decision, we must assume that $\bar{z}$ is from the first pattern when $D(\bar{z}) > 0$, otherwise from the second pattern.

Since the true correlation matrices and the mean values of the parameters are unknown in real practical problems, their sample analogs must be used instead. In our problem, we obtained the correlation matrix R by taking the weighted sum of the sample correlation matrices of the two patterns. The weights were made approximately proportional to the numbers of the vectors in each pattern (0.4 for the first pattern and 0.6 for the second). For more stable inversion of matrix R, its diagonal elements were increased by a factor of 1.1-1.2 (see [14]).

First, the discriminant function was calculated for 41 parameters (five parameters – Nos. 4, 14, 27, 40, and 47 – were discarded). A check on the same instruction material from which the coefficients of the discriminant function were calculated gave $\gamma = 0.13$. Then from these 41 parameters we selected the 26 best, i.e., the parameters that had the greatest difference between the mean values for the different patterns. These parameters were Nos. 6, 7, 9-12, 15, 17-21, 24-26, 31, 33, 36, 39, 42-46, 48, and 49. For this set of parameters we obtained $\gamma = 0.15$. Then the number of parameters was reduced to 14: Nos. 6, 7, 9, 11, 12, 15, 17, 20, 21, 33, 36, 42, 45, and 48. This gave $\gamma = 0.22$. Then the three parameters 20, 21, and 42 were tested, which gave $\gamma = 0.25$. Note that the best single parameter for discriminant recognition is No. 42, which gives $\gamma = 0.27$.

It should be pointed out that since γ was calculated from the same material with which instruction was performed, there is a danger that γ will be somewhat increased with new control material.

Comparison of the results of recognition by our algorithm with the results of the discriminant function indicates that when a great number of parameters are used, the discriminant function gives a somewhat lower γ than our algorithm (0.13 as opposed to 0.15). With a large number of parameters, however, it is difficult to understand through which parameters and by what means the discriminant function distinguishes the patterns. However, for a small number of parameters, the discriminant function gives appreciably worse results. For example, our algorithm is clearly preferable for a pair or a triplet of parameters. It should be stated that the γ values cited by us are only estimates of the true probability of an incorrect decision, and that these were obtained with the same material with which instruction was performed. A definitive comparison of the effectiveness of these two methods of pattern recognition could only be made with a great deal of additional control material. Unfortunately we had only a very small amount of control material at our disposal: 13 vectors for the first pattern and 12 vectors for the second. The check results are as follows: with the set of all four triplets, four earthquakes out of 24 were incorrectly classified (one from the first pattern and three from the second). This gives an estimate for γ_0 – the probability of an incorrect decision – of $4/24 \approx 0.17$, which is close to the value obtained with our instruction material: $0.15 + 0.06 = 0.21$ (the 0.06 is added to allow for selecting the best combinations from among others during instruction). The 90% confidence interval for γ_0 (see [9], p. 46) is (0.09-0.28). The best triplet, (11, 36, 42), gave five incorrect decisions (21%). Note that the discriminant function gave six incorrect decisions for 14 parameters.

5. Conclusion

The results obtained here show that the recognition of earthquakes whose foci are in the crust and immediately below the crust from the records of one seismic station (even under the requirement of promptness) is a quite practicable problem. This recognition can be performed with simple algorithms, using a small number of parameters (two, three, or more), and the probability of an incorrect answer is 20-25%. It is evidently not hopeless to attempt a quantitative estimate of focal depth from the dynamic parameters of one record (or, better yet, a set of records): see [13]. In this case the focal depth could be sought as a function (linear in the simplest case) of those parameters of the record that were found to be useful in our recognition problem.

The algorithm proposed here does the following:

1. Discards parameters that give little information for recognition, thereby reducing the material for further analysis;

2. Isolates the parameters and combinations of parameters in pairs and triplets that are useful in our recognition problem;

3. Gives for the isolated combinations of parameters the frequency γ of incorrect decisions, which with allowance for the sorting corrections

provides an estimate of the probability of an incorrect decision. An alternative method of estimating this probability is to check the algorithm on additional control material.

4. Provides information on which nonlinear functions of pairs or triplets of parameters it would be useful to form for pattern recognition and for the quantitative estimation of focal depth. This information can be obtained by examining two-dimensional graphs (for pairs) or series of two-dimensional graphs (for triplets, one of the parameters on each graph is fixed) on which the best pairs (or triplets) have been plotted.

LITERATURE CITED

1. Pod"yapol'skii, G. S. (1968), "Excitation of a long gravitational wave in the ocean by a seismic source," Izv. Akad. Nauk SSSR, Ser. Fizika Zemli, No. 1, pp. 7-24.
2. Tarakanov, R. Z., and N. V. Levin (1966), "Multi-asthenosphere model of the earth's upper mantle from seismological data," paper presented at the Session of the Scientific Council at Yuzhno-Sakhalinsk, 18-22 September 1965.
3. Bongard, M. M. (1967), Problems of Recognition, Moscow, Nauka.
4. Lunts, A. L., and V. L. Brailovskii (1967), "Evaluating criteria obtained in statistical decision rules," Izv. Akad. Nauk SSSR, Tekhn. Kibern., No. 3, pp. 99-110.
5. Cramér, H. (1951), Mathematical Methods of Statistics, Princeton, Princeton University Press.
6. Parzen, E. (1962), "On estimation of a probability density function and mode," Ann. Math. Stat., 33:1065-1078.
7. Aizerman, M. A., E. M. Braverman, and L. I. Rozonoer (1964), "Theoretical fundamentals of the method of potential functions in the theory of computer learning," Avtomatika i Telemekhanika, vol. 25, p. 917.
8. Braverman, E. M. (1966), "The method of potential functions in the theory of computer learning," Avtomatika i Telemekhanika, 27:100.
9. van der Waerden, B. L. (1957), Mathematische Statistik, Berlin, Springer Verlag.
10. Humbel, E.(1962), Statistics of Extremes, New York, Columbia University Press.
11. Pisarenko, V. F., and T. G. Rautian (1966), "Statistical classification by several criteria," Computational Seismology, No. 2, Moscow, Nauka, pp. 150-182.
12. Anderson, T. (1959), An Introduction to Multivariate Statistical Analysis, New York, John Wiley and Sons.
13. Nersesov, I. L., V. F. Pisarenko, T. G. Rautian, N. A. Smirnova, and R. I. Khalturin (1966), "Methods of recognition theory for statistical classification of shallow and deep earthquakes," in: Seismicity and Internal Structure of the Earth in the Far East, and Tsunami Problems, Yuzhno-Sakhalinsk.
14. Pisarenko, V. F. (1966), "Linear problems for inexactly specified random processes," Problemy Peredachi Informatsii, 2:56-57.

LOCAL STATISTICS OF EARTHQUAKE CATALOGS†

V. I. Keilis-Borok, V. M. Podgaetskaya,
and A. G. Prozorov

In analyzing earthquake catalogs one encounters the problem of distinguishing earthquakes whose occurrence is related to previous earthquakes. This paper investigates the first part of the problem: proof of the existence of interrelated earthquakes (not simply in the trivial case of aftershocks). The study boils down to a comparison of the statistical distributions of certain statistics for an actual earthquake catalog and for a randomized catalog of earthquakes which occur randomly and independently of one another. The catalog of random earthquakes was obtained by random shuffling of an actual catalog.

The method was worked out using catalogs of earthquakes of Central Asia [1].

1. Main Definitions

1.1. We introduce the following symbols:

- t – the origin time of the earthquake;
- λ, φ, h – the longitude, latitude, and depth of the hypocenter; we shall frequently combine these into a vector $\mathbf{x}$;
- z – the vector of the hypocenter coordinates;
- M – the earthquake magnitude;
- g – the measure of earthquake energy or magnitude;
- i – a subscript indicating the earthquake index in order of increasing time t;
- N – the number of earthquakes in the catalog;
- T – the time period the catalog covers ($T \geq t_N - t_1$).

The earthquake catalog will be represented as a flow or sequence of events $(t, \mathbf{x}, g)$ that are naturally ordered in time.

The statistics that we shall consider are the following:

- τ – the time interval between two successive earthquakes;
- d – the distance between their hypocenters;
- r – the distance between their epicenters;
- $\hat{\tau}(d)$ – the time interval from some *initial* earthquake to the first of subsequent earthquakes that occur at a distance of not more than d from the initial one;
- $\hat{d}(\tau)$ – the distance between the hypocenters of some *initial* earthquake and the first of subsequent earthquakes that occur in the time interval τ after the initial one, if any such occur;
- $\hat{r}(\tau)$ – the distance between the epicenters, as above, considered as a function of τ.

1.2. Random catalog. We shall define a *flow of independent events* a sequence of events $(t, \mathbf{x}, g)$ in which

a) the events that occur in different regions of the space $\{t, \mathbf{x}, g\}$ are statistically independent;

b) the probability of occurrence of one event in a small volume $\Delta V = \Delta t \times \Delta \mathbf{x} \times \Delta g$ is $\Lambda(t, \mathbf{x}, g) \Delta t \Delta \mathbf{x} \Delta g$; and

c) the probability of occurrence of two or more events is $0(\Delta V)$.

The parameter $\Lambda(t, \mathbf{x}, g)$ is the *intensity of the flow*.

† Translated from Vychislitel'naya Seismologiya, Vol. 5, Moscow, Nauka (1971), pp. 55-79.

We assume that the actual catalog could, in a first approximation, be considered random, i.e., as a flow of independent events that is completely defined by its intensity $\Lambda(t, \mathbf{x}, g)$. In order to find out in what way the actual catalog differs from the random one, we must construct such a random catalog with the same intensity as that of the actual catalog.

We construct the random catalog by randomizing the actual catalog.

We assume

$$\Lambda(t, \mathbf{x}, g) = f(\mathbf{x}, g)\,\lambda(t), \tag{1}$$

$$\sum_g \int_G f(\mathbf{x}, g)\,d\mathbf{x} = 1, \tag{2}$$

where G is the seismic region under study.

Let us consider the case of specific g values. Then we take as the estimate $f(\mathbf{x}|g)$

$$f^*(\mathbf{x}|g) = \frac{n(\mathbf{x}|g)}{N}, \tag{3}$$

where $n(\mathbf{x}|g)$ is the number of earthquakes in the actual catalog that have coordinates $\mathbf{x}$ and magnitude g. Two different estimation methods of estimation of $\lambda(t)$ will be used:

$$\lambda^*(t) = N/T = \text{const} \tag{4a}$$

or

$$\int_{t_1}^{t_2} \lambda^*(\tau)\,d\tau = n(t_1, t_2), \tag{4b}$$

where $n(t_1, t_2)$ is the number of earthquakes in the actual catalog for the time interval from t_1 to t_2.

Let us explain the physical meaning of (1)-(4); the meaning of (3) and (4) will also be clear from the algorithm given below for constructing random catalogs.

In equation (1) we assume that in the random catalog the intensity of the earthquakes can vary with time but will vary identically for the entire region under study. Nothing, by the way, hinders the construction of a random catalog in which the intensity is in the same manner dependent upon time only within individual parts of the entire region. If these parts are sufficiently small, however, the interrelation of the earthquakes will be interpreted as a change in intensity Λ, i.e., the local effect will be included in the random catalog.

In equation (3) we state that the density of the statistical distribution of the hypocenters in space is the realization of this density as given in the actual catalog.

In equation (4a) we assume that the intensity of the flow of earthquakes does not change with respect to time.

Equation (4b) means that we do not make this assumption but, as in (3), assume only that the distribution density of the earthquake moments coincides in time with the realization of this density as given in the actual catalog.

The algorithm for constructing random catalogs is as follows.

We write superscripts "r" and "a" on all symbols introduced above if they pertain to the random and actual catalogs, respectively. In accordance with (3), the coordinates of the hypocenter in the random catalog have the form

$$\mathbf{x}_i^r = \mathbf{x}_k^a, \quad g_i^r = g_k^a, \tag{5}$$

where k is a random number that is uniformly distributed on the interval 1-N. Thus the hypocenter and intensity of earthquakes in the random catalog are obtained by shuffling the actual catalog. In accordance with (4a),

$$t_i^r \doteq t_{i-1}^r + \xi_i \tag{6a}$$

where ξ is a random value with probability density $\lambda e^{-\lambda\xi}$, and t_0 is arbitrary.

In accordance with (4b),

$$t_i^r = t_i^a. \tag{6b}$$

Both catalogs – with constant (4a and 6a) and time-variable (4b and 6b) intensities – were in most cases constructed with replacement: the hypocenters ($\mathbf{x}_k^a$, g_k^a) that had been used once were not eliminated in the sampling of further earthquakes, in order to leave the spatial distribution of intensity $f^*(\mathbf{x}|g)$ unchanged.

1.3. Theoretical Distribution of Statistics. The distribution of τ for a flow of independent events with intensity λ constant in time is, as is well known, exponential:

$$P(\tau < \tau_0) = 1 - e^{-\lambda\tau_0}. \tag{7}$$

Theorem. For a random catalog with λ constant in time, the distributions of the statistics d, $\hat{\tau}(d)$, and $\hat{d}(\tau)$ are given by the equations

$$P(d < d_0) = \int_{|\mathbf{x}-\mathbf{y}|\leqslant d} f(\mathbf{x}) f(\mathbf{y})\,d\mathbf{x}\,d\mathbf{y}. \tag{8}$$

$$P(\hat{\tau}(d) < \tau) = 1 - \int_G f(\mathbf{x}) \exp[-\lambda F(d/\mathbf{x})\tau]\,d\mathbf{x}, \tag{9}$$

where

$$F(d \mid \mathbf{x}) = \int\limits_{|\mathbf{y}-\mathbf{x}| \leqslant d} f(\mathbf{y})\,d\mathbf{y}, \tag{10}$$

$$P(\hat{d}(\tau) < d) = (1 - e^{-\lambda\tau})^{-1}\, p(\hat{\tau}(d) < \tau). \tag{11}$$

Equation (8) is also valid for an intensity having the form (1).

Proof. Equation (8) is obvious. To prove equation (9), let us consider the conditional distribution of $\hat{\tau}(d)$, with the condition that the initial earthquake occurred at **x**. It is exponential with the parameter $\lambda F(d|\mathbf{x})$, since the events (t, **y**, g) in the region $\|\mathbf{y}-\mathbf{x}\| \leq d$ also form a flow of independent events not with parameter λ but with $\lambda F(d|\mathbf{x})$.

The unconditional distribution is obtained by averaging the conditional one with weight $f(x, y)$ over the region G, which gives (9).

The probability of an event $\{\hat{\tau}(d) < \tau\}$ can be calculated by another method: by the total probability formula

$$P(\hat{\tau}(d) < \tau) = P(\hat{\tau}(d) < \tau \mid A)\, P(A), \tag{12}$$

where A is the event that at least one earthquake occurred within time τ in region G. But from the definition of $\hat{d}(\tau)$,

$$P(\hat{\tau}(d) < \tau \mid A) = P(\hat{d}(\tau) < d), \tag{13}$$

and $\mathscr{P}(A) = 1 - e^{-\lambda\tau}$. Hence (11) follows, and the theorem is proved.

2. Method for Detecting Interrelated Earthquakes

2.1. If we compare the distributions of statistics τ, d, $\hat{\tau}$, and $\hat{d}$ for the random and actual catalogs, we can study the effect of the initial earthquake on subsequent earthquakes. In fact, the fictitious earthquakes in our random catalog must not affect one another. And if the distribution of the statistics for the actual catalog is different from that for the random one, this can be explained by the effect of a real initial earthquake on subsequent earthquakes (of course, only if the random catalog exactly corresponds to the given actual; if not, the resulting errors must be estimated).

Let $P_a(d)$ and $P_r(d)$ be probability density functions of d† for the actual and random catalogs, respectively.

We shall call the effect positive or negative, respectively, when the occurrence of the initial earthquake increases or decreases the probability of a subsequent earthquake. We cannot simply assume that for a given d the effect is positive if $P_a(d) > P_r(d)$ and vice versa if the effect of the initial earthquake for some d values distorts the ratio of P_a and P_r for other d, because of normalization. These distortions can be eliminated if physical considerations allow us to assume that the effect of the initial earthquake must not extend further than some threshold A. Then we have

$$P_r(d) = kP_a(d) \text{ for } d \geqslant A. \tag{14}$$

To estimate k, we can use the formula

$$k = \frac{\int_A^\infty P_r(s)\,ds}{\int_A^\infty P_a(s)\,ds}. \tag{15}$$

The difference

$$R(d) = kP_a(d) - P_r(d) \tag{16}$$

characterizes the effect of the initial earthquake without distortion. If $R > 0$, the effect is positive, and vice versa.

Let us rewrite formula (16) as

$$\bar{R}(d) = \frac{1}{r} R(d) = \left[\frac{1+r}{r} P_a(d) - \frac{1}{r} P_r(d)\right], \tag{17}$$

where

$$c = k - 1 = \int_0^\infty R(s)\,ds. \tag{18}$$

Since the catalogs are limited in extent, P_a and P_r are known only approximately, and all of the results should be interpreted statistically at some confidence level. The hypothesis on the value of A can be verified by testing the fit of the conditional distributions $P_a(d|d \geq A)$ and $P_r(d|d \geq A)$; some statistical tests can be used without resorting to this hypothesis.

The following three tests are used in this paper.

1. A. N. Kolmogorov's test [2] for the maximum absolute value of the difference between empirical distribution functions is used to prove a significant difference at confidence level α between the random and actual distributions (the existence of an earthquake interaction).

† All further discussion in 2.1 is valid for any statistic τ, d, $\hat{\tau}$, or $\hat{d}$; for simplicity, we shall mention only statistic d.

2. The binomial test [2] is used to compare the probabilities of occurrence of a subsequent earthquake in two intervals: $d_1 \le d \le d_2$ and $d_3 \le d \le d_4$. We let

$$p = \frac{p'}{p' + q'}, \tag{19}$$

where

$$p' = \int_{d_1}^{d_2} P_d(r)\,dr, \quad q' = \int_{d_3}^{d_4} P_d(r)\,dr. \tag{20}$$

A significant difference between P_r and P_a is obviously proof of a difference between the random and actual distributions. Here it can be asserted that the number of earthquakes that occur after the initial one in the actual catalog will be different from that in the random one, on at least one of these intervals.

3. The following method is used to prove the existence of interaction between hypocenters that are close to one another ($d \le d_0$). Such earthquakes, if they are independent, have a Poisson distribution with the parameter

$$\lambda = NP\,(\hat{d} \leqslant d_0). \tag{21}$$

From histograms of $P_r(\hat{d})$ and $P_a(\hat{d})$ we find the sample values of $P(\hat{d} \le d)$, and using the tables in [2] we obtain the confidence intervals for λ_r and λ_a, respectively. If we divide by N_r and N_a correspondingly, we find the confidence intervals for $P_r(\hat{d})$ and $P_a(\hat{d})$. If they do not overlap, then at the given confidence level it can be asserted that the effect in question exists; it is positive if $\lambda_a > \lambda_r$ and negative if $\lambda_a < \lambda_r$.

In conclusion, note that in the preceding discussion the values of d or $\hat{d}$ must be defined so that they are independent in the random catalog. The d values that are defined as $\|\mathbf{x}_1 - \mathbf{x}_2\|$, $\|\mathbf{x}_2 - \mathbf{x}_3\|$, etc., are pairwise dependent, just as are the $\hat{d}$ values that are defined by overlapping groups of earthquakes.

To avoid error, it is better to define d as the distance from each odd-numbered hypocenter to the next even-numbered one. Similarly, it is better to define $\hat{d}$ for nonoverlapping time intervals $k\tau_n \le t \le (k+1)\tau_n$; $k = 1, 2, \ldots$.

Here, unfortunately, the sample size of statistic d is reduced by half and the size of statistic $\hat{d}$ by even more.

This dependence was insignificant in practice in the cases we considered.

2.2. Effectiveness of Statistics When Epicentral Region Is Increased.

We must have a rather large volume of observations to study the weak effect of the initial earthquake on subsequent earthquakes. Since the observation time is limited, we must use observations over a large area, and be satisfied with the characteristics of the effect as averaged over this area.

An increase in the size of epicentral region affects the effectivenesses of the statistics introduced here in different ways. The statistic τ becomes ineffective, since the intensity Λ is increased and $P(\tau)$ is concentrated about $\tau = 0$. The effectiveness of statistic d is not increased, since distant earthquakes "split up" interacting pairs of earthquakes. The effectivenesses of $\hat{\tau}(d)$ and $\hat{d}(\tau)$ increase, and if the region is increased without limit these statistics allow us to study weak interactions. Although this assertion is likely to be valid in general, we shall restrict ourselves to proving it under the following assumptions.

Theorem. Let the seismic region G be a plane circle of radius R. The intensity $\Lambda(t, \mathbf{x}, g)$ is constant

$$\Lambda(t, \mathbf{x}, g) = \Lambda = \lambda\pi R^2; \tag{22}$$

The initial earthquake creates, on the average, n_a aftershocks with density $P_a(r)$, spatial distribution function $F_a(r)$, and time intensity $\lambda_a(t)$. Functions $P_a(r)$ and $\lambda_a(t)$ are finite and continuous;

$$\int_0^\infty \lambda_a(t)\,dt = n_a < 1.$$

Then the following assertions are valid:

(i) For any τ_0 and α, a large radius R_0 can be found such that the events $\{\tau < \tau_0\}$ for the random and actual catalogs are statistically indistinguishable at the confidence level α for any $R > R_0$.

(ii) For sufficiently large R_0, the statistical distinguishability of the events $\{r < r_0\}$ for the random and actual catalogs does not depend on R for $R > R_0$.

(iii) For some τ_0 and d_0 and any α, a radius R_0 is found such that the events $\{\hat{\tau}(d_0) < \tau_0\}$ for the random and actual catalogs are distinguishable at confidence level α for regions with any $R > R_0$.

(iv) An assertion similar to (iii) is valid for the events $\{\hat{d}(\tau_0) < r_0\}$.

Proof. (i) If in the random catalog the intensity is Λ, then in the actual catalog the intensity of independent earthquakes (not aftershocks) must be $\Lambda' = \Lambda(1 - n_a)$, so that together with the aftershocks, it equals Λ:

$$\Lambda'(1+n_a+n_a^2+\ldots)=\Lambda.$$

Hence it is easy to see that the following inequality holds in the actual catalog:

$$1-\exp(-\Lambda'\tau_0)\leqslant P_a(\tau<\tau_0)\leqslant 1. \qquad (23)$$

Since in the random catalog, according to equation (7),

$$P_r(\tau<\tau_0)=1-e^{-\Lambda\tau_0},$$

then

$$|P_a(\tau<\tau_0)-P_r(\tau<\tau_0)|\leqslant e^{-\Lambda\tau_0}+e^{-\Lambda'\tau_0}\leqslant$$
$$\leqslant 2\exp(-\lambda\pi\tau_0 R^2) \qquad (24)$$

from which assertion (i) follows.

(ii) For large R, ignoring the boundary effect, we can represent equation (8) as

$$F(r\,|\,\mathbf{x})=\int\limits_{\|\mathbf{y}-\mathbf{x}\|\leqslant r} f(\mathbf{y})\,d\mathbf{y}\cong\frac{r^2}{R^2}. \qquad (25)$$

Inasmuch as (25) is not a function of **x**, we have

$$P_r(r<r_0)\cong\frac{r_0^2}{R^2}. \qquad (26)$$

For the actual catalog the probability that the next earthquake will occur not further than d from the initial one, assuming that the initial earthquake occurred at point **x**, is written for large R as:

$$F(r_0\,|\,\mathbf{x})\cong\frac{\lambda'\pi R^2}{\lambda'\pi R^2+F_a(r_0)\lambda_a(0)}\frac{r_0^2}{R^2}+$$
$$+\frac{\lambda_a(0)}{\lambda'\pi R^2+F_a(r_0)\lambda_a(0)}F_a(r_0)\cong\frac{1}{R^2}\left(r_0^2+\frac{\lambda_a(0)F_a(r_0)}{\lambda'\pi}\right). \qquad (27)$$

Similarly to (26), we have by simplifying the last expression

$$P_a(r<r_0)\cong\frac{1}{R^2}\left(r_0^2+\frac{\lambda_a(0)F_a(r_0)}{\lambda'\pi}\right). \qquad (28)$$

Thus, in the interval 0-r that interests us, the number of earthquakes in the random catalog is

$$N_r(r_0)=\lambda\pi r_0^2, \qquad (29)$$

while in the actual catalog it is

$$N_a(r_0)=\lambda\pi r_0^2+\lambda_a(0)F_a(r_0). \qquad (30)$$

These numbers are independent of the region, and therefore their statistical distinguishability does not depend upon R (for sufficiently large R). Thus assertion (ii) is proved.

We select τ_0 such that $\lambda_a(\tau)=0$ when $\tau\geq\tau_0$, and r_0 such that $P_a(r_0)\equiv 0$ when $r>r_0$. Substituting (22) into (9), we obtain

$$P_r(\hat{\tau}(r_0)<\tau_0)=1-\int\limits_G\frac{1}{\pi R^2}e^{-\lambda\pi R^2\frac{r_0^2}{R^2}\tau_0}\,d\mathbf{x}\cong$$
$$\cong 1-e^{-\lambda\pi r_0^2\tau_0}, \qquad (31)$$

and substitution of (27) into (9) gives

$$P_a(\hat{\tau}(r_0)<\tau_0)=1-\exp(-\lambda\pi r_0^2\tau_0)-n_a \qquad (32)$$

Probabilities (31) and (32) are independent of R for sufficiently large R. As the size of the region G is increased, the significance of the statistical results increases as $\sqrt{\lambda\pi R^2\tau_0}=c\sqrt{R}$, from which assertion (iii) follows.

(iv) The proof is similar to that of (iii). It is sufficient to note that

$$(1-e^{-\Lambda\tau})=1-e^{-\lambda\pi R^2\tau}\to 1 \quad\text{for}\quad R\to\infty.$$

2.3. Distortions Introduced by Errors in Epicenter Coordinates.

It often happens that $P_a(r)>P_r(r)$ in some vicinity of $r=0$, so that the point of intersection of P_a and P_r can be taken as an estimate of the radius of the effect. This estimate will be exaggerated due to errors in determining the epicentral coordinates. Let us examine this exaggeration in somewhat greater detail.

Let the error of determination of the epicenter $P_0(\mathbf{x})$ be normal and independent of the coordinates of the point on the plane. We have

$$P_0(\|\mathbf{x}\|)\,dr=\frac{r}{\sigma^2}\exp\left(-\frac{r^2}{2\sigma^2}\right)dr, \qquad (33)$$

where $\|\mathbf{x}\|=r$. Just as before, we shall assume that the function $\overline{R}(r)$, which describes the effect of the initial earthquake on subsequent ones, is finite: $\overline{R}(r)\equiv 0$ when $r\geq A$ (see equation 17). In the presence of epicenter error, equation (17) gives a different function:

$$\overline{R}_0(\mathbf{z})=\int\limits_G\overline{R}(|\mathbf{x}|)P_0(|\mathbf{z}-\mathbf{x}|)\,d\mathbf{x}=$$
$$=\int\limits_G[kP_a(|\mathbf{x}|)-P_r(|\mathbf{x}|)]P_0(|\mathbf{z}-\mathbf{x}|)\,d\mathbf{x},$$

$$\overline{R}_0(r)\,dr=\int\limits_0^{2\pi}[R_0(\mathbf{z})\,r\,dr]\,d\varphi. \qquad (34)$$

TABLE 1. Coordinates of Vertices of Epicentral Regions

Region number	Region	Vertex coordinates					
		1	2	3	4	5	6
1	Surkhob River	39°21 70.13	39°45 70.63	39°81 73.96	39°42 74.09	38°96 70.24	—
2	Darvaz	38.72 70.00	38.92 70.28	39.19 71.48	39.04 71.62	38.38 70.37	—
3	Eastern Tien Shan	42.74 72.34	43.66 81.37	41.59 81.13	40.34 75.06	—	—
4	Fergana	40.05 69.28	41.59 71.97	41.38 73.19	40.26 74.56	40.00 73.28	40°00 69.35

The function $\bar{R}_0(r)$ will be approximately normal under the following not too limiting assumptions:

1) the error $P_0(\|x\|)$ has the form of (33);

2) the effect $\bar{R}(\|x\|)$ has the form of (33); in particular, it is a point effect when $\sigma = 0$, i.e., when $\bar{R}(r) = \delta(r)$;

3) region G is sufficiently large so that the integrals over it can be considered integrals in infinite limits.

Under these assumptions the function $\bar{R}_0(r)$ has the form of (33), and for the variance of $\bar{R}_0$ we have

$$\sigma(\bar{R}_0) = \sigma(\bar{R}) + \sigma(P_0). \tag{35}$$

This equation shows that if an effect exists ($c \neq 0$ in formula (15)), then for the statistic r the characteristic $\delta(\bar{R})$ of the radius of the effect is, in principle, $\sigma(P_0)$ – the mean-square error of the epicenter determination. If, for example, the effect is concentrated in an infinitely small vicinity of the epicenter ($\sigma = 0$), we estimate the radius of the effect in $\sigma_{\bar{R}}$ – the epicenter error.

The variance $\sigma(\bar{R}_0)$ can be estimated by the maximum likelihood method as follows:

$$\sigma^*(\bar{R}_0) = \sqrt{\frac{\sum_{i=1}^{n} a_i(r_i) r_i^2}{2\sum_{i=1}^{n} a_i(r_i)}}, \tag{36}$$

where $a_i(r_i) = R^*(r_i)$ is the sample value of the function $\bar{R}$ for a given r_i; $\sigma^*(R_0)$ will be the apparent mean-square radius of the earthquake effect.

3. Interrelationship of Earthquakes of Central Asia

3.1. Data. Here, we shall study the catalog of earthquakes of eastern Central Asia for 1930-1956 given in Chart 12 of [1]. The coordinates of the epicenters are rounded to 0.°1 and the focal depth is given to the nearest 10 km; the depth is generally given only for foci below the crust. Earthquake magnitudes in [1] are divided into five groups beginning with $M \approx 3.0$ which are represented with varying completeness for the different time intervals. The earthquakes of group V (those with the smallest magnitude) are covered particularly unevenly. Therefore, the statistical characteristics were studied for different periods: the interrelationship of the earthquakes of group I, and also that between groups IV and V was examined for the period of 1952-1956 – the period of the most complete and relatively most uniform coverage. The earthquake statistics for the other groups were taken from the entire interval of 1930-1956.

3.2. Epicentral Regions. Calculations were made for all of chart 12 of [1], which corresponds to the region from approximately 36° to 45°N. lat. and from 65° to 81° E. long. Below, this region is referred to by the number "0." In addition, the regions shown in Fig. 1 and Table 1 were studied individually.

Region 1 is a narrow strip along the Surkhob River, which contains a deep fault zone on the boundary between Pamir and Tien Shan dividing the foot of the Gissaro-Alai ridge (folded and metamorphized Paleozoic rocks) from a region of a Meso-Cenozoic depression (loose strata).† During the Quaternary the depressed region began to be lifted at a rate that considerably exceeded the rate of growth of the Gissaro-Alai. A zone of sharply contrasting vertical movements accompanied by lateral shifts was created. Region 2 is a narrow band of contact of the Pre-Darvaz depression and the Darvaz uplift, which are separated by the deep Darvaz fault. The dislocated sediments of the Meso-

† The neotectonic characteristics of the regions were taken by E. Ya. Rantsman, whom the authors thank.

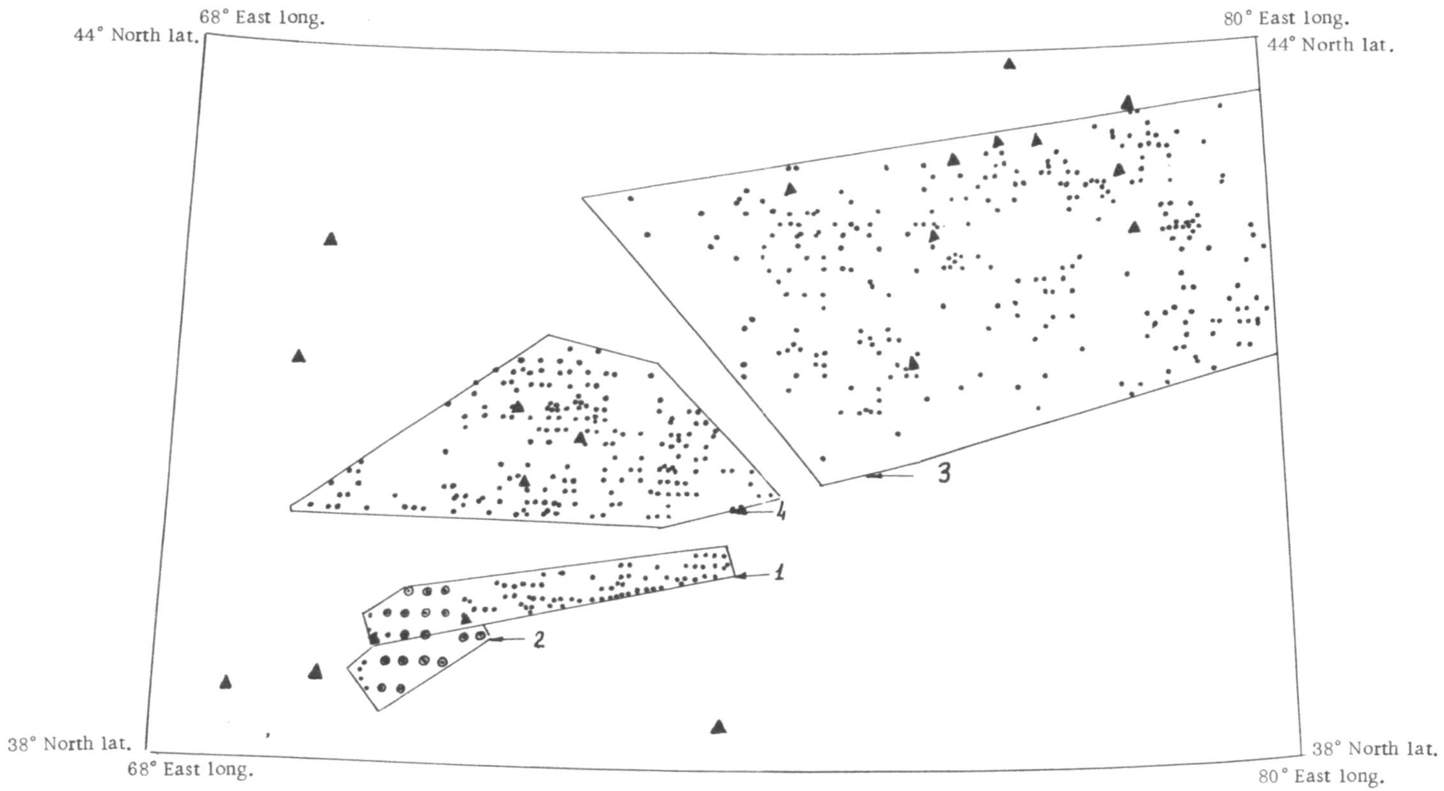

Fig. 1. Map of epicentral regions: 1) Surkhob River; 2) Darvaz Range; 3) Eastern Tien Shan; 4) Fergana. The points indicate the epicenters of earthquakes of group V, ⊙ denote places of multiple epicenters, and ▲ are seismic stations. See [1] for a more detailed map.

TABLE 2. Summary of Calculated Bar Graphs

Variant number	Figure	Type of statistic	Threshold	No. of the epicentral region	H, km	g_i	g_{i+k}	Increment† Δ	Intervals of histogram 0-B	Probability catalog	s_1		s_2		No. of earthquakes	
											a	r	a	r	a	r
1	2	3	4	5	6	7	8	9	10	11	12		13		14	
1a	2a	d	—	0	<75	V	V	50	1500	1	0	0	—	—	2833	5831
1b	2b	d	—	0	<75	V	V	5	50	1	0.931	0.958	—	—	2833	5831
2a	3a	$\hat{d}$	1	0	<75	V	V	50	1500	1	0	0	0.215	0.233	2827	5830
2b	3b	$\hat{d}$	1	0	<75	V	V	5	50	1	0.587	0,707	0,215	0.233	2827	5830
3	3c	$\hat{d}$	1	0	>75	V	V	5	100	1	0,310	0.335	0,244	0.220	2477	8765
4	—	$\hat{d}$	1	0	<75	IV	IV + V	5	100	1	0.489	0,679	0.125	0.181	88	271
5	—	$\hat{d}$	1	0	>75	IV	IV + V	5	100	1	0,194	0.337	0.278	0.221	36	113
6	—	$\hat{d}$	7	0	<75	III	III + IV	25	300	1	0.068	0.195	0,559	0.693	59	261
7	—	$\hat{d}$	7	0	All	II	II + III	25	300	1	0.000	0.033	0.806	0,953	31	302
8a	4	$\hat{\tau}$	5	0	<75	V	V	10	100	1*	0.803	0,888	—	—	2809	3672
										2*		0,861				3730
8b	4	$\hat{\tau}$	5	0	<75	V	V	1	10	1*	0.927	0.990	—	—	2809	3672
										2*		0.985				3730
9a	5a	$\hat{\tau}$	16	0	<75	V	V	5	50	1*	0,600	0.693	—	—	2286	3753
9b	5b	$\hat{\tau}$	16	0	<75	V	V	1	10	1*	0.813	0.910	—	—	2286	3753
10	5c	$\hat{\tau}$	50	0	<75	V	V	1	10	1*	0,566	0.628	—	—	3019	3826
11a	6b	$\hat{\tau}$	5	0	>75	V	V	10	100	1*	0,852	0.857	—	—	2265	2606
11b	6a	$\hat{\tau}$	5	0	>75	V	V	1	10	1*	0.978	0.983	—	—	2265	2606
12	7a	$\hat{\tau}$	5	1	<75	V	V	10	100	1*	0,638	0.741	—	—	456	1054
13	7b	$\hat{\tau}$	5	2	<75	V	V	10	100	1*	0.667	0,756	—	—	213	541
14	7c	$\hat{\tau}$	5	3	<75	V	V	0	100	1*	0.911	0,963	—	—	224	590
15	7d	$\hat{\tau}$	5	4	<75	V	V	0	100	1*	0.976	0,969	—	—	127	356

†Value of d in km, of τ in days.

zoic and Cenozoic and the metamorphized strata of the Paleozoic make contact along the fault. The total range of recent movements here is the greatest in Central Asia (up to 13 km). The region of depression was uplifted during the Quaternary. Region 3 is eastern Tien Shan – this region is remarkable, since the orogenesis here followed the formation of a platform. It is composed in the northern part of intrusive (granitoid) and metamorphized rocks of the Caledonian orogenesis. Sedimentary strata of the middle and upper Paleozoic with folding of the Hercynian phase predominate in the southern part of Eastern Tien Shan. Orogenic movements began in the Oligocene but reached a great range only in the late Pliocene. Displacements along deep fault zones occurred during the Quaternary in the zone of collision of the Tien Shan and the Kazakh platform (northern part of Eastern Tien Shan). Region 4 is the Fergana intermontane basin, which is made up of loose deposits of the Mesozoic and Cenozoic, and its zone of collision with the Altai and Fergana ridges, which are composed of dislocated sedimentary Paleozoic rocks. The zone of contact with the Altai ridge is represented by the deep Yuzhno-Fergana fault.

3.3. Aftershocks were eliminated by eye. Bunches of earthquakes of a given intensity group occurring after stronger earthquakes were not considered. Since the conclusions obtained below are not changed substantially when aftershocks are eliminated, the problem of the strict definition of aftershocks is not considered here.

3.4. A summary of the calculated variants is given in Table 2.

Column 1 gives the variant number, 2 the number of the figure in which the calculated histograms are given, 3 the statistic calculated, 4 the threshold d in determining $\hat{\tau}(d)$ or the threshold τ in determining $\hat{d}(\tau)$, 5 the number of the epicentral region from Table 1 and Fig. 1, 6 the depth interval, 7 the intensity group of the initial earthquake g_i, 8 the intensity group of the subsequent earthquakes g_{i+k}, 9-13 the characteristics of the calculated histograms (these were computed with a step $\Delta(g)$ on the interval 0-B (10)), 11 the type of random catalog (events outside the interval were also calculated and taken into account in normalization; the relative number of such events is denoted by s_1(12)), 13 the relative number of initial earthquakes s_2 after which there were no subsequent earthquakes in time τ (for statistic $\hat{d}(\tau)$), and 14 the number of initial earthquakes N.

Columns 12, 13, and 14 each contain two numbers: the first is for the actual catalog and the second is for the random catalog.

As a rule the focal depth is not indicated for normal earthquakes. In a number of cases, therefore, we consider not d but r. In calculating d we assumed h = 20 km for the earthquakes for which h was not given in [1].

The calculated histograms are shown in Table 6.

Figures 2-9 give graphs of R (see formula (16)); the broken lines show the shapes of the graphs before aftershocks were eliminated.

3.5. Effect of Initial Earthquake on Subsequent Earthquakes (Existence). The earthquakes are mainly independent, since R is very small. An interrelationship between earthquakes still exists, however. For normal earthquakes, all of the histograms are significantly different for the random and actual catalogs. This is demonstrated by A. N. Kolmogorov's test (Table 3). Only for the earthquakes of magnitude group II was the volume of statistics insufficient for this test, so we had to resort to the Poisson test (Table 5). For deep earthquakes, the existence was proved by the Poisson test only for the greatest volume of statistics – for the earthquakes of group V (Table 5). In all cases there was a significant deviation from zero and the actual histograms exceeded the random, i.e., the effect was positive. The significant difference between P_a and P_r for variants 4-7 reflects the trivial fact of the effect of an initial earthquake on weaker ones, i.e., the existence of aftershocks. The objection may arise that the effect considered here might also be due to unidentified aftershocks: that subsequent earthquakes are weaker than the initial ones, although the effect is masked by errors in magnitude. We considered this objection and rejected it after direct study of the relation between the magnitudes of coupled initial and subsequent earthquakes.

3.6. Distances to Which the Effects Extend. It makes sense to study these distances only for the weakest earthquakes of group V; for stronger earthquakes, the volume of statistical data is too small.

We can use both d and $\hat{d}$ to study the effect of normal earthquakes. The fact that the histograms of d and $\hat{d}$ are not monotonic is explained in part by the discreteness of the hypocenter coordinates; which affects P_r and P_a in the same way and is therefore not important for what follows. Assuming that the effect does not extend further than A = 200 km, then we can find R(d) and R($\hat{d}$) (see 2.1), shown in Figs. 2 and 3. Table 3 shows the confidence level of fit of the conditional distributions for the actual and random catalogs for d ≥ A or $\hat{d}$ ≥ A. The fit is quite satisfactory, which confirms the hypothesis in 2.2. Comparison of Figs. 2 and 3 shows (see also the confidence level of difference in Table 3), that the sensitivity of statistic $\hat{d}$ is more sensitive than that of d, in agreement with 2.2. Since minimization underlies the definition of statistics includes a minimization, however, this reduces the dimensions of the effect in space. Using statistic $\hat{d}$, we can show a significant effect only at a distance of 25-50 km (Table 4). Statistic d, however, indicates a significant effect of an initial earthquake on subsequent earthquakes at a distance of 50-100 km; the confidence level is given in Table 4. We recall that the above examples of effect are exaggerated due to the error in epicenter determination (according to [1], they can be up to 50 or more km) (see 2.3). For small d, however, the errors in d do not include the regional errors of the travel time curve – the main source of epi-

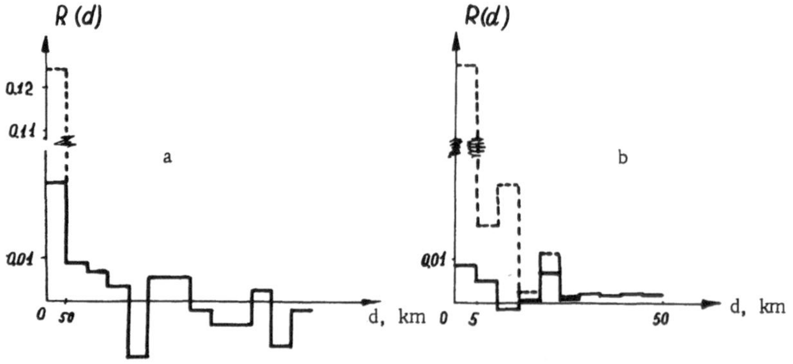

Fig. 2. R(d) for normal earthquakes (here and in Figs. 3-9 graphs for earthquakes of group V are given; the broken line shows R before elimination of aftershocks).

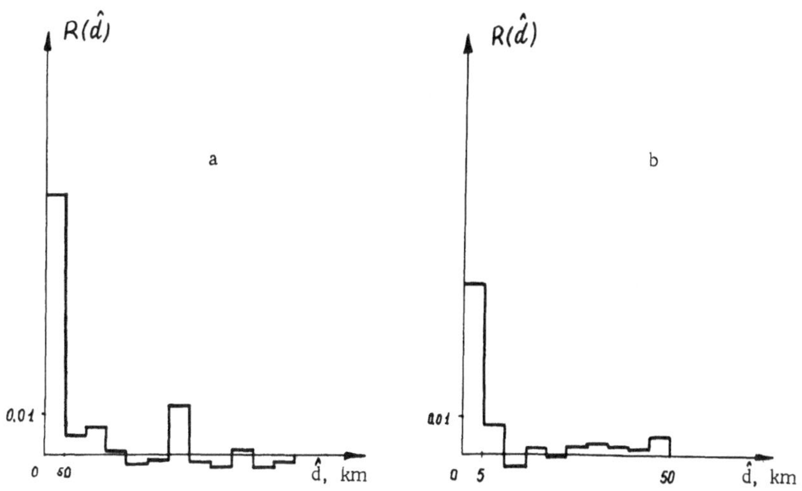

Fig. 3. R(d̂(1 day)) for normal earthquakes.

TABLE 3. A. N. Kolmogorov's Test

Number of variant	Length of histogram	$\max \lvert F_a - F_r \rvert$	$\sqrt{\frac{N_a N_r}{N_a + N_r}}$	Confidence level of agreement
1a	—	0.041	43.7	0.0040
2a	—	0.053	43.7	0.0000
4	$d < 100$ km	0.232	8.1	0.0020
6	$\hat{d} < 300$ km	0.333	6.8	0.0000
8a	$\hat{\tau} < 100$ days	0.089	39.9	0.0000
9a	$\hat{\tau} < 50$ days	0.102	37.7	0.0000
10	$\hat{\tau} < 10$ days	0.062	41.1	0.0000
1a	$d > 200$ km	0.013	43.7	0.902
2a	$\hat{d} > 200$ km	0.011	43.7	0.975
3	$\hat{d} > 5$ km	0.0075	42	0.999
8a	$\hat{\tau} > 50$ days	0.0059	39.9	0.770
11a	—	0.0095	34.8	0.999
10	5 days $\geqslant \tau \geqslant$ 10 days	0.076	15.5	0.124

TABLE 4. Binomial Test

Number of variant	Intervals d and τ	Confidence interval		Confidence level of difference
		for P_a	for P_r	
3	$d \leqslant 5$ km $\tau \leqslant 1$ days	0.0039—0.0137	0.0011—0.0038	0.98
7	$d \leqslant 25$ km $\tau \leqslant 7$ days	0.0313—0.6150	0.0000—0.0252	0.998
11b, 3	$d \leqslant 5$ km $\tau \leqslant 1$ days	0.0226—0.0682	0.0016—0.0225	0.99

TABLE 5. Poisson Test

Number of variant	Interval		Confidence level of difference
	First	Second	
1a	50—200 km	200—1500 km	0,999
1a	100—150 km	150—∞ km	0,90
2a	25—50 km	50—∞ km	0,996
8a	10—50 days	50—∞ days	0,9999
8a	20—50 days	50—∞ days	0,9999
8a	30—50 days	50—∞ days	0,997
9a	30—40 days	40—∞ days	0,9998
10	3—5 days	5—∞ days	0,997
8c	50—100 days	100—∞ days	0,95
8c	20—40 days	40 days	0,998
12	0—1 days	>1 days	0,9999
13	0—1 days	>1 days	0,9999
14	0—1 days	>1 days	0,9999
15	0—1 days	>1 days	0,9929
14	30—100 days	>100 days	0,9999
15	5—75 days	>75 days	0,9981
12—13	0—1 days, 21 var	0—1 days, 22 var	0,9861
14—15	0—1 days, 23 var	0—1 days, 24 var	0,94

Fig. 4. R(d̂(1 day)) for deep earthquakes.

center error – so that the errors in d are considerably smaller than the absolute epicenter errors.

The broken line in Fig. 2 shows R(d) before the elimination of aftershocks; they attenuate quickly (practically by 30 km). This unexpected result – that the effect of stronger earthquakes extends less far than that of weaker earthquakes – implies that for weak earthquakes we are not dealing simply

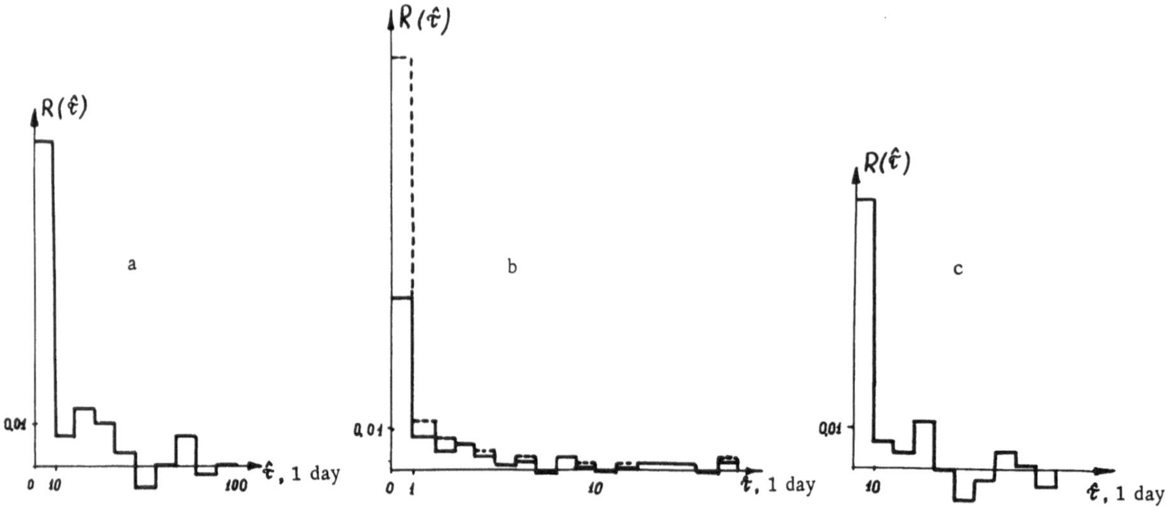

Fig. 5. R(τ̂(5 km)) for normal earthquakes; the average intensity of earthquakes in random catalog was constant in time in a and b, and varied in time in c.

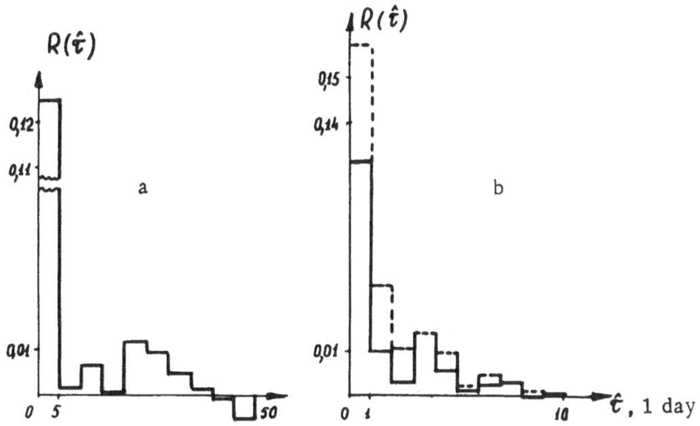

Fig. 6. $R(\hat{\tau}(16\text{ km}))$ for normal earthquakes ($h \leq 75$ km).

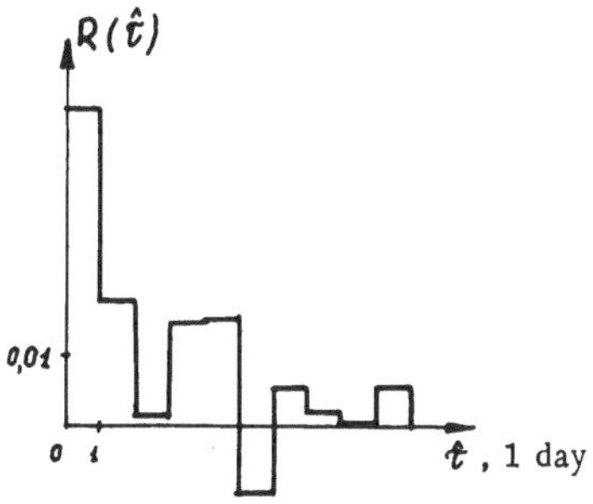

Fig. 7. $R(\hat{\tau}(50\text{ km}))$ for normal earthquakes ($h \leq 75$ km).

with the effect of one earthquake on another but with some common cause of earthquakes (for example, a temporary increase in activity in some region at a given time).

The existence of an effect at $d \leq 5$ km and $\tau \leq 1$ day has been proved for the earthquakes of group V. This effect is significantly less than that of surface earthquakes (see Figs. 3b and 4 and Table 5).

Note that the graphs of $R(\hat{d})$, while correctly reflecting the effect of the initial earthquake, sometimes create an incorrect representation of the statistical significance of one or another deviation from zero. For example, R(325 km) in Fig. 3a and R(22.4 km) in Fig. 4 seem to contradict common sense. An increase in $\hat{d}$, however, increases the share of random earthquakes that can occur at that distance and, along with this, increases the size of allowable random fluctuations.

3.7. The interval of the effect in time, too, could only be studied for earthquakes of group V. In this case, to avoid distortions of the activity Λ, we must use the random catalog without replacement (see 1.2). We assume that the threshold A of the extent of the effect is equal to 50 days for surface earthquakes, then according to A. N. Kolmogorov's test the conditional distributions have a good fit when $\hat{\tau} \geq A$ (Table 3). Figures 5, 6, and 7 show $R(\hat{\tau})$ for $r_n = 5$ km, $r_n = 16$ km, and $r_n = 50$ km.

When $r_n = 5$ km, i.e., practically at the epicenter, the positive effect lasts for not less than 30 days; the confidence level is given in Table 4. In a circle of radius 16 km, the effect of the initial earthquake is greater than at the epicenter (compare Figs. 6 and 5), and a similar result can be shown about the duration of the effect (see Table 4). When the radius is increased to 50 km, the sensitivity of statistic $\hat{\tau}$ is greatly reduced. In particular, even at $\hat{\tau} > 5$ days, A. N. Kolmogorov's test (Table 3) shows an insignificant difference, with confirmation of point (i) of the theorem of 2.2. The statistics $\hat{\tau}$ ($r_n = \infty$) and τ coincide. For an interval of 3-5 days, however, the initial earthquake has an effect in a circle of radius $r_n = 50$ km. As was pointed out above, the effect of weak earthquakes does not necessarily imply their direct interaction but it can mean that they are related only by some common cause. In Fig. 5c, a change in activity with time, uniform for all of Central Asia (if such a change occurred), was included in the random catalog of type 2 (see 1.2). The positive effect was somewhat reduced, but the conclusions about its extent were unchanged (see Table 4). A significant negative effect appeared in the interval of 50-100 days (i.e., a decrease in the probability of appearance of a subsequent earthquake in that interval). The confidence level is given in Table 4. The variation of activity over all of Central Asia can also be checked by examining the numbers of normal earthquakes for

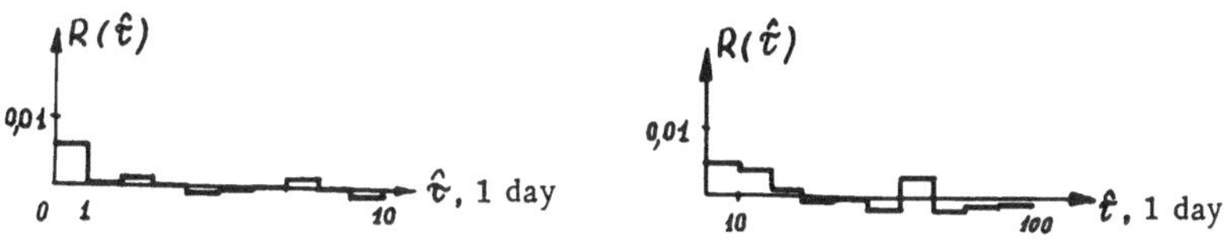

a b

Fig. 8. $R(\hat{\tau}(5\text{ km}))$ for intermediate earthquakes ($h > 75$ km).

TABLE 6. Histograms of Local Statistics

Version 1a

α	r
0,0692	0.0420
0,0711	0.0632
0.0803	0,0749
0.0782	0.0840
0,1021	0.1156
0.0965	0.1003
0.0842	0.0804
0.0679	0,0710
0,0430	0.0506
0.0490	0.0456
0.0433	0.0511
0.0373	0.0427
0.0342	0.0331
0.0270	0.0273
0.0246	0.0269
0,0239	0.0226
0.0207	0,0171
0.0126	0,0130
0,0067	0.0103
0.0074	0.0079
0.0067	0.0063
0,0021	0.0058
0.0042	0.0031
0,0025	0.0021
0,0007	0.0009
0,0007	0.0007
0,0007	0.0008
0,0004	0.0003
0,0000	0.0000
0.0000	0,0000

Version 1b

0.01310	0.00137
0.00742	0.00171
0,00283	0.00309
0,00247	0.00171
0,00920	0,00480
0,00811	0,00635
0,00882	0,00429
0,00494	0.00514
0,00918	0,00755
0.00600	0.00600

Version 2a

0.1149	0.0604
0,0808	0.0825
0.0846	0.0847
0.0780	0.0835
0.0907	0.1003
0,0680	0.0750
0.0614	0.0545
0.0405	0.0458
0.0233	0.0281
0.0242	0.0250
0.0211	0.0252
0.0187	0.0225
0.0169	0.0154
0.0135	0.0137
0.0120	0.0113
0.0088	0.0117
0.0063	0.0069
0.0046	0.0060
0.0032	0,0039
0.0025	0.0026
0.0032	0.0022
0.0004	0,0029
0.0011	0.0014
0.0021	0.0005
0.0007	0.0002
0.0000	0.0002
0.0000	0.0007
0.0000	0.0002
0.0000	0.0000
0.0000	0.0000

Version 2b

0.04020	0,0017
0.00958	0.0027
0.00332	0.0050
0.00406	0.0022
0,00700	0.0081
0.01105	0.0093
0.00960	0.0074
0.00774	0.0060
0.01130	0.0099
0.01180	0.0081

Version 3

0.0077	0.0022
0.0040	0.0034
0.0137	0.0136
0.0085	0.0089
0,0202	0.0280
0.0214	0.0208
0.0275	0.0233
0.0234	0.0245
0.0279	0.0279
0.0335	0.0324
0.0347	0.0288
0.0226	0.0281
0.0279	0.0302
0.0262	0.0262
0.0275	0.0284
0.0267	0 0241
0.0315	0.0262
0.0206	0,0226
0.0262	0.0244
0.0198	0.0216

Version 4

0.1591	0.0295
0.0682	0.0900
0.0568	0.0074
0,0227	0.0037
0.0113	0.0074
0.0227	0.0111
0.0113	0,0037
0.0000	0.0037
0.0113	0.0111
0.0000	0.0000
0.0000	0.0074
0.0000	0.0000
0.0000	0.0111
0.0000	0.0037
0.0000	0.0111
0.0000	0,0074
0,0000	0.0111
0.0000	0.0000
0.0227	0.0074

Version 5

0.0278	0.0088
0.0000	0.0000
0.0278	0.0088
0.0278	0,0000
0,0556	0.0265
0.0556	0,0000
0.0000	0,0265
0.0000	0,0353
0.0278	0.0176
0.0556	0.0265
0.0833	0.0442
0.0000	0.0353
0.0000	0.0619
0.0000	0.0000
0.0556	0,0442
0.0000	0.0884
0.0000	0.0000
0.0000	0.0176
0.0833	0.0265
0.0278	0.0530

Version 6

0.3390	0,0114
0.0169	0,0114
0.0000	0.0077
0.0169	0.0077
0.0000	0.0038
0.0000	0.0000
0,0000	0,0038
0.0000	0.0038
0.0000	0.0153
0,0000	0.0191
0,0000	0.0191
0.0000	0.0077

Version 7

0,1935	0,0000
0.0000	0.0331
0.0000	0.0000
0,0000	0.0000
0,0000	0.0000
0.0000	0.0000
0,0000	0.0000
0,0000	0,0000
0,0000	0.0000
0.0000	0.0000
0.0000	0.0662
0.0000	0.0331

Version 8a†

0.0725	0.0103	0.0155
0.0185	0.0128	0.0137
0.0218	0,0106	0.0201
0.0208	0.0125	0.0115
0.0143	0.0130	0.0158
0.0060	0.0117	0.0142
0,0087	0,0084	0.0126
0.0136	0.0079	0.0110
0.0102	0.0138	0.0105
0.0091	0.0103	0.0142

Version 8b

0.0370	0,0005	0.0024
0.0083	0.0008	0.0019
0.0045	0.0005	0.0011
0.0064	0.0011	0.0011
0.0049	0,0019	0.0032
0,0015	0.0005	0.0016
0.0038	0.0016	0.0011
0.0007	0.0016	0,0013
0.0030	0.0008	0,0014
0.0023	0.0008	0.0005

Version 9a

0.1480	0,0461
0.0391	0.0434
0.0361	0.0346
0.0297	0.0336
0.0349	0,0277
0.0316	0.0269
0,0252	0.0245
0,0222	0.0245
0.0191	0.0232
0.0143	0.0221

Version 9b

0,0526	0,0088
0.0151	0.0075
0.0117	0.0104
0,0173	0,0077
0.0150	0.0117
0,0094	0.0096
0.0079	0.0069
0.0098	0.0085
0,0060	0.0075
0.0101	0.0109

Version 10

0.0907	0.0562
0.0610	0.0499
0.0484	0.0520
0.0486	0.0392
0.0463	0.0363
0.0319	0,0461
0.0311	0.0290
0.0271	0.0280
0.0224	0.0240
0.0247	0.0217

Version 11a

0.0221	0.0173
0.0190	0.0150
0.0172	0.0161
0.0141	0.0150
0.0128	0.0134
0,0146	0.0165
0,0146	0.0115
0.0124	0.0142
0;0115	0.0130
0.0097	0.0111

Version 11b

0.00795	0.00192
0.00309	0.00192
0.00176	0.00192
0.00177	0.00192
0.00044	0.00115
0.00088	0.00153
0.00177	0.00153
0.00177	0.00077
0.00088	0.00115
0.00177	0.00345

Version 12*

0.0307	**0.0009**
0.0772	0.0265
0.0372	0.0360
0.0460	0.0331
0.0525	0.0313
0.0394	0,0341
0.0197	0.0284
0.0153	0.0199
0.0262	0.0161
0.0306	0.0218
0.0175	0.0132

Version 13

0.0704	**0.0000**
0.0995	0.0185
0.0234	0.0221
0.0328	0.0166
0.0422	0.0350
0.0328	0.0166
0,0094	0.0295
0.0187	0.0240
0.0375	0.0314
0.0141	0.0203
0.0234	0.0277

Version 14

0.0446	**0.0000**
0.0712	0,0000
0.0045	0,0034
0.0134	0,0034
0.0000	0,0034
0.0000	0.0068
0.0000	0.0017
0.0000	0.0034
0.0000	0.0085
0.0000	0.0017
0.0000	0.0051

Version 15

0.0137	**0.0000**
0.0137	0.0028
0,0000	0.0000
0.0000	0,0028
0,0000	0.0056
0.0000	0.0000
0.0000	0.0056
0.0081	0.0084
0.0000	0.0000
0.0000	0.0056
0.0000	0.0000

*The first numbers in versions 12-15 denote the significance of the histogram during the first day.

†In versions 8a and 8b, statistics are calculated for the random catalog with average intensity (4a) for c_1 and with variable intensity (4b) for c_2.

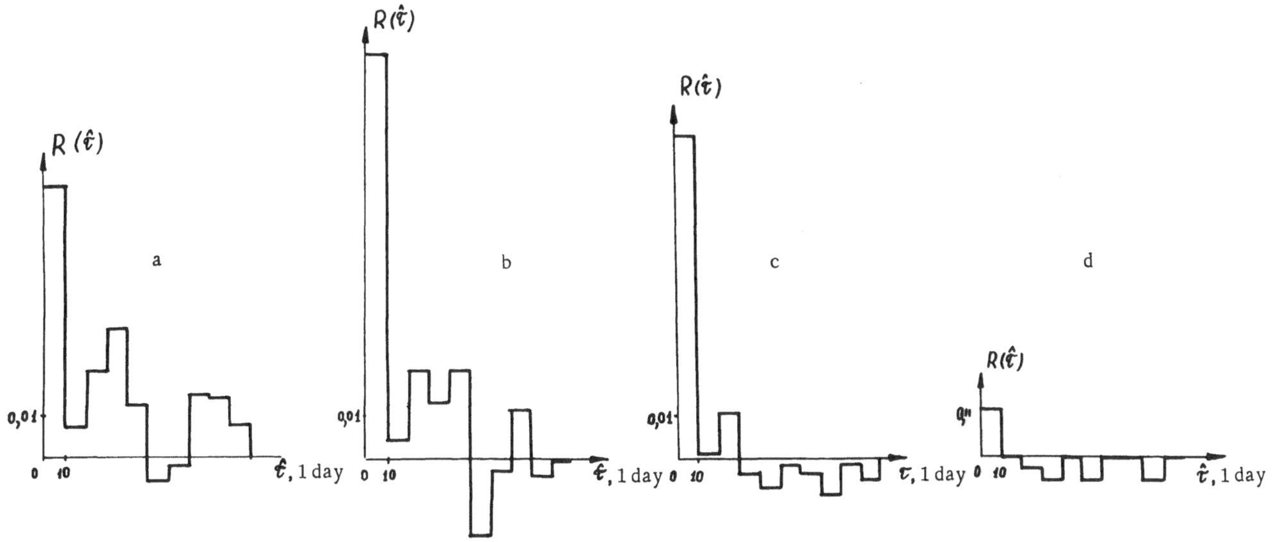

Fig. 9. R($\hat{\tau}$ (5 km)) for normal earthquakes for individual regions indicated in Fig. 1: a) Surkhob River, b) Darvaz, c) Eastern Tien Shan, d) Fergan.

different years. For example, in 1952 there were 620 and in 1956 460 normal earthquakes of group V. If it is assumed that earthquakes occur absolutely randomly with the same intensity, these numbers must obey a Poisson distribution with the same parameter. Since the mean-square deviation is estimated to be $\sigma \leq \sqrt{620} \approx 25$, and according to the central limit theorem [3], the Poisson distribution is close to normal, the confidence level of the difference in earthquake intensities in 1952 and 1956 is about 0.999.

A positive effect in deep earthquakes exists only in the hypocenter during the first day after the initial earthquake.

The excess in Fig. 8b in the interval of 0-20 days is insignificant according to the tests we used (see Table 3, for example).

3.8. The statistics studied above were averaged over all of eastern Central Asia. Let us compare the properties of these statistics for the individual regions. Figure 9 shows graphs for four regions of Central Asia which were isolated using neotectonic considerations (see 3.2). In all of these regions, the binomial test proved at a 99% confidence level the positive effect of an initial earthquake of group V on subsequent earthquakes (see Table 4). The Surkhob River and Darvaz regions are similar in the mean density of epicenters of group V. For example, the fraction of epicenters for which in 100 days there was no subsequent earthquake after the initial one agrees accurately with the random fluctuations (column 12, Table 2). The eastern Tien-Shan and Fergana regions have similar properties. The positive effect in the first day (Table 6 and Fig. 8a-d) was significantly different in these regions. In the Darvaz region it was greater than in the Surkhob River region, and in the Tien-Shan it was greater than in the Fergana region. The confidence level is comparatively low, however, especially in the latter case. The result must therefore be refined with a greater volume of statistical data (note that only the data for the five best-studied years were used). In the interval of 30-100 days for the eastern Tien Shan region and in the interval of 5-75 days for the Fergana region, the binomial test indicates a significant negative effect of the initial earthquake on subsequent earthquakes (Table 4).

LITERATURE CITED

1. Atlas of Earthquakes in the USSR: Moscow (1962).
2. Yanko, Ya. (1961), Tables of Mathematical Statistics, Moscow.
3. Cramér, H. (1951), Mathematical Methods of Statistics, Princeton, Princeton University Press.
4. Lukk, A. A. (1968), "Sequence of aftershocks of the Dzhuren deep-focus earthquake of 14 March 1965," Izv. Akad. Nauk SSSR, Ser. Fizika Zemli, No. 5.